Fundamentals
of College Algebra

Fundamentals
of College Algebra

Carl B. Allendoerfer
Professor of Mathematics, University of Washington

Cletus O. Oakley
Professor and Department Head, Emeritus
Department of Mathematics, Haverford College

McGraw-Hill Book Company *New York/St. Louis/San Francisco/Toronto/London/Sydney*

FUNDAMENTALS OF COLLEGE ALGEBRA

Library of Congress Catalog Card Number 66-22291

01369

1234567890QB7321069876

Preface

The purpose of this book is to present a modern treatment of a collection of topics which have traditionally been brought together under the heading College Algebra. This title is actually a misnomer, for only about half of this material can properly be called "Algebra", but tradition is so strong that we are keeping the customary name. The book assumes that the students have completed Intermediate Algebra and are preparing to enter a standard course in Analytic Geometry and Calculus. A knowledge of Trigonometry is not assumed (except in Sec. 10.11), but, of course, students who complete this book must also study Trigonometry before they enter a course in Calculus.

Chapters 1 to 11 are essentially identical with the Second Edition of the authors' "Fundamentals of Freshman Mathematics". Chapter 12 on Probability is taken from the Second Edition of the authors' "Principles of Mathematics". In addition, we have included new material dealing with applications to business and economics. The subjects covered are simple applications of matrices, consumption matrices, elementary linear programming, supply and demand functions, elasticity of demand, expectation, and 2×2 games.

In writing this book we have been influenced by the recommendations of many groups who have been working toward the reform of the mathematical curriculum and, in particular, by those of the Committee on the Undergraduate Program in Mathematics of the Mathematical Association of America and of the Commission on Mathematics of the College Entrance Examination Board. The book, however, represents our own ideas for a modern course of instruction and is not written to follow any specific outline suggested by such official bodies.

The book contains slightly more material than can be covered in a three-hour semester course (or a five-hour quarter course), and thus the instructor is in a position to select those topics which he feels are most suitable for his students. The following are examples of how the book can be used.

1　*Twelfth-grade Mathematics in High School.* This material is suitable for a fourth semester of Algebra. Chapters 3, 4, and 5 are review material and can be taken very rapidly. The inclusion of a unit on Probability is in keeping with many modern curricular recommendations.

2　*College Algebra in a College or University.* The book is intended to be used in an introductory course for those students who have entered with three years of high school mathematics, or whose preparation is otherwise unsuitable for them to enter a course in Analytic Geometry and Calculus. As such it fits the recommendations of CUPM for Course 0 in the recently proposed General College Curriculum. If the class is weak, Chapters 3, 4, and 5 must be covered, and Chapter 12 and parts of Chapter 11 will necessarily be omitted. If the class is strong, Chapters 3, 4, and 5 can be omitted and the remainder of the book completed.

The attention of prospective users is called to the following special features of the book.

1 Chapter 1 includes an introduction to sets and logic. The set theory is used throughout the rest of the book, but the logic may be omitted or postponed at the option of the instructor.

2 The discussion of mathematical induction in Chapter 2 is unusually careful. Problem material includes many types of situations other than the traditional summation of finite series.

3 Matrix and vector algebra is introduced in Chapter 7. These topics are becoming of great importance in the physical and social sciences and need to be part of the precalculus curriculum. Determinants are also introduced, together with their most important properties.

4 The solution of inequalities in one and two variables is stressed. The treatment is based upon the principle of continuity, which makes the arguments seem almost transparent.

5 Exponential and logarithmic tables are included in the Appendix, so that the book is completely self-contained.

6 The applications to business and economics make the book suitable for a first course in mathematics for students in a College of Business Administration. A great deal of pains has been devoted to the problems in these sections in the hope that they are in the spirit of actual problems in the world and are still simple enough for classroom use. No other text combines the standard material of College Algebra with so many applications of these kinds. We are very indebted to James H. Allendoerfer for his assistance in the collection of the materials for these portions of the book.

7 The problem sets are large and are well graded. The more difficult problems have been marked with asterisks (*). Answers are given for the odd-numbered problems. The even-numbered problems essentially duplicate the odd-numbered ones and can be used for review and alternative assignments. Answers to the even-numbered problems are published in the Instructor's Manual.

8 Color has been used to improve the clarity of the illustrations. Each figure is accompanied by a full legend so that illustrations stand by themselves without reference to the text material.

9   Color has also been used to emphasize important statements in the text.   Colored marginal notes are included to help the student locate essential ideas which might otherwise be missed.

"It is hoped that the book is relatively free of errors, but each author blames the other for those that may be discovered".

CARL B. ALLENDOERFER
CLETUS O. OAKLEY

To the Student

As you approach the study of this book, there are surely many questions about it which may occur to you. You may wonder how it differs from mathematics books which you have previously studied, why we have chosen to include the topics which are given, what you are expected to gain from its study, or even what is in it for you. This foreword is written to answer questions of this type and to help you start your studies in the right frame of mind.

In the first place, we mean for you to read the book. Some students treat books of this kind as a collection of problems and never read the explanations; if you do this, you will miss many of our main objectives. We are trying to teach you the language of mathematics and to explain to you some of its most important ideas. Solving the problems is important, but understanding the ideas is far more essential.

Since mathematics is a technical subject, we shall need to introduce you to a new terminology and to use unfamiliar words such as *set, union, intersection, distributive law, matrix, vector, function, event, probability,* and many others. Each of these involves an important idea and requires a definition. You must read these definitions and make sure that you understand them before you proceed. To help you with this, we frequently follow definitions or other important ideas by Exercise A, B, C, etc. You should answer these as you read; they are generally easy and will help to point up subtle points which you might well have missed.

When you understand the language, you should then wish to use it for something which we hope will be meaningful to you. We shall begin by stating a general type of problem which we wish to solve. Examples of such problems are: Can we solve a certain type of equation? If so, how do we do it? How can we tell what the graph of a certain equation will look like without plotting a large number of points? How do we use mathematics to express relationships between certain physical quantities? How do we define and measure the probability of an event? What do we mean by a hypothesis to be tested? Before proceeding to read the text, sit back and try to answer these questions yourself. If you can answer them without the help of the book, so much the better. Then read the book to see how well you did.

The book is organized into five main topics:

1 In Chaps. 1 and 2 we discuss the foundations of the subject, methods of reasoning, and the meaning of our number system.

2   In Chaps. 3 to 5 we present an algebra review. This may seem familiar to you, but it is more sophisticated than the usual beginning algebra book. We hope that after working through these chapters, you will have a deeper understanding of what you have been doing all these years.

3   Chapters 6 to 8 introduce you to various methods for solving equations and inequalities and to the ideas of vectors and matrices which arose from the problem of solving simultaneous linear equations.

4   In Chap. 9 we begin the study of relationships between two variables. Here we introduce the ideas of functions and relations which permeate the applications of mathematics to the physical and social sciences. You cannot manage to do your job, however, by just knowing about functions in general; you need to know a lot about the collections of functions called "elementary functions" which play the most prominent role in college mathematics. These are the algebraic functions (Chap. 10), the exponential and logarithmic functions (Chap. 11), and the trigonometric functions which you have studied elsewhere. The list of elementary functions is completed by the hyperbolic functions. We cannot explain these until you have studied some calculus.

5   The last topic (Chap. 12) is that of probability. There are many events in daily life whose outcome cannot be predicted with certainty and which can only be approached by an understanding of the odds involved. This theory was invented to serve the needs of professional gamblers, but now it is fundamental in many familiar situations such as public opinion polls, quality control in industrial processes, economic forecasting, genetics, treatment of disease, and military strategy. Our chapter introduces you to the basic ideas of probability theory which you will apply later in your work in engineering, biological, or social science.

We hope that by studying this book you will have acquired a mastery of the language and essential ideas of elementary mathematics. When you are finished with it, you should be prepared to go ahead with the more advanced ideas and techniques of calculus, differential equations, and modern algebra which are the keystones of our modern scientific and engineering developments.

Contents

Appendix   381

Fundamentals
of College Algebra

MATHEMATICAL METHOD

1.1 Introduction

When you became interested in the study of science or engineering, you were undoubtedly told that you would have to learn a good deal of mathematics in order to be successful in your career. The basic ideas and relationships in the physical sciences have been expressed in mathematical terms for a very long time, and in recent years the use of mathematics in the biological and social sciences has increased tremendously. As a potential scientist you are studying this book to learn some of the mathematics that you will need in your own field, but before you can appreciate the mathematics to come you need to know something of the nature of mathematics and its relationship to science. Let us, then, consider various ways in which mathematics is used by practicing scientists.

Substitution in Formulas You are surely acquainted with such formulas as

$A = \pi r^2$ for the area A of a circle of radius r, or
$s = 16t^2$ for the distance s in feet through which a body falls from rest in a time t measured in seconds

For a specific value of r (say 2 in.) you now calculate A, and for a given time interval (say 3 sec) you calculate s. Only the simplest processes of algebra and arithmetic are needed in formulas like these, but not all the formulas of science are so elementary. You have probably met formulas which are more complicated algebraic expressions, but even these do not cover all the needs of science. At the very beginning you will meet other types of functions such as the trigonometric, logarithmic, and exponential functions; and before you have gone very far you must be able to handle the symbolism of the calculus. In all these cases, there is a common idea: you are given an expression which relates one variable to a number of others. For specific values of these last variables you are asked to determine the value of the first variable by following a prescribed set of rules. To do so, you must learn the rules and develop speed and accuracy in your calculations. All of this, however, is basically mechanical and requires no real thought. Indeed, all these calculations can be performed with great ease by modern high-speed calculating machines. Since these machines are expensive and are not particularly efficient when used on simple problems, you must still learn to calculate. When you do so, however, you must remember that you are acting as a machine and not as a human being. The real reasons why a scientist must know mathematics lie, therefore, in other directions.

Definitions of Scientific Terms   When you study any new subject, your first task is to learn the meanings of the special, technical terms which are introduced. This is especially important in science where words are used with very precise meanings. Since much of science is based on measurement and is therefore quantitative in character, these technical definitions are best expressed in terms of mathematics. Although you doubtless have an intuitive idea of the meaning of "the area of a circle", we cannot define this precisely without the use of the calculus. In the same way such familiar notions as "velocity" and "acceleration" remain vague until they are expressed in mathematical terms. Indeed, these basic concepts of physics cannot be defined until we have reached the final chapters of this book. After the terms of science are defined, statements about them must be made; and these generally appear as mathematical equations. Since you will, therefore, have to know mathematics in order to understand what your science teachers are saying, mathematics may well be thought of as the "language of science". If you do not understand the language, you cannot learn science.

Analysis of Scientific Situations   You will often be confronted by a set of scientific statements and asked to draw a conclusion from them. In most such situations, you will need to rely upon your mathematics, and frequently you will have to do some real thinking. In algebra you have already met problems of this type and have called them "word problems" or "story problems". You will remember that they are often more difficult than routine manipulations, but success in solving these kinds of problems is of the utmost importance to a scientist. Let us consider a very familiar example of such a problem.

Illustration 1   Mary is now four years older than John, and in three years her age will be twice his age. What are their ages now?

Solution

Let $\quad x =$ John's age now $\qquad x + 3 =$ John's age in 3 years
$\qquad x + 4 =$ Mary's age now $\qquad x + 7 =$ Mary's age in 3 years

Then $x + 7 = 2(x + 3)$; and finally $x = 1$. Therefore John is now one, and Mary is five.

If we had a large number of such problems to solve, we would have generalized the above example to read as follows.

Illustration 2 Mary is now a years older than John, and in b years her age will be c times his age.   What are their ages now?

Solution

Let $\quad x =$ John's age now $\qquad\quad x + b =$ John's age in b years
$\qquad x + a =$ Mary's age now $\qquad x + a + b =$ Mary's age in b years

Then $$x + a + b = c(x + b)$$

Or, solving, $$x = \frac{a + b - bc}{c - 1}$$

This is now a formula into which we can substitute to solve all age problems of this type, and you may wonder why you were asked to learn how to solve age problems when you might just as well have substituted in the formula.   The answer is that you must learn how to analyze problems and not merely to substitute in formulas.   Age problems like those above are actually only one type of age problems, and the kinds of scientific problems that arise are tremendous in number and diversity.   You must develop your mental powers to be able to deal with these.

You, as a student, may, however, take a different view, and many engineers are especially prone to do so.   You may say: "Aren't the really important problems solved in general terms, so that all that I must do is to substitute in a formula?   My handbook is full of formulas, and I see no reason why I should learn how to obtain them."   The answer to this depends really upon what kind of an engineer you wish to be, and is in three parts:

1   If you are doing a particular job, such as the stress analysis of a bridge, by all means use the handbook formulas and get the answers as rapidly as possible.   If you are a routine engineer (essentially a "slave") this will be the limit of your capabilities, and when you are in difficulty you will have to call for help.

2   If, however, you are a better engineer, you will know the derivations of the formulas and hence will understand the assumptions that were made at the outset.   You will be able to avoid using the formulas in situations for which they were not designed, and your suspension bridges will not collapse in a moderate wind.

3 If, finally, you are a creative engineer, who is in the forefront of progress toward new designs, you will not only be able to understand the derivations of existing formulas but you will also be able to derive new ones to fit new situations.

This book is, in spirit, dedicated to the education of creative scientists and engineers and is intended to provide them with the mathematics which they will need. We shall insist that you learn how to calculate, but we shall also emphasize your understanding of the basic ideas and theoretical developments of the subject.

1.2   Abstract Nature of Mathematics

Although we have seen that mathematics is an essential tool for science, it is an entirely different kind of subject. Science is closely tied to the physical world, but mathematics is completely abstract. Many people shudder at the thought of anything abstract and consequently may have a mental block against mathematics. Actually, there is nothing so terrifying about the abstractness of mathematics once its true nature is understood. In order to assist this understanding, let us describe the essentials of a mathematical structure or theory.

As with any new subject, we begin mathematics by discussing the new, technical terms which we must introduce. Our intuition tells us that each of these should have a definition, but sooner or later we will find that our definitions are going in circles. To take a simple example, we may define:

Point: the common part of two intersecting *lines.*
Line: the figure traced by a *point* which moves along the shortest path between two *points.*

Here we have defined *point* in terms of *line* and *line* in terms of *point,* and so we have shed no real light on the nature of either *point* or *line.*

The situation is somewhat similar to that which we would encounter if we tried to learn a foreign language, say French, by using only an ordinary French dictionary—not a French-English dictionary. We look up a particular French word and find it described in more French words and we find ourselves no further ahead. Without a knowledge of a certain amount of French, a French dictionary is useless. In mathematics we have a similar difficulty.

undefined words

The only way to avoid circular definitions in mathematics, or any other subject, is to take a small number of words as *undefined.* All other mathematical words will be defined in terms of these with the understanding that our definitions may also contain common English words ("is", "and", "the", etc.) which have no special mathematical meanings. It is not easy to decide which words should be left undefined and which should be defined in terms of the undefined words. Many choices can be made, and the final decision is largely based upon considerations of simplicity and elegance.

Illustration 1 Let us suppose that *point, line,* and *between* are undefined. Then we may define:

> *Line segment:* that portion of a line contained between two given points on a line.

The words in this definition other than *point, line,* and *between* are without special meanings and thus may be used freely.

Our use of undefined words is the first phase of our abstraction of mathematics from physical reality. The penciled line on our paper and the chalk line on our blackboard are physical realities, but *line,* the undefined mathematical concept, is something quite apart from them. In geometry we make statements about a *line* (which we shall call *axioms*) which correspond to observed properties of our physical lines, but if you insist on asking: "What is a *line?*" we must give you the somewhat disturbing answer: "We don't know; it isn't defined."

Once we have built up our vocabulary from undefined words and other words defined in terms of them, we are ready to make statements about these new terms. These statements will be ordinary declarative sentences which are so precisely stated that they are either true or false. We will exclude sentences which are ambiguous or which can be called true or false only after qualifications are imposed on them.

The following are acceptable statements:

All triangles are isosceles.
$6 = 2 + 5$
The number 4 is a perfect square.

axioms

Our task, now, is to decide which of our statements are true and which are false. In order to give meaning to this task, we must first establish a frame of reference on which our later reasoning will be based. At the very beginning we must choose a few statements which we will call "true" by assumption; such statements are called "axioms". These axioms are statements about the technical words in our vocabulary and are completely abstract in character. They are not statements about the properties of the physical world. You must have heard that an "axiom is a self-evident truth", but axioms can be any statements at all, evident or not. Since mathematical theories can begin with any set of axioms at all, they are infinite in their variety; some of them are interesting and useful, others merely interesting, and still others only curiosities of little apparent value. The choice of a set of axioms which leads to an interesting and useful theory requires great skill and judgment, but for the most part such sets of axioms are obtained as models of the real world. We look about us, and from what we see we construct an abstract model in which our undefined words correspond to the most important objects that we have identified, and in which our axioms correspond to the basic properties of these objects. The mathematics which you will use as a scientist is entirely based on axioms which were derived in this fashion.

From our set of axioms (which we have assumed to be true) we now proceed to establish the truth or falsehood of other statements which arise. We must agree upon some rules of procedure, which we call the "Laws of Logic", and by means of these rules we seek to determine whether a given statement is true or false. We shall not dwell upon these Laws of Logic here, but if you are interested you can read about them in the References given at the end of the chapter. Except for a few tricky places which we will discuss below, you can rely upon your own good sense and previous experience in logical thinking. Whenever doubts arise, however, you must refer back to the full treatment of these logical principles.

When we have shown that the truth of a given statement follows logically from the assumed truth of our axioms, we call this statement a *theorems* "theorem" and say that "we have proved it". The truth of a theorem, therefore, is relative to a given set of axioms; absolute truth has no meaning when applied to mathematical statements. The main business of a mathematician is the invention of new theorems and the construction of proofs for them. The discovery of a new theorem depends upon deep intuition and intelligent guessing, and the process of making such a discovery is very much like that of creative effort in any field. After our intuition has led us to believe that a certain statement is true, we must still prove it; and this is where our use of logical deduction comes in.

Our abstract mathematical system, then, consists of four parts:

1 Undefined words
2 Defined words
3 Axioms, i.e., statements which are assumed to be true
4 Theorems, i.e., statements which are proved to be true

Since we shall need to have a good understanding of the nature of proof, we will devote the rest of this chapter to a discussion of various problems which you will meet in mathematical proofs.

1.3 Sets

Before we can proceed we must introduce a concept which pervades all mathematics; namely that of a *set*.

We think of a set as a collection of objects: pencils, trees, numbers, points, etc. The individual components of the set are called its *elements*. As an example, consider the set consisting of the four boys named John, Joe, Jerry, Jim. This set has four elements. Sets may be of any size. We may think of the set of all particles of sand on a beach; this has a finite number of elements, but this number is certainly very large. A set, however, may have infinitely many elements. An example of an infinite set is the set of all positive integers: 1, 2, 3, 4, 5, ... Indeed a set may contain no elements, in which case we call it the *empty* set, or the *null* set.

We can describe sets in this way, but *set* is a primitive notion which cannot be defined. Hence we take *set* and *element* to be undefined. The statement: "*p* is an element of a set *P*" is similarly an undefined relationship.

Examples and Notation In the list below we give some typical examples of sets occurring in mathematics and indicate the notations appropriate for these. Note that we regularly use curly brackets { } to represent a set; but there are exceptions to this as we shall see in items 1, 4, and 5 below.

1 $\emptyset$, the empty, or null, set containing no elements. For example, $\emptyset$ is the set of all men living now who were born in 1600 A.D.

2 {3}, the finite set, of which 3 is the only element. Note that this is quite different from the real number 3.

3 {2, 7, 15, 36}, a finite set of four elements.

4 $X = \{x \mid x \text{ is a real number}\}$, the set of all real numbers. This expression should be read: "The set X is the set of numbers x such that x is a real number", the vertical line standing for "such that".

5 $X \times Y = \{(x,y) \mid x \text{ and } y \text{ are real numbers}\}$, the set of all ordered pairs (x,y) of real numbers (see Sec. 2.12). This set is sometimes called the *Cartesian Product* of X and Y.

6 $\{x \mid x \text{ is a positive integer}\}$, the infinite set of all positive integers. We shall often write this as the set $\{1, 2, 3, 4, 5, \ldots\}$.

7 $\{x \mid x \text{ is an even positive integer}\}$, the infinite set of all even positive integers. We shall often write this as the set $\{2, 4, 6, 8, 10, \ldots\}$.

8 $\{R \mid R \text{ voted for Abraham Lincoln for President}\}$, the set of all people who voted for Lincoln.

9 $\{T \mid \text{Triangle } T \text{ is isosceles and lies in a given plane}\}$, the set of all isosceles triangles which lie in a given plane.

10 $\{L \mid L \text{ is a line parallel to line } M\}$, the set of all lines parallel to a given line M.

There are several types of relations between sets which we shall need in the future. One of these is the notion of *identity*.

Definition Two sets are said to be *identical* if and only if every element of each is an element of the other. When A and B are identical, we write $A = B$.

Illustration 1

a The sets $\{1, 7, 10, 12\}$ and $\{12, 7, 1, 10\}$ are identical. Note that order in which the elements are written is not relevant.

b The sets $\{x \mid x \text{ is an even prime number}\}$ and $\{2\}$ are identical. These are just two different ways of defining the same set.

c The sets {Carl, Cletus} and {Dorothy, Louise} are not identical even though each contains two elements.

When A and B are not identical, it is still possible that every element of A is an element of B. In such a case we say that A is a *proper subset* of B.

Definition   A set A is a proper subset of a set B if and only if (1) every element of A is an element of B and (2) A and B are not identical.

subsets

We write this relationship $A \subset B$, read "A is a proper subset of B".

Illustration 2

a   If $A = \{1, 4, 6\}$ and $B = \{1, 4, 6, 15\}$, then $A \subset B$.
b   If $A = \{x \,|\, x$ is an even positive integer$\}$ and $B = \{x \,|\, x$ is a positive integer$\}$, then $A \subset B$.

If we are not certain whether or not A and B are identical, but if we know that every element of A is an element of B, we say that A is a *subset* of B.

Definition   A set A is a subset of B if and only if every element of A is an element of B.

We write this relationship $A \subseteq B$, read "A is a subset of B".

Illustration 3   Let A be the set of rainy days in January of this year and B be the set of all days in January of this year. Then surely $A \subseteq B$. We cannot know the truth of $A \subset B$ unless we specify the place to which we are referring and investigate the records of the weather bureau.

Remarks

1   If we know that $A \subset B$ we can conclude that $A \subseteq B$; but given $A \subseteq B$, we need more information before deciding whether or not $A \subset B$ is true.
2   By convention the empty set $\emptyset$ is a subset of every set, and is a proper subset of every set except itself.

The next notion is that of a 1 to 1 correspondence between two sets. Suppose that in your classroom there are 30 seats and 30 students (all present). When the students all sit down, one to a chair, they set up a correspondence between the set of students and the set of seats. Since there is one seat for each student and one student for each seat, this is called a "1 to 1 correspondence".

Definition   Two sets $A = \{a_1, a_2, \ldots\}$ and $B = \{b_1, b_2, \ldots\}$ are said to be in 1 *to* 1 *correspondence* when there exists a pairing of the a's and the b's

1 to 1 correspondence

such that each a corresponds to one and only one b and each b corresponds to one and only one a.

Illustration 4 Establish a 1 to 1 correspondence between the set of numbers $\{1, 2, \ldots, 26\}$ and the set of letters of the alphabet $\{a, b, \ldots, z\}$.
We make the pairing

$$
\begin{array}{cccc}
1 & 2 & \cdots & 26 \\
a & b & \cdots & z
\end{array}
$$

However, there are many other possible pairings, such as

$$
\begin{array}{ccccc}
2 & 3 & \cdots & 26 & 1 \\
a & b & \cdots & y & z
\end{array}
$$

Illustration 5 Establish a 1 to 1 correspondence between the set $\{1, 2, 3, 4, 5, \ldots\}$ and the set $\{2, 4, 6, 8, 10, \ldots\}$.
Let n represent an element of $\{1, 2, 3, 4, 5, \ldots\}$. Then the pairing $n \leftrightarrow 2n$ gives the required correspondence, examples of which are

$$
\begin{array}{cccccccc}
1 & 2 & \cdots & 50 & \cdots & 100 & \cdots \\
2 & 4 & \cdots & 100 & \cdots & 200 & \cdots
\end{array}
$$

Exercises

A Establish a 1 to 1 correspondence between the sets $\{$John, Joe, Jerry, Jim$\}$ and $\{$Mildred, Marcia, Ruth, Sandra$\}$.

B Establish a 1 to 1 correspondence between the sets $\{1, 2, 3, 4, 5, \ldots\}$ and $\{3, 6, 9, 12, 15, \ldots\}$.

Problems 1.3

1 Which pairs of the following sets are identical?
 a $\{$Robert, Richard, Ralph$\}$ **b** $\{1, 2, 3\}$
 c $\{3, 2, 1\}$ **d** $\{a, b, c\}$
 e $\{$Richard, Ralph, Robert$\}$ **f** $\{b, a, s\}$

2 Which pairs of the following sets are identical?
 a $\{$January, February, March, April$\}$
 b $\{1, 2, 3, 4\}$
 c $\{$algebra, geometry, analysis, probability$\}$
 d $\{$March, April, geometry, 3$\}$
 e $\{3, 2, 1, 4\}$
 f $\{$April, 3, March, geometry$\}$

In Probs. 3 to 8 list all subsets of the given set. Which of these are proper subsets?

3 $\{4, 7\}$ **4** $\{a, b\}$
5 $\{2, 4, 6\}$ **6** $\{a, b, c\}$
7 $\{a, b, c, d\}$ **8** $\{1, 2, 3, 4\}$
*9 Count the number of subsets in each of Probs. 3 to 8 that you have worked. Now guess a general formula for the number of subsets of a given, finite set. Now prove that your guess is correct.
10 Prove that if $A \subseteq B$ and $B \subseteq A$, then A and B are identical.

Problems 11 to 21 refer to the following situation: A four-sided die (a tetrahedron) has its faces marked with the numbers 1, 2, 3, 4, respectively. Two such dice are thrown and the numbers on the resulting bottom faces are recorded by symbols such as (2,1)—meaning that the number on the bottom face of the first die is 2 and that of the second die is 1. We call symbols like (2,1) the *outcomes* of throws of our two dice. What is the set of all possible outcomes of throws of our two dice such that the two numbers:

11 *A*: Are both even   **12** *B*: Are both odd
13 *C*: Are equal   **14** *D*: Have sum equal to 5
15 *E*: Have sum equal to 9   **16** *F*: Have an even sum
17 *G*: Have the first number 1 and the second number odd
18 *H*: Have the first number even and the second number 2
19 *I*: Have sum less than 12
20 *J*: Are unrestricted
21 Write all relationships of the forms $A = B$, $A \subset B$, which are true for the above sets: $A, \ldots, J$.

In Probs. 22 to 30 establish a 1 to 1 correspondence between the given two sets whenever this is possible.

22 The set of negative integers; the set of positive integers
23 $\{2, 4, 6, 8, 10, \ldots\}$; $\{3, 6, 9, 12, 15, \ldots\}$
24 The set of married men; the set of married women
25 The set of Chevrolets in operation; the set of Fords in operation
26 The set of all students in your university; the set of mathematics professors in your university
27 The set of all French words; the set of all English words
28 The set of positions on a baseball team; the set of players of one team as listed in their batting order
29 The set of elective offices in your state government; the set of elected officials
30 The set of all integers; the set of positive integers

1.4 Open Sentences

We said above that a *statement* is a sentence which can meaningly be called true or false. In addition to statements, mathematicians frequently write assertions such as:

1 $x + 1 = 3$
2 $x^2 - 5x + 6 = 0$
3 $x^2 - 4 = (x - 2)(x + 2)$
4 $x^2 = 4$ and $x + 1 = 6$

It is not possible to say whether these are true or false, for we are not told the value of x. In the examples above, the assertion is:

1   True for $x = 2$, false for other values of x
2   True for $x = 2$ or 3, false for other values of x
3   True for all x, false for no x
4   True for no x, false for all x

Such assertions are called *open sentences*.

Definition   An open sentence is an assertion containing a variable x, which becomes a statement when x is given a specified value.

Remarks

1   Open sentences may contain two or more variables, but for simplicity we begin by considering the case of a single variable.
2   This definition is faulty in that it refers to the "variable" x which so far has not been defined.   We shall repair this deficiency in a moment.

Exercise

A   Write several open sentences containing two or more variables.

When we confront the problem of assigning values to x, we first must decide what values of x are permissible.   That is, we must agree upon a set of numbers, geometric figures, people, etc., which will be the subject of our discourse.   This set is called the *universal set* (or replacement set), U.   The understanding is that the symbol x in our open sentence may be replaced by any element of U.   If x appears more than once in a given open sentence, it must be replaced by the same element of U each time that it appears.   When we make such a replacement, we say that "we have given x a value".   These ideas permit us to define a *variable*.

variable   **Definition**   A variable is a symbol in an assertion which may be replaced by any element of a given universal set U.

Variables are commonly, but not exclusively, represented by letters at the end of the alphabet: x, y, z.
In contrast to a variable, there is the notion of a *constant*.

Definition   A constant is a fixed element of a given set.

Frequently this given set is a set of numbers, and a constant is represented by the desired numeral such as 2, ½, π, etc.   In other numerical cases the constant is represented by a letter in the first part of the alphabet such as a, b, c, d.   Here we mean that a stands for a fixed, but arbitrary element of the given set.
When we are given an open sentence with variable x, we can consider the set of those elements of U (i.e., values of x) whose substitution for x

converts the open sentence into a true statement. We call this the *truth set* of the given open sentence.

Definition The truth set of an open sentence is the set of elements of the universal set U whose substitution for x converts the open sentence into a true statement.

truth set

Remarks

1 We shall write truth sets in the form

$$\{x \mid x^2 - 5x + 6 = 0\}$$

where $x^2 - 5x + 6 = 0$ is the given open sentence. Generally the universal set U is evident from the context, but in case of doubt we write expressions such as

$$\{x(\text{integers}) \mid x^2 - 5x + 6 = 0\}$$

which implies that U is the set of integers.

2 For more formal work, we shall use notations, such as p_x, q_x, r_x, etc., to represent open sentences with variable x. The truth set of p_x will be written as

$$P \qquad \text{or} \qquad \{x \mid p_x\}$$

3 If the open sentence is an equation which is true for *all* values of x, we call it an *identity*. In such a case we can write

$$P = U \qquad \text{or} \qquad \{x \mid p_x\} = U$$

or in particular

$$\{x \mid x^2 - 4 = (x - 2)(x + 2)\} = U$$

4 The process commonly called "solving an equation" is the same as "finding the truth set of an open sentence which is an equation".

The steps in many mathematical processes amount to replacing an open sentence by one *equivalent* to it. For example we proceed from

$$2x + 1 = 5$$

to

$$2x = 4$$

to

$$x = 2$$

equivalent open sentences

At each step we have replaced the previous open sentence by a simpler one which has the same truth set. Hence we define equivalent open sentences as follows:

Definition Two open sentences p_x and q_x are equivalent if and only if their truth sets are identical, i.e., if $P = Q$. When p_x and q_x are equivalent, we write $p_x \leftrightarrow q_x$.

Illustration 1

a   The open sentences

$$p_x: \quad x^2 + 5x = -6$$
and
$$q_x: \quad x^2 + 5x + 6 = 0$$

are equivalent since $P = Q = \{-2, -3\}$.

b   The open sentences

$$p_x: \quad x = 2$$
and
$$q_x: \quad x^2 = 4$$

are not equivalent, since $P = \{2\}$ and $Q = \{-2, 2\}$.

Later in this chapter we shall need to consider the *negation* of an open sentence, p_x. This is defined as follows.

Definition   The *negation* of an open sentence p_x is the open sentence "It is false that p_x" or any open sentence equivalent to this. We use the symbol $\sim p_x$ to represent the negation of p_x.

negations

In simple cases the negation can be formed by inserting a *not* in an appropriate place, as suggested in Illustration 2.

Illustration 2   Let U be the set of integers. Then negations are formed as follows:

Open sentence	Negation
$x^2 - 4 = 0$	$x^2 - 4 \neq 0$
x is an even integer	x is not an even integer
	or
	x is an odd integer

Remarks

1   It is evident that the truth set of $\sim p_x$ consists of those elements of U which are not elements of the truth set of p_x. Thus the truth set of $\sim(x^2 = 4)$, where U is the set of integers, is the set of all integers different from 2 and -2.

2   The formation of the negation of more complicated open sentences can be quite tricky. For example, the negation of

$$x^2 = 4 \quad \text{and} \quad x + 1 = 6$$
is
$$x^2 \neq 4 \quad \text{or} \quad x + 1 \neq 6$$

Similarly the negation of

$$x^2 = 9 \quad \text{or} \quad x + 2 = 7$$
is
$$x^2 \neq 9 \quad \text{and} \quad x + 2 \neq 7$$

1.5   Set Operations

The example just given of the way of finding the truth set of $\sim p_x$ suggests a more general situation. Let us suppose that we have decided on a universal set U, and have before us a subset of U, namely A. Then we can form the set consisting of those elements of U that are not in A. We call this the *complement* of A (relative to U) and write it as A'.

complement

Definition   Given a universal set U and a subset $A \subseteq U$, then the *complement* of A (written A') is the set of elements of U which are not elements of A (Fig. 1.1).

A' is the complement of A relative to U.

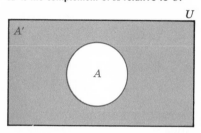

Figure 1.1
The complement of A relative to U consists of those elements of U which are not in A.

Illustration 1

a   Let
    $\qquad U = \{1, 2, 3, 4, 5, 6, 7, 8, 9, 10\}$
    and
    $\qquad A = \{1, 3, 5, 7, 9\}$
    Then
    $\qquad A' = \{2, 4, 6, 8, 10\}$

b   Let U be the set of positive integers and A be the set of even positive integers. Then A' is the set of odd positive integers.

c   If $A = U$, then $A' = \emptyset$.

d   If $A = \emptyset$, then $A' = U$.

e $\{x \mid \sim p_x\} = \{x \mid p_x\}'$

Exercise

A   Show that $(A')' = A$.

If we are given two sets, A and B, there are two important ways of operating with these to produce new sets. The first is called the *union* of A and B.

Definition   The union of A and B, written $A \cup B$, is the set of elements which belong to either A or B or to both A and B (Fig. 1.2).

union

Illustration 2

a   Let
    $\qquad A = \{1, 3, 5\} \qquad \text{and} \qquad B = \{2, 4, 6\}$
    Then
    $\qquad A \cup B = \{1, 2, 3, 4, 5, 6\}$

Union of A and B

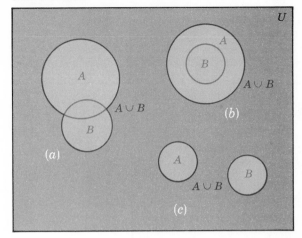

Figure 1.2
The union of A and B consists
of those elements which are
in A or in B or in both A and B.

b Let $\qquad A = \{1, 2, 6\} \qquad$ and $\qquad B = \{1, 2, 3\}$
Then $\qquad\qquad A \cup B = \{1, 2, 3, 6\}$

c Let A be the set of even positive integers and B be the set of odd positive integers. Then $A \cup B$ is the set of all positive integers.

d Let $\qquad A = \{x \mid x^2 = 9\} = \{-3, 3\}$
$\qquad\qquad B = \{x \mid (x^2 - 4x + 3 = 0)\} = \{1, 3\}$
Then $\quad A \cup B = \{x \mid (x^2 = 9) \text{ or } (x^2 - 4x + 3 = 0)\} = \{-3, 1, 3\}$

When "or" is used to combine two open sentences as above, we interpret it to mean "either or both" in the same sense as the legal phrase "and/or".

e If P and Q are the truth sets of p_x and q_x, respectively, then $P \cup Q = \{x \mid p_x \text{ or } q_x\}$.

The second operation of this kind is called *intersection*.

Definition   The intersection of A and B, written $A \cap B$, is the set of elements
intersection   that belong to both A and B (Fig. 1.3).

Illustration 3

a Let $\qquad A = \{1, 3, 5\} \qquad$ and $\qquad B = \{2, 4, 6\}$
Then $\qquad\qquad A \cap B = \emptyset$

b Let $\qquad A = \{1, 2, 6\} \qquad$ and $\qquad B = \{1, 2, 3\}$
Then $\qquad\qquad A \cap B = \{1, 2\}$

c Let $\qquad A = \{x \mid x^2 = 9\} = \{-3, 3\}$
$\qquad\qquad B = \{x \mid x^2 - 4x + 3 = 0\} = \{1, 3\}$
Then $\quad A \cap B = \{x \mid (x^2 = 9) \text{ and } (x^2 - 4x + 3 = 0)\} = \{3\}$

d If P and Q are the truth sets of p_x and q_x, respectively,

$$P \cap Q = \{x \mid p_x \text{ and } q_x\}$$

Intersection of A and B

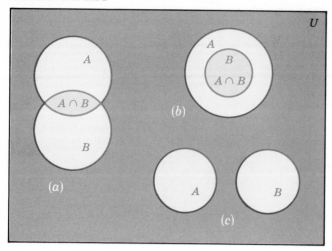

Figure 1.3   The intersection of A and B consists of those elements which are in both A and B. In (c), $A \cap B = \emptyset$.

That is, $P \cap Q$ is the set of those elements in U (values of x) for which *both p_x and q_x are true.*

Problems 1.5

In Probs. 1 to 6 let $U = \{1, 2, 3, 4, 5, 6, 7, 8, 9, 10\}$; $A = \{1, 2, 5, 7, 9\}$; $B = \{2, 3, 4, 6, 8\}$; and $C = \{5, 6, 9, 10\}$. Find simpler expressions for the following sets.

1 $A \cap B$ **2** $A \cup B$
3 $(A \cup B)'$ **4** $(A \cap B)'$
5 $(A \cup B) \cap C$ **6** $(A \cap B) \cup (C')$

In Probs. 7 to 18 the sets $A, \ldots, J$ are those defined in Probs. 11 to 20 in Sec. 1.3. Find simpler expressions for the following sets.

7 $A \cap C$ **8** $B \cap C$
9 $D \cup E$ **10** $G \cup H$
11 $A \cap A$ **12** $B \cap G$
13 F', assuming $U = I$ **14** J', assuming $U = I$
15 $(A \cup G) \cap C$ **16** $(B \cup H) \cap C$
17   Verify that $A \cap (B \cup C) = (A \cap B) \cup (A \cap C)$.
18   Verify that $A \cup (B \cap C) = (A \cup B) \cap (A \cup C)$.
19   For any sets A and B where $A \subset B$, find $A \cap B$ and $A \cup B$.
20   For any set A, find $A \cap A'$ and $A \cup A'$.

In Probs. 21 to 30 one of the following relations is true: $A \subset B$, $A = B$, $B \subset A$. Write the correct relation in each case. The universal set in Probs. 21 to 28 is the set of all integers.

	A	B
21	$\{x \mid 2x + 4 = x - 5\}$	$\{x \mid 3x + 5 = 2x - 4\}$
22	$\{x \mid x^2 + 4 = -2x + 2\}$	$\{x \mid x^2 + 2x = -2\}$
23	$\{x \mid (x + 1)(x - 2) = 0\}$	$\{x \mid x + 1 = 0\}$
24	$\{x \mid x - 2 = 0\}$	$\{x \mid (x - 2)x = 0\}$
25	$\{x \mid (x - 2)(x - 3) = 0\}$	$\{x \mid x - 2 = 0\} \cup \{x \mid x - 3 = 0\}$
26	$\{x \mid x - 2 = 0\}$	$\{x \mid x$ is a positive even integer$\}$ $\cap \{x \mid x$ is a prime$\}$
27	$\{x \mid x + 2 = 4\}$	$\{x \mid (x + 2)^2 = 16\}$
28	$\{x \mid x^2 = 25\}$	$\{x \mid x + 1 = 6\}$
29	$\{T(\text{triangles in a plane}) \mid T$ is equilateral$\}$	$\{T(\text{triangles in a plane}) \mid T$ is isosceles$\}$
30	$\{L(\text{lines in 3-space}) \mid L$ does not intersect given line $M\}$	$\{L(\text{lines in 3-space}) \mid L$ is parallel to line $M\}$

31 Which of the following pairs of open sentences are equivalent? The universal set is the set of integers.

- **a** $3x - 5 = x + 7$ $x - 3 = 3$
- **b** $x^2 = 4$ $x = 2$
- **c** $x^2 = 9$ $x = 3$ or $x = -3$
- **d** $(x - 1)(x - 5) = 0$ $x^2 - 6x + 5 = 0$
- **e** $(x - 2)(x + 2) = x^2 - 4$ $(x + 2)(x + 2) = x^2 + 4$

32 Which of the following pairs of open sentences are equivalent? The universal set U is the set of integers.

- **a** $5x + 1 = 5x - 1$ $2x = 1$
- **b** $(x + 1)(2x - 1) = 0$ $x + 1 = 0$
- **c** $(x - 4)(x + 4) = x^2 - 16$ x is an integer.
- **d** $x = 0$ x is greater than -1 and x is less than $+1$.
- **e** $4x + 3 = 0$ x is not an integer.

In Probs. 33 to 38 write the negation of the given open sentence, and find the truth set of this negation, assuming that $U = \{1, 2, 3, 4, 5\}$.

33 $2x = 4$ **34** $x + 4 = 7$
35 $(x - 3)(x - 4) = 0$ **36** $(x - 1)(x - 5) = 0$
37 $x^2 = 9$ and $x + 2 = 4$ **38** $x^2 = 16$ or $x - 1 = 4$

1.6 Implications

The great bulk of mathematical theorems are assertions of the form: "If ..., then ..." where the dots are to be filled with either statements or open sentences. Such assertions are called *implications;* a few typical examples are:

If x is an odd integer, then x^2 is an odd integer.
If $x^2 = 1$, then $(x + 1)(x - 1) = 0$.
If triangle x is equilateral, then x is equiangular.

In these examples, as in almost all interesting cases, the dots above are replaced by open sentences, and the implication can be written in the standard form

$$\text{If } p_x, \text{ then } q_x.$$

or more briefly

$$p_x \rightarrow q_x$$

where p_x and q_x are open sentences.

There is, moreover, a subtlety here which is often ignored. In the example above:

If x is an odd integer, then x^2 is an odd integer.

we really mean:

For *all* integers x: If x is an odd integer, then x^2 is an odd integer.

In other words, our implications are *general* statements which are to be true for *all* values of the variable involved. In order to be complete, we should write our implications in the form

$$\forall_x(p_x \rightarrow q_x)$$

which is read

$$\text{For all } x: \text{ If } p_x, \text{ then } q_x.$$

Notice that this is no longer an open sentence, but is a statement which is true or false. We are at liberty to define the circumstances under which it is true and do so as follows.

Definition   The statement $\forall_x(p_x \rightarrow q_x)$ is *true* if and only if $P \subseteq Q$, where P and Q are the truth sets of p_x and q_x, respectively.

true
implications
This definition may seem arbitrary, but it is motivated by intuition obtained from special cases such as those given below. Observe that we can make any definition we please, and that no proof is required. Nevertheless, if our mathematics is to be useful in our daily lives, our definitions should not be in conflict with our intuition.

Illustration 1   It seems reasonable to agree to the truth of

For all x: If $x = 2$, then $x^2 = 4$.

Let p_x be $x = 2$; q_x be $x^2 = 4$. Then $P = \{2\}$ and $Q = \{-2, 2\}$, and $P \subseteq Q$. In this case, therefore, the above definition agrees with common sense.

Illustration 2   It does not seem reasonable to assert the truth of

For all x: If $x^2 = 4$, then $(x - 2)(x - 3) = 0$.

In particular, when $x = -2$ this becomes

If $4 = 4$, then $(-4)(-5) = 0$.

which is certainly false. Let us examine the truth sets. Let p_x be $x^2 = 4$ and q_x be $(x - 2)(x - 3)$. Then $P = \{-2, 2\}$ and $Q = \{2, 3\}$. In this case the statement $P \subseteq Q$ is false; and again our intuition gives the same result as the above definition.

This situation can be illustrated by the following diagrams. Let the interior of the rectangle represent the universal set U. Let P be represented by the points in the interior of the circle labeled P (see Fig. 1.4), and let Q be represented by the points in the interior of the circle labeled Q. Of the many relative positions in which circles P and Q may lie, there are just two that correspond to the statement $P \subseteq Q$. These are drawn in Fig. 1.5. This helps us formulate a test that will tell us when $\forall_x(p_x \rightarrow q_x)$ is true.

Graphical representation of sets

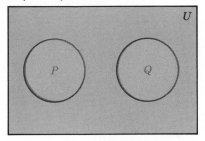

Possible positions of P and Q when $P \subseteq Q$

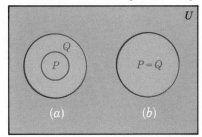

Figure 1.4 Sets are represented by the interiors of circles.

Figure 1.5 If $P \subseteq Q$, then P must either be inside Q or be identical with Q.

Rule

1 If q_x is true for each x for which p_x is true, then $\forall_x(p_x \rightarrow q_x)$ is true.

2 If there is at least one value of x for which p_x is true and q_x is false, then $\forall_x(p_x \rightarrow q_x)$ is false.

Let us apply this to the implications in Illustrations 1 and 2.

Illustration 3

a For all x: If $x = 2$, then $x^2 = 4$. There is only one value for which $x = 2$ is true, namely 2. When we substitute this for x in $x^2 = 4$, we obtain $4 = 4$, which is true. Hence the implication is true (Fig. 1.6).

b For all x: If $x^2 = 4$, then $(x - 2)(x - 3) = 0$. There are two values of x for which $x^2 = 4$ is true, namely 2 and -2. If we substitute 2 for x in $(x - 2)(x - 3) = 0$, we obtain $(2 - 2)(2 - 3) = 0(-1) = 0$, which is true. If, however, we substitute -2 for x in $(x - 2)(x - 3) = 0$, we obtain $(-2 - 2)(-2 - 3) = (-4)(-5) = 0$, which is false. Hence the implication is false (Fig. 1.7).

$P = \{x \mid x = 2\}$; $Q = \{x \mid x^2 = 4\}$; $P \subseteq Q$

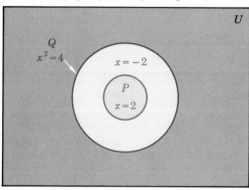

Figure 1.6
Since the set where $x = 2$ is included in the set where $x^2 = 4$, the implication "For all x: If $x = 2$, then $x^2 = 4$" is true.

In the implication $p_x \to q_x$ it is entirely possible that there is no value of x for which p_x is true; i.e., $P = \emptyset$. Since $\emptyset$ is a subset of any set, it follows that $P \subseteq Q$ and that $p_x \to q_x$ is true. (See Illustration 4.)

$P = \{x \mid x^2 = 4\}$; $Q = \{x \mid (x - 2)(x - 3) = 0\}$; $P \not\subseteq Q$

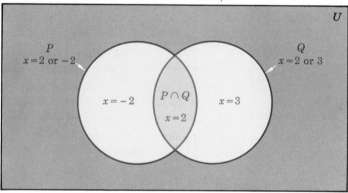

Figure 1.7 Since the set where $x^2 = 4$ is not contained in the set where $x = 2$ or 3, the implication "For all x: If $x^2 = 4$, then $(x - 2)(x - 3) = 0$" is false.

Illustration 4 Consider the following implication:

For all odd integers x: If $x = 4$, then $x^2 = 6$.

Since no odd integer equals 4, $P = \emptyset$, and the implication is true.

1.7   Derived Implications

Associated with the implication $\forall_x(p_x \to q_x)$ there are three other implications with which it is often confused:

converse   *Converse:* $\forall_x(q_x \to p_x)$

inverse   *Inverse:* $\forall_x(\sim p_x \to \sim q_x)$

contrapositive   *Contrapositive:* $\forall_x(\sim q_x \to \sim p_x)$

In terms of the truth sets, P and Q, of p_x and q_x, these become:

> *Given implication:* $P \subseteq Q$
> *Converse:* $Q \subseteq P$
> *Inverse:* $P' \subseteq Q'$
> *Contrapositive:* $Q' \subseteq P'$

We must now examine how the truth of each of these is related to the truth of each of the others. The case when $P = Q$ is trivial, for in this case all four statements are true.

Theorem 1 Let $\forall_x(p_x \to q_x)$ be a given implication where p_x is equivalent to q_x (i.e., $P = Q$). Then the given implication, its converse, its inverse, and its contrapositive are all true.

Let us now examine the case where $P \neq Q$. First let us assume that $P \subset Q$ is true and see what can be said about the truth of $Q \subset P, P' \subset Q'$, and $Q' \subset P'$. Our assumption tells us that Fig. 1.8*a* represents the true

$P \subseteq Q$ **and its derived implications**

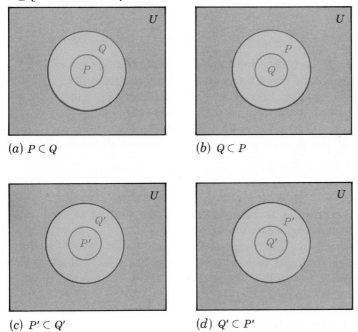

(a) $P \subset Q$ (b) $Q \subset P$

(c) $P' \subset Q'$ (d) $Q' \subset P'$

Figure 1.8 These diagrams, where $P \neq Q$, help us to discover the relative truth or falsehood of an implication, its converse, its inverse, and its contrapositive.

situation. Then clearly $Q \subset P$ is false. Moreover, P' (the outside of P) contains Q' (the outside of Q); so $P' \subset Q'$ is false and $Q' \subset P'$ is true.

By similar reasoning with Fig. 1.8*b*, *c*, and *d* we can draw the following conclusions.

Comparative truth of an implication and its derived implications when $P \neq Q$

Given implication	Converse	Inverse	Contrapositive
Assumed true	Proved false	Proved false	Proved true
Proved false	*Assumed true*	Proved true	Proved false
Proved false	Proved true	*Assumed true*	Proved false
Proved true	Proved false	Proved false	*Assumed true*

We summarize the most important conclusions from this table in Theorem 2.

Theorem 2 Let $\forall_x(p_x \rightarrow q_x)$ be a given implication where p_x is not equivalent to q_x (i.e., $P \neq Q$). Then

implication
equivalent
to its con-
trapositive

1 If either the given implication or its contrapositive is true, both are true.
2 If either the converse or the inverse is true, both are true.
3 The given implication and its converse are not both true.
4 The given implication and its inverse are not both true.

Exercises

A What can be said about the converse and the contrapositive?
B What can be said about the inverse and the contrapositive?
C By examining Fig. 1.8*b*, *c*, and *d*, complete the argument which establishes the last three lines of the above table.
D Consider the situation where $P \subseteq Q$ and $Q \subseteq P$ are both false. Draw an appropriate figure, and show that in this case all four implications are false.

Remarks

1 This theorem warns us against confusing the truth of an implication with that of its converse or inverse.
2 It also tells us that the contrapositive is just another way of stating a given implication. If we have trouble proving the truth of an implication, we may shift to the contrapositive. If the contrapositive is true, so is the implication.

Illustration 1 The contrapositive of

For all x: If x is odd, then x^2 is odd.

is

For all x: If x^2 is not odd, then x is not odd.

or better

For all x: If x^2 is even, then x is even.

Illustration 2   The contrapositive of

> For all pairs of lines L and M: If L is parallel to M, then L and M are coplanar.

is

> For all pairs of lines L and M: If L and M are not coplanar, then L is not parallel to M.

Exercise

E   Write the converse and inverse of the implications of Illustrations 1 and 2.   Are these true or false?

The idea of a contrapositive can be extended to more complicated implications such as

$$\forall_x[(p_x \text{ and } q_x) \rightarrow r_x]$$

The contrapositive is $\forall_x[(\sim r_x) \rightarrow \sim(p_x \text{ and } q_x)]$, which can be rewritten

(1) $$\forall_x[(\sim r_x) \rightarrow (\sim p_x \text{ or } \sim q_x)]$$

There are also two partial contrapositives, each of which is true if and only if the given implication is true, namely:

(2) $$\forall_x[(\sim r_x \text{ and } q_x) \rightarrow (\sim p_x)]$$
(3) $$\forall_x[(p_x \text{ and } \sim r_x) \rightarrow (\sim q_x)]$$

Illustration 3   The three contrapositives of

> For all x: If x is even and x is a prime, then $x = 2$.

are

1   For all x: If $x \neq 2$, then x is odd or x is not a prime.
2   For all x: If $x \neq 2$ and x is a prime, then x is odd.
3   For all x: If x is even and $x \neq 2$, then x is not a prime.

1.8   Alternative Expressions for Implications

Mathematics frequently expresses implications in language different from that used above, and consequently you must learn to recognize implications even when they are disguised in a fashion which may seem confusing at first.   In the expressions that follow we have omitted the symbol $\forall_x$ and the corresponding phrase "For all x"; for hereafter these will be understood even though they do not appear explicitly.   This omission is the usual practice in mathematics.

Definition The following six statements all carry the same meaning.

$$p_x \to q_x$$
$$P \subseteq Q$$

sufficient p_x is sufficient for q_x.

necessary q_x is necessary for p_x.

if If p_x, then q_x.

only if Only if q_x, then p_x.

The last four statements can be understood by referring to Fig. 1.8a. From this we see that:

For x to be in Q, it is sufficient that it be in P.
For x to be in P, it is necessary that it be in Q.
If x is in P, then x is in Q.
Only if x is in Q is it in P.

Illustration 1 The implication, "If a polygon is a square, then it is a rectangle", can be rewritten in the following ways:

The fact that a polygon is a square is a sufficient condition that it is a rectangle.
The fact that a polygon is a rectangle is a necessary condition that it is a square.
A polygon is a square only if it is a rectangle.

The first two of these may be rephrased:

A sufficient condition that a polygon is a rectangle is that it is a square.
A necessary condition that a polygon is a square is that it is a rectangle.

Remark From the definition above it follows that the converse of an implication can be obtained in any of the following ways, depending on how the given implication is phrased:

Interchange p_x and q_x.
Interchange P and Q.
Replace "necessary" by "sufficient".
Replace "sufficient" by "necessary".
Replace "if" by "only if".
Replace "only if" by "if".

In a similar fashion, the equivalence $p_x \leftrightarrow q_x$ can be expressed in a number of alternative ways.

Definition The following six statements all carry the same meaning:

$$p_x \leftrightarrow q_x$$
$$P = Q$$

p_x is necessary and sufficient for q_x.

q_x is necessary and sufficient for p_x.
p_x if and only if q_x.
q_x if and only if p_x.

Illustration 2   The equivalence, "A triangle is equilateral if and only if it is equiangular", can be rewritten in the following ways:

A triangle is equiangular if and only if it is equilateral.
A necessary and sufficient condition that a triangle is equilateral is that it is equiangular.
A necessary and sufficient condition that a triangle is equiangular is that it is equilateral.

Problems 1.8

In Probs. 1 to 6 determine the truth or falsehood of the given implication by finding P and Q and checking the truth of $P \subseteq Q$.   The universal set is the set of integers.

1   For all x: If $x^2 = 9$, then $x = 3$.
2   For all x: If $(x + 1)(x - 3) = 0$, then $x = 3$.
3   For all x: If $3x - 2 = x + 4$; then $3x = 9$.
4   For all x: If $x + 5 = 7$; then $x^2 = 4$.
5   For all x: If $4x^2 = 9$; then $x = 5$.
6   For all x: If $2x + 1 = 4$; then $x = \frac{3}{2}$.

In Probs. 7 to 12 state the converse, inverse, and contrapositive of the given implication.

7   For all integers x: If x is divisible by 3, then $2x$ is divisible by 6.
8   For all integers x: If $x \neq 0$, then x^2 is greater than zero.
9   For all triangles T: If T is equilateral, then T is isosceles.
10   For all quadrilaterals Q: If Q is a parallelogram, then the diagonals of Q bisect each other.
11   For all pairs of lines L and M: If L and M are parallel, then L and M do not intersect.
12   For all pairs of triangles T and S: If T is congruent to S, then T is similar to S.
13   Write the contrapositive of the inverse of $\forall_x(p_x \to q_x)$.
14   Write the converse of the contrapositive of $\forall_x(p_x \to q_x)$.
15   Write the inverse of the converse of $\forall_x(p_x \to q_x)$.
16   Write the contrapositive of the inverse of $\forall_x(p_x \to q_x)$.

In Probs. 17 to 20 write the contrapositive and the two (or more) partial contrapositives of the given implication.

17   For all integers x: If x is positive and $x^2 = 9$, then $x = 3$.
18   For all integers x: If x is not divisible by 3 and x is not of the form $3n + 1$ (where n is an integer), then x is of the form $3m + 2$ (where m is an integer).

19 For all triples of lines L, M, and N:
If L and M do not intersect, and
$\qquad$ L and N do not intersect, and
$\qquad$ L and M are coplanar, and
$\qquad$ L and N are coplanar, and
$\qquad$ M and N do intersect,
then M and N are identical.

20 For all pairs of triangles ABC and $A'B'C'$:
If AB is congruent to $A'B'$, and
$\qquad$ AC is congruent to $A'C'$, and
$\qquad$ angle C is congruent to angle C',
then ABC is congruent to $A'B'C'$.

In Probs. 21 to 24 consider the implication $\forall_x[(p_x \text{ and } q_x) \to r_x]$, and let P, Q, R be the truth sets of p_x, q_x, and r_x, respectively. By definition this implication is true if and only if $(P \cap Q) \subseteq R$.

21 Draw a circle diagram illustrating $(P \cap Q) \subset R$, where $P \neq Q$ and $(P \cap Q) \neq R$.

22 Draw a diagram illustrating the partial contrapositive $(R' \cap Q) \subset P'$, and show that this is equivalent to that drawn for Prob. 21. Assume $R' \neq Q$ and $(R' \cap Q) \neq P$.

23 Draw a diagram illustrating the partial contrapositive $(P \cap R') \subset Q'$, and show that this is equivalent to that drawn for Prob. 21. Assume $P \neq R'$ and $(P \cap R') \neq Q$.

24 Draw a circle diagram illustrating the contrapositive $R' \subset (P' \cup Q')$, and show that this is equivalent to that drawn for Prob. 21. Assume $P \neq Q$ and $R' \neq (P' \cup Q')$.

In Probs. 25 to 30 write the given implication using the "sufficient condition" language.

25 If the base angles of a triangle are equal, the triangle is isosceles.

26 If two triangles are congruent, their corresponding altitudes are equal.

27 If two lines are perpendicular to the same line, they are parallel.

28 If two spherical triangles have their corresponding angles equal, they are congruent.

29 If $3x + 2 = x + 4$, then $x = 1$.

30 If $x^2 = 0$, then $x = 0$.

In Probs. 31 to 36 write the given implication, using the "necessary condition" language.

31 If a triangle is inscribed in a semicircle, then it is a right triangle.

32 If $x = 3$, then $x^2 = 9$.

33 If a body is in static equilibrium, the vector sum of all forces acting on it is zero.

34 If a body is in static equilibrium, the vector sum of the moments of all forces acting on it is zero.

35 If two forces are in equilibrium, they are equal, opposite, and collinear.

36 If three nonparallel forces are in equilibrium, their lines of action are concurrent.

In Probs. 37 to 42 write the given implication using the phrase "only if".

37 The implication of Prob. 31   **38** The implication of Prob. 32
39 The implication of Prob. 33   **40** The implication of Prob. 34
41 The implication of Prob. 35   **42** The implication of Prob. 36

In Probs. 43 to 48 write the converse of the given implication using "necessary" and then "sufficient" language. Give two answers to each problem.

43 The implication of Prob. 25   **44** The implication of Prob. 26
45 The implication of Prob. 27   **46** The implication of Prob. 28
47 The implication of Prob. 29   **48** The implication of Prob. 30

In Probs. 49 to 54 write the converse of the given implication using the phrase "only if".

49 The implication of Prob. 31   **50** The implication of Prob. 32
51 The implication of Prob. 33   **52** The implication of Prob. 34
53 The implication of Prob. 35   **54** The implication of Prob. 36

In Probs. 55 to 58 write the given equivalence in "necessary and sufficient" language.

55 Two lines are parallel if and only if they are equidistant.

56 An integer is even if and only if it is divisible by 2.

57 Three concurrent forces are in equilibrium if and only if their vector sum is zero.

58 A lever is balanced if and only if the algebraic sum of all moments about its fulcrum is zero.

59 A man promised his girl: "I will marry you only if I get a job." He got the job and refused to marry her. She sued for breach of promise. Can she logically win her suit? Why?

1.9 Direct Proof

We recall that the procedure for proving the truth of $p_x \rightarrow q_x$ [remember this means $\forall_x(p_x \rightarrow q_x)$] was to show that q_x is true for each x for which p_x is true, or that $P \subseteq Q$. In the examples of Sec. 1.6 this was easy, for there were only finitely many values of x for which p_x was true. Usually, however, there are infinitely many such values of x, and so we must turn to more sophisticated methods.

In the method known as direct proof, we are supposed to know the truth of a number of implications, say:

$$p_x \to q_x \qquad q_x \to r_x \quad \text{and} \quad r_x \to s_x$$

Then we wish to conclude the truth of

$$p_x \to s_x$$

Three given implications: $p_x \to q_x$; $q_x \to r_x$; $r_x \to s_x$

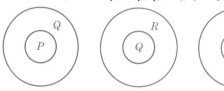

Figure 1.9
An implication is represented by a pair of concentric circles.

The argument is immediate, for we are, in fact, given the diagrams shown in Fig. 1.9. From these we derive Fig. 1.10. Thus P is contained in S, and $p_x \to s_x$ is true.

The chain of implications leading to $p_x \to s_x$

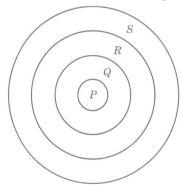

Figure 1.10
A chain of implications is represented by a collection of concentric circles.

There are endless variations of this pattern, and great skill is required in arranging the work so that the proof is valid. One common trick, for instance, is to replace an implication by its contrapositive if this will simplify matters.

Illustration 1 Given: $(p_x \to \sim q_x)$
                              $(r_x \to q_x)$
Prove: $(p_x \to \sim r_x)$

Proof Since $(r_x \to q_x)$ is equivalent to $(\sim q_x \to \sim r_x)$ we have: $(p_x \to \sim q_x)$; $(\sim q_x \to \sim r_x)$. Hence we conclude that $(p_x \to \sim r_x)$.

Illustration 2 As a specific example, consider the following proof of

If x is odd, then x^2 is odd. [x is an integer.]

This amounts to the following:

Given:

    a If x is odd, then $x = 2a + 1$ where a is an integer.

    b If $x = 2a + 1$ where a is an integer, then x is odd.

    c The rules of algebra.

Prove: If x is odd, then x^2 is odd.

Proof

1 If x is odd, then $x = 2a + 1$. [Given]

2 If $x = 2a + 1$, then $x^2 = 4a^2 + 4a + 1$. [Algebra]

3 $4a^2 + 4a + 1 = 2(2a^2 + 2a) + 1 = 2b + 1$, where b is an integer. [Algebra]

4 If $x = 2a + 1$, then $x^2 = 2b + 1$. [Substitution]

5 If $x^2 = 2b + 1$, then x^2 is odd. [Given]

6 If x is odd, then x^2 is odd. [From 1, 4, and 5]

1.10   Indirect Proof

If you have difficulty in constructing a direct proof, you can sometimes make progress by using other tactics. The method of indirect proof is based upon the fact that $p_x \rightarrow q_x$ [remember that this means $\forall_x(p_x \rightarrow q_x)$] must be either true or false. If we can show that it is not false, then it must be true. We proceed by assuming that it is false, combining this assumption with other known facts, and (if we are successful) arriving at a contradiction. Since contradictions are impossible in correct thinking, we must have made a mistake somewhere. Our only dubious statement was the assumption that $p_x \rightarrow q_x$ is false. Hence this must be in error, and $p_x \rightarrow q_x$ must be true.

In practice, then, we must form the statement "$p_x \rightarrow q_x$ is false". It is at this point that most errors occur, for students forget what we said in Sec. 1.6. Our discussion there amounts to the following:

The implication $p_x \rightarrow q_x$ is false if and only if there exists at least one value of x for which p_x is true *and* q_x is false (i.e., $\sim q_x$ is true). We proceed by making precisely this assumption.

Illustration 1 Suppose that we are trying to construct indirect proofs of the following:

    a If x^2 is even, then x is even. [x is an integer.]

    b If x is not a perfect square, then x cannot be written in the form $x = a^2/b^2$, where a and b are integers. [x is an integer.]

    c If x^2 is greater than zero, then $x \neq 0$. [x is an integer.]

To start the indirect proof, we assume that for at least one value of x:

assume
$p_x \rightarrow q_x$
is false

    a x^2 is even *and* x is odd.

    b x is not a perfect square, *and* x can be written in the form $x = a^2/b^2$, where a and b are integers.

    c x^2 is greater than zero, *and* $x = 0$.

Illustration 2   In other situations the variable x may represent something other than a number.   For example, consider:

a   If a *pair* of lines are cut by a transversal so that a pair of alternate interior angles are equal, then the lines are parallel.   [x represents a *pair* of lines.]

b   If *triangle ABC* has $a \neq b$, then triangle ABC has angle $A \neq$ angle B.   [x represents triangle ABC.]

Standard notation for a triangle

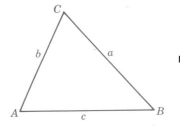

Figure 1.11

To start the indirect proof, we assume that:

a   There exists at least one pair of lines, such that the lines are cut by a transversal so that alternate interior angles are equal, *and* the lines are not parallel.

b   There exists at least one triangle ABC such that $a \neq b$ *and* angle $A =$ angle B.

Indirect proof also can be applied to statements which are not implications.   To begin the indirect proof, we assume the negation of the given statement.

Illustration 3   Consider constructing indirect proofs of the following:

a   There are infinitely many prime numbers.

b   The integer 2 cannot be written in the form $2 = a^2/b^2$, where a and b are integers.

To start the indirect proof, we assume that:

a   There are only finitely many prime numbers.

b   The integer 2 can be written in the form $2 = a^2/b^2$, where a and b are integers.

find a contradiction   The remainder of an indirect proof consists in establishing a contradiction based upon this assumption and other known facts.   There is no standard way of doing this, but the following illustrations should give the idea.

Illustration 4   Construct an indirect proof of

If x^2 is even, then x is even.   [x is an integer.]

We assume that there is at least one integer x for which x^2 is even *and x* is odd.

From Sec. 1.9, Illustration 2, we know that if x is odd, then x^2 is odd. Hence it is impossible to have x odd and x^2 even. This is the desired contradiction.

Illustration 5   Construct an indirect proof of the following: If a pair of lines are cut by a transversal so that a pair of alternate interior angles are equal, then the lines are parallel.

We assume that there is at least one pair of lines such that the lines are cut by a transversal so that alternate interior angles are equal, *and* the lines are not parallel. That is: assume that $\angle 1 = \angle 2$ and that AB and CD

Assumption at beginning of indirect proof

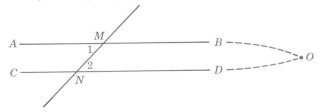

Figure 1.12   The figure assumes that angle 1 = angle 2 and that lines AB and CD meet at O.

meet at O. Now in triangle MON, $\angle 1$ is an exterior angle, and hence it must be greater than the interior angle $\angle 2$. This contradicts the assumption that $\angle 1 = \angle 2$.

1.11   Other Methods of Proof

Use of the Contrapositive   When we are trying to prove the truth of an implication $p_x \rightarrow q_x$, we can just as well prove the contrapositive, $\sim q_x \rightarrow \sim p_x$. For we have seen that an implication and its contrapositive are equivalent. Sometimes the contrapositive is easier to prove, and then we should choose this method. Often there are great similarities between indirect proof and the proof of the contrapositive.

Illustration 1   Let us consider the theorem of Sec. 1.10, Illustration 5. The contrapositive of the stated implication is: If two lines are not parallel, the alternate interior angles obtained by cutting these lines by a transversal are not equal.

We establish this by the precise argument used in Illustration 5 above.

Proof by Enumeration   Suppose we wish to prove that for *all x, p_x* is true, under the assumption that the variable x refers to a *finite* universal set U. We can then give a complete proof by checking the truth of p_x for each x that is an element of U.

Illustration 2   Let U be the set $\{0, 1\}$, and let "addition", $\oplus$, be defined by the following table.

$\oplus$	0	1
0	0	1
1	1	0

Prove that for all x in U, $x \oplus 1 = 1 \oplus x$.   To do so we let x take the values 0 and 1 in turn.

When $x = 0$, we have

$$0 \oplus 1 = 1 \qquad 1 \oplus 0 = 1$$

Hence
$$0 \oplus 1 = 1 \oplus 0$$

When $x = 1$, both sides of $x + 1 = 1 + x$ become $1 + 1$; so they are equal.   Therefore, for all x in U,

$$x + 1 = 1 + x$$

Remark   This method applies only when U is finite.   If U is infinite, we could never finish the job of checking each value of x; and this method fails.

Proof of Existence   Before you spend a lot of time and money (on a high-speed computer, say) trying to solve a problem, it is a good idea of determine in advance that the problem actually does have a solution.   You have probably never seen problems that do not have solutions, for most textbooks and teachers consider it to be bad form to ask students to do something which is impossible.   In actual practice, however, such problems may arise and it is a good idea to know how to recognize them.   A very simple example of such a problem is the following:

Find all the integers x which satisfy the equation

$$7x + 5 = 2x + 9$$

In order to reassure you that you are working on problems that do have solutions, mathematicians have developed a number of "existence theorems".   These are statements of the following form:

There exists a number x which has a given property.

An important example of such a theorem is this one:

If a and b are any real numbers such that $a \neq 0$, there exists a real number x which satisfies the equation $ax + b = 0$.

The best way of proving such a theorem is to exhibit a number x with the required property.   The proof of the above theorem amounts to checking that $x = -(b/a)$ satisfies the given equation.

Although there are other forms of existence proofs, a constructive proof of this kind is considered to be of greater merit, and this method is used widely in establishing the existence of solutions of various types of equations.

1.12   Methods of Disproof

If you have tried unsuccessfully to prove a conjectured theorem, you may well spend some time trying to disprove it.   There are two standard methods for disproving such statements.

Disproof by Contradiction   In this case we assume that the given statement is true and then proceed to derive consequences from it.   If we succeed in arriving at a consequence which contradicts a known theorem, we have shown that the given statement is false.

Illustration 1   Disprove the statement: "The square of every odd number is even."

Of course, this immediately contradicts our previous result (Sec. 1.9, Illustration 2) that the square of every odd number is odd.   But let us disprove it from first principles.   Since every odd number can be written in the form $2a + 1$, where a is an integer, and since every even number can be written in the form $2b$, where b is an integer, the given statement implies that

$$(2a + 1)^2 = 2b \qquad \text{for some } a \text{ and } b$$

or
$$4a^2 + 4a + 1 = 2b$$

Both sides are supposed to represent the same integer, but the left-hand side *is not* divisible by 2, while the right-hand side *is* divisible by 2.   This is surely a contradiction, and so the given statement is false.

Disproof by Counterexample   This method is effective in disproving statements of the form

For *all* values of x, a certain statement involving x is true.

An example is the following:

For all values of x, $x^2 + 16 = (x + 4)(x - 4)$.

counter-
example

In order to disprove such an assertion, we proceed to find a "counterexample".   In other words, we look for *one* value of x for which the statement is false; and since the statement was supposed to be true for *all* values of x, this single counterexample is the end of the matter.   In the above example, $x = 0$ does the job.

Illustration 2   Disprove the statement: "The square of every odd number is even."

All that we have to do is to find a single odd number whose square is odd. Since $3^2 = 9$, we have established the disproof.

Problems 1.12

In Probs. 1 to 10 construct an indirect proof of the given implication. You may use any theorems of geometry which you learned in high school.

1 If triangle ABC (see Fig. 1.11) has $a \neq b$, then triangle ABC has $\angle A \neq \angle B$.

2 If triangle ABC (see Fig. 1.11) has $\angle A \neq \angle B$, then triangle ABC has $a \neq b$.

3 If quadrilateral $ABCD$ has unequal diagonals, then quadrilateral $ABCD$ is not a rectangle.

4 If point P is not on line L, then there is no more than one line through P perpendicular to L.

5 If point P is on line L, then in a plane through L there is no more than one line through P perpendicular to L.

6 In triangles ABC and $A'B'C'$; if $b = b'$, $c = c'$, $\angle A \neq \angle A'$, then $a \neq a'$.

7 Two distinct circles can have at most two distinct points in common.

8 Two distinct planes can have at most a line in common.

9 Define a tangent to a circle at point P as the line perpendicular to the radius OP at P. Prove that P is the only point which is on both the circle and the tangent.

10 Define a tangent to a circle as a line which has one and only one point in common with the circle, namely point P. Prove that this tangent at P is perpendicular to the radius OP.

In Probs. 11 to 14 find the error in the given direct "proofs".

11 Given: **1** If $x = 5$, then $x^2 = 25$.
        2 If $x^2 \neq 25$, then $x \neq -5$.
Prove: **3** If $x = 5$, then $x = -5$.
Proof: **4** If $x = 5$, then $x^2 = 25$.   [Given]
        5 If $x^2 = 25$, then $x = -5$.   [Contrapositive of (2)]
        6 If $x = 5$, then $x = -5$.   [From (4) and (5)]

12 Given: **1** If $x = 3$, then $x^2 - 5x + 6 = 0$.
        2 If $x = 4$, then $x^2 - 5x + 6 \neq 0$.
Prove: **3** If $x \neq 4$, then $x = 3$.
Proof: **4** If $x \neq 4$, then $x^2 - 5x + 6 = 0$.   [Contrapositive of (2)]
        5 If $x^2 - 5x + 6 = 0$, then $x = 3$.   [From (1), given]
        6 If $x \neq 4$, then $x = 3$.   [From (4) and (5)]

13 Given: **1** If two triangles have all three pairs of corresponding angles equal, then they are similar.
        2 If two triangles are congruent, then they are similar.
Prove: **3** If two triangles have all three pairs of corresponding angles equal, then they are congruent.
Proof: **4** Immediate from (1) and (2)

14 Given: **1** If integer x is even, then x^2 is even.
Prove: **2** If integer x is odd, then x^2 is odd.
Proof: **3** (2) is the contrapositive of (1).

In Probs. 15 to 22 disprove the given statement.

15 The sum of two even integers is odd.
16 The product of two odd integers is a perfect square.
17 $(x + 4)^2 = x^2 + 16$
18 $(x + 2)^3 = x^3 + 8$
19 Two triangles are congruent if two sides and the angle opposite one of these sides in one triangle are equal, respectively, to the corresponding parts of the other triangle.
20 If two triangles are similar, they have the same area.
21 The sum of the exterior angles of any triangle is equal to $180°$.

Angles 1, 2, and 3 are exterior angles.

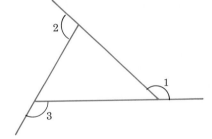

22 Any two medians of a triangle bisect each other.
23 Give an indirect proof of the theorem: "There exist an infinite number of primes." If you are unable to do so, consult Courant and Robbins, "What Is Mathematics?" page 22. This theorem is due to Euclid.

1.13 Mathematical Models

By this time you should have begun to understand what we mean by saying that mathematics is abstract. Mathematical proof is a process of reasoning by given rules from a set of axioms (which are assumed to be true) to a valid conclusion, which we call a "theorem". Because of the abstract character of mathematics, we cannot expect to prove anything about our physical world by purely mathematical means.

Scientists, however, spend their lives uncovering the secrets of nature, and engineers put these discoveries to work for the benefit of our society. You may quite properly wonder how an abstract subject like mathematics has become such an important tool for scientists and engineers. The key to this matter is the concept of a "mathematical model" of nature. The first step in the study of any branch of science is that of observing nature. When enough facts have been collected, the scientist begins to organize them into some pattern. In quantitative sciences like astronomy, chemistry, and physics this pattern is expressed in terms of mathematics. The undefined terms of the abstract mathematics (points, lines, etc.) represent physical objects; refined abstract concepts (velocity, acceleration, force, etc.) are then defined to correspond to intuitive ideas which seem important to the scientist. Then mathematical equations involving these concepts are used as axioms to describe the observed behavior of nature. All of these, taken together, constitute our mathematical model. This model, of course, is only a picture of nature; it differs from nature just as a model of an aircraft differs from the real plane itself. But just as a great deal can be learned about a plane from a model which is studied in a wind tunnel, we can use our mathematical model to help us understand nature. From our axioms, we can deduce theorems, which are true only in our abstract sense. Nevertheless, if our model is well constructed, these theorems will correspond to observable properties of nature which we may well not have suspected in advance. At the very worst, these theorems are intelligent guesses about nature and serve as guides for our experimental work. At their best, when the model is a good one as is the case in most physical sciences, our mathematical results can almost be identified with physical truth. In those portions of science which you are likely to be studying along with this book, this correspondence is so close that you may not realize the difference between mathematics and nature itself. It is our hope that the study of this chapter will have helped you to appreciate this important distinction.

References

Carroll, Lewis: "Logical Nonsense", Putnam, New York (1934).

Courant, Richard, and Herbert Robbins: "What Is Mathematics?" Oxford, New York (1941).

Stabler, E. R.: "An Introduction to Mathematical Thought", Addison-Wesley, Reading, Mass. (1953).

Stoll, Robert R.: "Sets, Logic, and Axiomatic Theories", Freeman, San Francisco (1961).

Suppes, Patrick: "Introduction to Logic", Van Nostrand, Princeton, N.J. (1957).

Tarski, Alfred: "Introduction to Logic", Oxford, New York (1946).

THE NUMBER SYSTEM

2.1   Introduction

Since numbers are basic ideas in mathematics, we shall devote this chapter to a discussion of the most important properties of our number system. We do not give a complete account of this subject, and you are likely to study it in more detail when you take more advanced courses in mathematics. Numerous suggestions for further reading are given at the end of the chapter.

Let us retrace briefly the development of numbers as it is usually presented in schools. As a young child you first learned to count, and thus became acquainted with the *natural numbers* 1, 2, 3, ... In your early study of arithmetic you learned how to add, subtract, multiply, and divide pairs of natural numbers. Although some divisions such as $6 \div 3 = 2$ were possible, it soon developed that new numbers had to be invented so as to give meaning to expressions like $7 \div 2$ and $3 \div 5$. To handle such situations, fractions were introduced, and the arithmetic of fractions was developed.

It should be noted that the invention of fractions was a major step in the development of mathematics. In the early days many strange practices were followed. The Babylonians considered only fractions whose denominators were 60, the Romans only those whose denominators were 12. The Egyptians insisted that the numerators must be 1, and wrote ⅓ + ¹⁄₁₅ instead of ⅖. Our modern notation dates from Leonardo of Pisa (also called Fibonacci), whose great work *Liber Abaci* was published in A.D. 1202.

Later on you became acquainted with zero and negative numbers such as -7, -3, $-\frac{5}{3}$, $-4\frac{1}{5}$, etc., and you learned how to calculate with these. The entire collection consisting of the positive and negative integers, zero, and the positive and negative fractions is called the system of *rational numbers*. The advantage of using this system in contrast to the system of purely positive numbers is that it is possible to subtract any rational number from any rational number. With only positive numbers available, $3 - 5$, for instance, is meaningless. It is interesting to note that it took many years before negative numbers were permanently established in mathematics. Although they were used to some extent by the early Chinese, Indians, and Arabs, it was not until the beginning of the seventeenth century that mathematicians accepted negative numbers on an even footing with positive numbers.

When you were introduced to *irrational* numbers such as $\sqrt{2}$ and π, you were told that these could not be expressed as ordinary fractions. Instead, they are written as infinite decimal expansions such as $1.4142\ldots$ and $3.1415\ldots$. The decimal expansions of the rational numbers are also infinite; for example,

$$\frac{1}{4} = 0.25000\cdots$$
$$\frac{1}{3} = 0.33333\cdots$$
$$2 = 2.00000\cdots$$
$$\frac{1}{7} = 0.142857142857\cdots$$

These, however, repeat after a certain point, whereas the irrationals do not have this property. The collection of all these, the rationals plus the irrationals, is called the system of *real* numbers. It is quite difficult to give a completely satisfactory definition of a real number, but for our present purposes the following will suffice.

Definition   A *real number* is a number which can be represented by an infinite decimal expansion.

real
number

If you wish a more subtle definition of a real number, read Courant and Robbins, "What Is Mathematics?" chap. 2.

Although a real number is a definite mathematical object, we can express such a number in a great variety of notations. For example, we can write 7 in the following ways:

$$\text{VII, } 111_{\text{two}} \text{ (base two), } 2\frac{1}{3}, 7.000\ldots, 9 - 2$$

The rational number usually written $\frac{1}{2}$ can also be written

$$\frac{2}{4}, \frac{8\pi}{16\pi}, 0.5000\cdots, \frac{\frac{1}{8}}{\frac{1}{4}}, \frac{1}{4} + \frac{1}{4}, \left(\frac{1}{\sqrt{2}}\right)^2, \sqrt{\frac{1}{4}}$$

For each real number it is customary to adopt a "simplest" expression which is commonly used to represent it (7 and $\frac{1}{2}$ in the examples above), but we shall not hesitate to use other representations when they are more convenient.

Since ½ and ¾ are merely names for the same number, we call them equal and write

$$\frac{1}{2} = \frac{2}{4}$$

More generally, we define equality for real numbers as follows.

Definition Two symbols, a and b, representing real numbers are equal if *equality* and only if they represent the same real number.

From this definition, Theorem 1 follows at once.

Theorem 1 If $a, b,$ and c represent real numbers and if $a = b$; then $a + c = b + c, a - c = b - c, ac = bc,$ and $a/c = b/c$ (provided $c \neq 0$).

For, since a and b represent the same real number, $a + c$ and $b + c$ represent the same real number. Hence $a + c = b + c$. The other cases follow the same argument.

Exercises

A Show that if $a = b$, then $b = a$.
B Show that if $a = b$ and $b = c$, then $a = c$.

2.2   Addition of Real Numbers

Addition is defined for *pairs* of real numbers such as $2 + 3 = 5$, $-3 + 2\frac{1}{2} = -\frac{1}{2}$, etc. Indeed, the sum of every pair of real numbers is defined as a third real number. We give this property the name "closure" and write the following law.

Closure Law of Addition The sum $a + b$ of any real numbers is a unique real number c.

This property of closure may seem so simple that you wonder why we mention it at all. Let us consider some situations where closure is not true.

Illustration 1

a The sum of two *odd* numbers is *not* an odd number.
b The sum of two *irrational* numbers is *not necessarily* irrational, for $(2 + \sqrt{3}) + (4 - \sqrt{3}) = 6$; i.e., the irrational numbers are not closed under addition.
c The sum of two *prime* numbers is *not necessarily* a prime, for $7 + 11 = 18$; i.e., the prime numbers are not closed under addition.

You are very familiar with the fact that the order of addition is not important. For instance, $2 + 4 = 4 + 2$, $-3 + \pi = \pi + (-3)$, etc. To describe this property, we say that addition is "commutative" and write the following law.

Commutative Law of Addition $a + b = b + a$

It is slightly more difficult to add three numbers such as $2 + 4 + 7$, for addition is defined for *pairs* of real numbers and not for *triples*. Normally, we first add $2 + 4 = 6$, and then add $6 + 7 = 13$. But we could just as well have added $4 + 7 = 11$ and then $2 + 11 = 13$. That is, $(2 + 4) + 7 = 2 + (4 + 7)$. To describe this property, we say that addition is "associative" and write the following law.

Associative Law of Addition $(a + b) + c = a + (b + c)$

Actually, the sum $a + b + c$ of three real numbers needs to be defined, for originally we knew only how to add two numbers, $a + b$. Therefore we make the following definition.

Definition $a + b + c$ is defined to the sum $(a + b) + c$.

We now prove a theorem which illustrates the fact that the sum of three real numbers is the same regardless of the order in which the addition is performed.

Theorem 2 $a + b + c = c + b + a$

Proof
$$\begin{aligned} a + b + c &= (a + b) + c && \text{[Definition]} \\ &= (b + a) + c && \text{[Commutative Law]} \\ &= c + (b + a) && \text{[Commutative Law]} \\ &= (c + b) + a && \text{[Associative Law]} \\ &= c + b + a && \text{[Definition]} \end{aligned}$$

In a similar fashion we can define the sum of four real numbers.

Definition $a + b + c + d$ is defined to be the sum

$$(a + b + c) + d$$

The number zero plays a special role in addition; the sum of zero and any real number a is a itself:

$$a + 0 = 0 + a = a$$

Since this leaves a identically as it was before the addition, we make the following definition:

Definition The real number *zero* is called the *identity element* in the addition of real numbers.

zero

This statement is equivalent to the statement: "For any real number a, $a + 0 = 0 + a = a$."

The set of real numbers is the union of three subsets, namely:

The set of positive real numbers
The set of negative real numbers
The set with the single element, *zero*, which is neither positive nor negative

If p is a positive real number and $-p$ is its negative, then $p + (-p) = 0$ and $(-p) + p = 0$. This enables us to conclude that, for every real number a, there is another real number ^-a, called its *additive inverse*, such that $a + {}^-a = 0$ and $^-a + a = 0$. For when a is positive so that $a = p$ (where p is positive), then $^-a = -p$, and when a is negative so that $a = -p$, then $^-a = p$. Finally, when $a = 0$, then $^-a = 0$. Putting all these together, we have the following definition.

Definition The *additive inverse* of a real number a is the real number ^-a having the property that

additive
inverse

$$a + {}^-a = {}^-a + a = 0$$

The use of the term *inverse* may be motivated as follows: We start at 0 and add a, thus obtaining a. We now wish to retrace our steps and return to 0; hence we must add ^-a to a. The operation of adding ^-a undoes the operation of adding a and thus is said to be the *inverse* operation.

Exercise

A Show that $^-(^-a) = a$.

We must further define the *difference* of two real numbers. Of course, this is familiar when a and b are both positive and $a > b$. Other cases, however, must be treated, and we include these in the definition below. We introduce the symbol $a - b$ to denote subtraction and define it as follows.

Definition Let a and b be real numbers. Then, by definition,

$$a - b = a + {}^-b$$

In other words, in order to subtract b from a, add ^-b to a.

You will notice that the symbol "$-$" is used in two distinct ways: (1) to represent a negative number such as -2, and (2) to represent the operation of subtraction as in $a - b$. It is further customary to give it a third meaning: (3) to represent the additive inverse ^-a by $-a$. In order for these three uses of the symbol "$-$" to be consistent, we must assume the conventions:

For all real numbers a and b:

$$-(-a) = a \quad \text{and} \quad a - b = a + (-b)$$

With this understanding we shall drop the notation ^-a and use minus signs freely without specifying the sense in which they are employed.

We shall have frequent occasion to refer to the absolute value of a real number a. This is written $|a|$ and is defined as follows.

Definition The *absolute value* of a real number a, $|a|$, is the real number

absolute value

such that:

a If a is positive or zero, then $|a| = a$.
b If a is negative, then $|a| = -a$.

Illustration 2 $\quad |5| = 5 \quad\quad |-6| = 6 \quad\quad |0| = 0$

2.3 Multiplication of Real Numbers

Now that the essential laws of addition are before us, the laws of multiplication are easy to learn; they are almost the same, with "product" written in the place of "sum".

Closure Law of Multiplication
The product $a \times b$ of any two real numbers is a unique real number c.

Commutative Law of Multiplication
$a \times b = b \times a$

Associative Law of Multiplication
$(a \times b) \times c = a \times (b \times c)$

We now ask: "What is the identity element for multiplication?" It should be the number b such that, for any a, $a \times b = a$. In other words, multiplication by b leaves a unchanged, just as in addition the addition of 0 to a leaves a unchanged. Clearly, the correct choice for the identity element is 1.

Definition The real number 1 is called the *identity element* in the multi-

multiplicative identity

plication of real numbers.

This statement is equivalent to the statement: "For any real number a, $a \times 1 = 1 \times a = a$."

If a is a real number different from zero, we know that

$$a \times (1/a) = (1/a) \times a = 1$$

This enables us to conclude that for every real number a ($\neq 0$), there is another real number a', called its *multiplicative inverse*, such that

$$a \times a' = a' \times a = 1$$

Definition The *multiplicative inverse* or a real number a ($\neq 0$) is the real number a' having the property that

multiplicative
inverse

$$a \times a' = a' \times a = 1$$

Exercise

A Show that $(a')' = a$ if $a \neq 0$.

Next let us define *division*. Just as the difference of a and b is defined to be the sum of a and the additive inverse of b, the quotient of a by b is defined to be the product of a and the multiplicative inverse of b.

Definition Let a and b be real numbers, and let $b \neq 0$. Then the *quotient* of a by b $\left(\text{written } a/b \text{ or } \dfrac{a}{b}\right)$ is defined to be

$$a/b = a \times b'$$

Note that division by zero is not defined.

Up to this point we have distinguished the quotient $1/a = 1 \times a'$ from the multiplicative inverse a', but there is no longer any reason for doing so. Hereafter we shall write $1/a$ for the multiplicative inverse of a with the understanding that the slant ($/$) is used in the two senses: (1) to represent the multiplicative inverse $1/a$, and (2) to represent the quotient a/b. Since these uses are compatible, this will cause no confusion.

Exercises

B Show that $b \times (a/b) = a$, where $b \neq 0$.
C Show that if $a/b = c$, then $a = b \times c$, where $b \neq 0$.

Remark The multiplicative and divisibility properties of zero are among the more troublesome parts of the study of the real numbers. In Sec. 2.4, Theorem 3, we shall show that for any real number a, $a \times 0 = 0$; but let us assume this result here. There are three situations to consider.

1 Let $a \neq 0$, then $0/a = 0$. For $0/a = 0 \times (1/a) = 0$ by the above definition and Theorem 3.
2 Let $a \neq 0$, then $a/0$ is *meaningless*. Suppose that we were to write $a/0 = c$, and that the result of Exercise C above still holds true. Then $a = c \times 0 = 0$, which cannot be the case since $a \neq 0$.

3 The symbol $0/0$ is *indeterminate*. Suppose that $0/0 = c$, or that $0 = c \times 0$. Since any real number c satisfies this equation, the value of c is indeterminate. Do not confuse $0/0$ with a/a $(a \neq 0)$, which is equal to 1.

In summary, we note that *zero may never appear in the denominator of a fraction:*

NEVER DIVIDE BY ZERO.

2.4   The Distributive Law

There is one final law; this connects multiplication and addition. You are used to writing $4(2 + 3) = (4 \times 2) + (4 \times 3)$; $2(x + y) = 2x + 2y$; etc. Or probably you did the reverse in factoring when you wrote $3x + 6y = 3(x + 2y)$. These are illustrations of the following law.

Distributive Law $a \times (b + c) = (a \times b) + (a \times c)$

This law is the basis for many familiar operations. For example, the usual way of multiplying 15×23 is

$$
\begin{array}{r}
15 \\
23 \\
\hline
45 \\
30 \\
\hline
345
\end{array}
$$

But this really amounts to the statement that

$$
\begin{aligned}
15 \times 23 &= 15 \times (20 + 3) \\
&= (15 \times 20) + (15 \times 3) \\
&= 300 + 45 \\
&= 345
\end{aligned}
$$

As a more complicated example, consider the following illustration.

Illustration 1   Show that $(a + b)(c + d) = ac + bc + ad + bd$.

$$
\begin{aligned}
(a + b)(c + d) &= (a + b)c + (a + b)d && \text{[Distributive Law]} \\
&= c(a + b) + d(a + b) && \text{[Commutative Law]} \\
&= (ca + cb) + (da + db) && \text{[Distributive Law]} \\
&= ca + cb + da + db && \text{[Property of Addition]} \\
&= ac + bc + ad + bd && \text{[Commutative Law]}
\end{aligned}
$$

The distributive law has a number of important consequences. The first of these states the multiplicative property of zero.

Theorem 3   Let a be any real number; then $a \times 0 = 0$.

Proof

$a \times 0 = 0$ **1** $\quad 0 = 0 + 0$ [Definition, Sec. 2.2]
2 $\quad a \times 0 = a \times (0 + 0)$ [Theorem 1]
3 $\quad a \times 0 = (a \times 0) + (a \times 0)$ [Distributive Law]
4 $\quad a \times 0 = a \times 0$ [Identity]

Subtracting (4) from (3), we obtain:

5 $\quad 0 = a \times 0$

From this theorem we conclude the following useful result.

Theorem 4 If a and b are two real numbers such that $ab = 0$, then $a = 0$, or $b = 0$.

Proof If $a = 0$, the theorem is immediately verified.
If $a \neq 0$, then $1/a$ is defined. Hence we may write

$$(1/a)(ab) = (1/a)(0)$$

Using the associative law for multiplication and Theorem 3, we find that $b = 0$, which proves the theorem.

This theorem has very many applications, especially in the solution of equations.

Illustration 2 Solve $x^2 - 5x + 6 = 0$.
By factoring we find that $(x - 2)(x - 3) = 0$.
From Theorem 4 we see that either

$$x - 2 = 0 \quad \text{and} \quad x = 2$$
or $\quad x - 3 = 0 \quad \text{and} \quad x = 3$

Hence 2 and 3 are solutions of the given equation.

THE PRODUCT OF TWO REAL NUMBERS IS ZERO IF AND ONLY IF AT LEAST ONE OF THE TWO FACTORS IS ZERO.

A second consequence of the distributive law is the set of rules for multiplying signed numbers. These are easily derived from the following theorem.

Theorem 5 For any real number a, $(-1) \times a = -a$.

Proof

1 $\quad 1 + (-1) = 0$ [Definition of additive inverse]
2 $\quad [1 \times a] + [(-1) \times a] = 0 \times a$ [Distributive law]
3 $\quad 0 \times a = 0$ [Theorem 3]
4 $\quad 1 \times a = a$ [Definition of 1]
5 $\quad a + [(-1) \times a] = 0$ [(2), (3), and (4)]

6 $a + (-a) = 0$ [Definition of additive inverse]
7 $a + (-1) \times a = a + (-a)$ [(5) and (6)]
8 $(-1) \times a = -a$ [Subtraction of a from both sides]

Corollary $(-1) \times (-1) = 1$

Put $a = -1$ in Theorem 5 and apply the convention that $-(-a) = a$. Now we can prove the usual rules as follows.

Theorem 6   Let p and q be any positive real numbers.   Then:

$p \times (-q)$
$= -(pq)$

a $p \times (-q) = -(pq)$
b $(-p) \times (-q) = pq$

$(-p)$
$\times (-q)$
$= pq$

Proof   Write $(-p) = (-1) \times p$; $(-q) = (-1) \times q$; and

$$-(pq) = (-1) \times p \times q$$

Then the identities of the theorem follow from the associative and commutative properties of multiplication and the corollary to Theorem 5.

2.5   Formal Properties of Real Numbers

In summary of Secs. 2.2 to 2.4, we state the following properties of the arithmetic of real numbers.   The letters a, b, c stand for arbitrary real numbers.

Addition

R1 $a + b$ is a unique real number   [Closure Law]
R2 $(a + b) + c = a + (b + c)$ [Associative Law]
R3 $a + 0 = 0 + a = a$ [Identity Law]
R4 $a + (-a) = (-a) + a = 0$ [Inverse Law]
R5 $a + b = b + a$ [Commutative Law]

Multiplication

R6 $a \times b$ is a unique real number   [Closure Law]
R7 $(a \times b) \times c = a \times (b \times c)$ [Associative Law]
R8 $a \times 1 = 1 \times a = a$ [Identity Law]
R9 $a \times \dfrac{1}{a} = \dfrac{1}{a} \times a = 1$ for $a \neq 0$ [Inverse Law]
R10 $a \times b = b \times a$ [Commutative Law]

Distributive Law

R11 $a \times (b + c) = (a \times b) + (a \times c)$

These 11 laws form the foundation of the entire subjects of arithmetic and ordinary algebra.   They should be carefully memorized.   In more

field

advanced mathematics they are taken to be the axioms of an abstract system called a *field*. Hence we may say that the real numbers form a field.

Problems 2.5

Addition

In Probs. 1 to 4 use the commutative and associative laws to establish the truth of the given statement. Model your proofs on that given for Theorem 2.

1 $4 + 2 + 7 = 2 + 7 + 4$ **2** $1 + 5 + 9 = 9 + 5 + 1$
3 $a + b + c = c + a + b$ **4** $a + b + c = b + c + a$
5 Define $a + b + c + d + e$.
6 Assuming that $a + b + c + d + e$ has been defined (Prob. 5), define $a + b + c + d + e + f$.
7 Find the additive inverse of each of the following: $4, -2, \pi, 0, -\sqrt{5}$.
8 Find the additive inverse of each of the following: $-6, 2, \frac{3}{4}, -0, 3$.
9 Find the absolute value of each of the following: $7, -3, 0, \frac{2}{3}, -\frac{1}{2}$.
10 Find the absolute value of each of the following: $-8, -0, \pi^2, \frac{3}{5}, -5$.

In Probs. 11 to 16 evaluate the given expression.

11 $[4 + (2 - 5)] - [6 - (5 - 2)]$
12 $[3 - (-2 + 4)] - [12 + (8 - 4)]$
13 $[7 + (1 - 8)] + [3 - (6 - 2)]$
14 $[(-12 + 4) + 7] + [-7 + (4 - 8)]$
15 $\{[(-2 + 4) - (12 + 6)] - [-15 + 8]\} - 24$
16 $\{[(5 - 9) - (6 - 12)] - [8 + 5]\} + 14$

Multiplication

17 Define $a \times b \times c$.
18 Assuming that $a \times b \times c$ is defined (Prob. 17), define $a \times b \times c \times d$.

In Probs. 19 to 22 use the commutative and associative laws to establish the truth of the given statement.

19 $4 \times 5 \times 7 = 7 \times 5 \times 4$ **20** $6 \times 2 \times 3 = 6 \times 3 \times 2$
21 $a \times b \times c = b \times a \times c$ **22** $a \times b \times c = c \times a \times b$
23 Find the multiplicative inverse of each of the following: $\frac{1}{3}, -4, -\frac{3}{4}, 1, 0$.
24 Find the multiplicative inverse of each of the following: $2, -\pi, \frac{1}{2}, \sqrt{5}, \frac{7}{3}$.

In Probs. 25 to 30 evaluate the given expression.

25 $(-2)[3(4 - 2) + 6] + 5[-2(-3 + 8) + 9]$
26 $5[-9(5 - 2) + 4] - 4[3(4 + 8) - 45]$
27 $3[8(-2 + 4) - 5(7 - 9)] + 5[(4 - 8)6 - (9 - 4)3]$

28 $-6[3(7 + 6) - 4(3 - 8)] - 4[(7 - 3)9 - 16]$
29 $4\{[-6(3 - 7) + 5(6 + 3)] - 16(2 - 3)\} - 11$
30 $-3\{[7(9 - 4) - 6(3 + 7)] + 5(-4 + 8)\} + 9$

Subtraction and Division

31   Does the commutative law hold for the subtraction of real numbers?
32   Does the commutative law hold for the division of real numbers?
33   Does the associative law hold for the subtraction of real numbers?
34   Does the associative law hold for the division of real numbers?
35   Is there an identity element for subtraction?   If so, what is it?
36   Is there an identity element for division?   If so, what is it?

Zero

37   What meaning is to be attached to each of the following?

$$\frac{4}{0} \qquad \frac{0}{4} \qquad \frac{4}{4} \qquad \frac{0}{\frac{1}{4}} \qquad \frac{0}{0}$$

38   What meaning is to be attached to each of the following?

$$\frac{2}{0} \qquad \frac{0}{2} \qquad \frac{2}{2} \qquad \frac{0}{\frac{1}{2}} \qquad \frac{0}{0}$$

39   For what real values of x are the following expressions meaningless?

$$\frac{2x - 1}{x + 1} \qquad \frac{5}{x} \qquad \frac{x + 4}{x - 2} \qquad \frac{0}{x^2 + 5} \qquad \frac{4}{x^2 - 5x + 4}$$

40   For what real values of x are the following expressions meaningless?

$$\frac{x}{3} \qquad \frac{3x + 2}{x - 1} \qquad \frac{x - 3}{x^2 + 8} \qquad \frac{3}{x} \qquad \frac{2}{x^2 + 4x + 4}$$

41   For what real values of x are the following expressions indeterminate?

$$\frac{x}{2x^2} \qquad \frac{x + 2}{2x + 4} \qquad \frac{1 + x^2}{3 + x^2} \qquad \frac{0}{x} \qquad \frac{x - 3}{x^2 - 9}$$

42   For what real values of x are the following expressions indeterminate?

$$\frac{2x + 6}{x + 3} \qquad \frac{0}{x^2} \qquad \frac{x^4}{2x^3} \qquad \frac{3 + x^2}{1 + x^4} \qquad \frac{2x + 5}{4x^2 - 25}$$

In Probs. 43 to 46 factor and solve for x.

43 $x^2 + 5x + 6 = 0$ **44** $x^2 - 8x + 12 = 0$
45 $x^2 - 4 = 1$ **46** $x^2 - 16 = 18$

Proofs

In Probs. 47 to 52 prove or disprove the given statement.   You may use R1 to R11 as given axioms.

47 $(a + b) \times c = (a \times c) + (b \times c)$
48 $a + (b \times c) = (a + b) \times (a + c)$
49 $a \div (b + c) = (a \div b) + (a \div c)$
50 $(a - b) + b = a$
51 If $a \neq 0$, $ax + b = 0$ has a unique solution.
52 To any real number a there corresponds a real number x such that $0x = a$.
53 Let "addiplication" be defined (with symbol $\odot$) as follows:

$$a \odot b = (a + b) + (a \times b)$$

Under addiplication are the real numbers closed? Is addiplication commutative; associative? Is there an identity; an addiplicative inverse?

In Probs. 54 to 57 state the reason for each step in the given proof.

54 If $a + b = c$, then $b = (-a) + c$.
Proof: **1** $a + b = c$
2 $(-a) + (a + b) = (-a) + c$
3 $[(-a) + a] + b = (-a) + c$
4 $0 + b = (-a) + c$
5 $b = (-a) + c$

55 If $ab = c$ and $a \neq 0$, then $b = \left(\dfrac{1}{a}\right) \times c$.

Proof: **1** $ab = c$

2 $\left(\dfrac{1}{a}\right) \times ab = \left(\dfrac{1}{a}\right) \times c$

3 $\left(\dfrac{1}{a} \times a\right)b = \left(\dfrac{1}{a}\right) \times c$

4 $(1)b = \left(\dfrac{1}{a}\right) \times c$

5 $b = \left(\dfrac{1}{a}\right) \times c$

56 If $a + c = b + c$, then $a = b$.
Proof: **1** $a + c = b + c$
2 $(a + c) + (-c) = (b + c) + (-c)$
3 $a + [c + (-c)] = b + [c + (-c)]$
4 $a + 0 = b + 0$
5 $a = b$

57 If $ac = bc$ and $c \neq 0$, then $a = b$.
Proof: **1** $ac = bc$

2 $ac\left(\dfrac{1}{c}\right) = bc\left(\dfrac{1}{c}\right)$

3 $a\left(c \times \dfrac{1}{c}\right) = b\left(c \times \dfrac{1}{c}\right)$

4 $a \times 1 = b \times 1$
5 $a = b$

2.6   The Natural Numbers

The counting numbers, 1, 2, 3, . . . , are called the *natural numbers* or the *positive integers*. These are special cases of the real numbers, but they do not have all of the properties R1 to R11. We leave it to you to verify that the natural numbers do satisfy R1, R2, R5, R6, R7, R8, R10, and R11.

Exercise

A   Choose $a = 2$, $b = 3$, $c = 5$, and for these natural numbers verify R1, R2, R5, R6, R7, R8, R10, and R11.

Let us look at the other laws. The natural numbers cannot satisfy R3 or R4, since R3 involves zero and R4 involves negative numbers and neither zero nor the negative numbers are natural numbers. The natural numbers cannot satisfy R9 since fractions of the form ½, ⅓, etc., are not natural numbers.

Exercises

B   Prove or disprove the statement: "For every pair of natural numbers a and b, there is a natural number x such that $a + x = b$."
C   Prove or disprove the statement: "For every pair of natural numbers a and b, where $b \neq 0$, there is a natural number x such that $bx = a$."

divisibility   The natural numbers, however, do have properties which are not shared by some of the real numbers. Many of these have to do with their divisibility. We recall the following definition.

Definition   A natural number b is called a divisor (or factor) of a natural number a if and only if there is a natural number x such that $a = bx$. In this case a is said to be divisible by b.

prime
number   Given a natural number, it is often important to make a list of its factors. Some numbers, called *primes*, have only two factors. We recall their definition.

Definition   A natural number is called *prime* if and only if it has no natural numbers as factors except itself and 1. For special reasons 1 is usually not considered prime.

Illustration 1   2, 3, 5, 7, 11, . . . are primes, whereas 4, 6, 8, 9, 10, . . . are not primes.

In factoring a natural number like 60, we may write

$$60 = 20 \times 3$$

and then factor these factors and continue factoring until only prime numbers are left as factors.   Thus

$$60 = 20 \times 3 = 4 \times 5 \times 3 = 2 \times 2 \times 5 \times 3$$

This can be carried out in other ways, such as

$$60 = 15 \times 4 = 5 \times 3 \times 2 \times 2$$

Notice that these two sets of prime factors of 60 are the same except for their order.   This illustrates a general property of the natural numbers which is stated as a theorem.

Theorem 7   Unique Factorization Theorem   A natural number $\neq 1$ can be expressed as a product of primes in a way which is unique except for the order of the factors.

We omit the proof of this theorem.   You can find it, for instance, in "Principles of Mathematics", 2d ed., by Allendoerfer and Oakley, page 98.

The full collection of theorems about the natural numbers is called *Number Theory*.   This is one of the most appealing branches of mathematics, and is generally the subject of an advanced college course.   A few of the many reference books on the subject are listed at the end of this chapter.

2.7   The Integers—Mathematical Induction

The integers consist of the natural numbers, zero, and the negatives of the natural numbers: ... $-3, -2, -1, 0, 1, 2, 3, \ldots$.   Often we call these respectively the "positive integers", "zero", and "negative integers".   The integers are thus special cases of the real numbers, but they fail to have all the properties R1 to R11.   We leave it to you to verify that the integers do satisfy all of R1 to R11 except R9.   From these properties we can infer the truth of a most important theorem for the integers, which was false for the natural numbers (Sec. 2.6, Exercise B).

Theorem 8   For every pair of integers a and b, there exists a unique integer x such that $a + x = b$.

Proof

1   Existence. $x = b - a$ is such an integer.
2   Unicity.   Suppose that x_1 and x_2 are two solutions of $a + x = b$. Then $a + x_1 = a + x_2$, or $x_1 = x_2$.

Exercise

A Prove or disprove the statement: "For every pair of integers a and b, where $b \neq 0$, there is an integer x such that $bx = a$."

The integers have another property which is essential for many parts of mathematics. This property permits us to use a method of proof called *Mathematical Induction* when we are concerned with theorems such as those given below. In the statements of these theorems we use the notation $a > b$, to mean "a is greater than b" and the notation $a \geq b$ to mean "a is greater than or equal to b". We defer the full treatment of such inequalities to Chap. 8.

1 For all integers $n \geq 2$, the product of n odd integers is odd; i.e., the product of any number of odd integers is odd.

2 For all positive integers n, and for any pair of integers x, y, where $x \neq y$, $x^n - y^n$ is divisible by $x - y$.

3 For all integers $n \geq 1$, $1 + 2 + \cdots + n = \dfrac{n(n+1)}{2}$.

4 For all integers $n \geq 2$, and all real numbers $a, b_1, \ldots, b_n$:

$$a(b_1 + \cdots + b_n) = ab_1 + \cdots + ab_n$$

5 For all integers $n \geq 1$, $n(n+1)(n+2)$ is divisible by 3; i.e., the product of every three consecutive positive integers is divisible by 3.

All these theorems have the common feature: their statement contains an open sentence p_n with variable n (Sec. 1.4), which is to be proved true for all n that are elements of a given infinite subset of the integers. Since we cannot prove such a theorem by verification in any finite number of cases, we must devise a general proof by some other means.

We shall illustrate the ideas behind such a proof in several informal examples.

Illustration 1 Prove: For all integers $n \geq 2$, the product of n odd integers is odd.

Let $\{a_1, a_2, \ldots, a_n\}$ be a set of n odd integers. We must show that $a_1 \times a_2 \times \cdots \times a_n$ is odd. First we note that we have proved earlier that the product of two odd integers is odd. Therefore $a_1 \times a_2$ is odd. Now $a_1 \times a_2 \times a_3 = (a_1 \times a_2) \times a_3$, which is the product of the odd integers $(a_1 \times a_2)$ and a_3. Since the product of two odd integers is odd, it follows that $a_1 \times a_2 \times a_3$ is odd. This argument can be repeated successively to show, in turn, that $a_1 \times a_2 \times a_3 \times a_4$ is odd, etc. Hence we are convinced of the truth of the theorem for all $n \geq 2$.

Illustration 2 Prove: For all integers $n \geq 2$ and all real numbers $a, b_1, \ldots, b_n$, it is true that $a(b_1 + \cdots + b_n) = ab_1 + \cdots + ab_n$.

First we recall (Secs. 2.2 and 2.4) that

$$a(b_1 + b_2) = ab_1 + ab_2 \qquad \text{[Distributive Law]}$$
$$b_1 + b_2 + b_3 = (b_1 + b_2) + b_3 \qquad \text{[Definition]}$$

Hence $a(b_1 + b_2 + b_3) = a[(b_1 + b_2) + b_3]$ [Theorem 1]
$\qquad\qquad\qquad = a(b_1 + b_2) + ab_3$ [Distributive Law]
$\qquad\qquad\qquad = (ab_1 + ab_2) + ab_3$ [Distributive Law]
$\qquad\qquad\qquad = ab_1 + ab_2 + ab_3$ [Definition]

Continuing in this fashion, we have

$$a(b_1 + b_2 + b_3 + b_4) = a(b_1 + b_2 + b_3) + ab_4$$
$$= (ab_1 + ab_2 + ab_3) + ab_4$$
$$= ab_1 + ab_2 + ab_3 + ab_4$$

Since the process can continue indefinitely, we have again convinced ourselves of the truth of the theorem.

The trouble with these informal arguments is that in the end we must say: "this continues indefinitely", "etc.", "and so on", or something of the kind. How do we really know that we shall not be blocked at some stage far in the future? Since there is no logical means of giving such an assurance, we must formulate an axiom which effectively says that no such blocks will appear. In order to formulate this axiom, let us analyze the informal arguments above. In each case we began with a true fact for $n = 2$, namely, in Illustration 1, that the product of two odd integers is odd and, in Illustration 2, that $a(b_1 + b_2) = ab_1 + ab_2$. We used this fact to derive a true statement for $n = 3$, $n = 4$, $n = 5$, etc., in turn, and concluded that the process would never stop. By analogy we can think of the positive integers as the rungs of an infinitely tall ladder based on the ground and reaching to the sky. The bottom rung is 1, the next 2, and so on. We wish to climb this ladder to any desired rung. To do so, there are two essential steps:

I   We must first get our foot on a low rung (the second rung in the above illustrations).

II   We must be able to climb from any rung to the next rung. Clearly, if we can do these two things, we can climb up as far as we please.

To formalize this idea, we now state our axiom.

Axiom of Mathematical Induction   Let a be an integer (positive, negative, or zero), and let A be the set of integers which are greater than or equal to a; that is, $A = \{n \mid n \geq a\}$.
If S is a subset of A with the two properties:

I S contains a.

II   For all integers k in A: if k belongs to S, then $k + 1$ belongs to S.

then the set S is equal to the set A.

In many applications of this axiom we have $a = 1$, so that A is the set of all positive integers.

When we use this axiom to prove theorems of the type under consideration, the set A and the open sentence p_n are given to us in the statement of the theorem. We choose S to be the subset of A consisting of those integers for which p_n is true. With these interpretations, we can reformulate the axiom as an operational procedure, which we call the Principle of Mathematical Induction.

Principle of Mathematical Induction

Let us be given a set of integers $A = \{n \mid n \geq a\}$ and a proposition of the form: For all n in A, p_n. We can prove the truth of this proposition by establishing the following:

I p_a is true. (We use the symbol p_a to denote the proposition obtained from the open sentence p_n by substituting a for n.)

II For all k in A, the implication $p_k \to p_{k+1}$ is true.

Let us now illustrate this method.

Illustration 3 (See Illustration 1.) Prove: For all integers $n \geq 2$, the product of n odd integers is odd.

We have: $\qquad a = 2 \qquad A = \{n \mid n \geq 2\}$

$\qquad p_n$: the product of n odd integers is odd.

Following the Principle of Mathematical Induction, we establish the two necessary facts:

I p_2 is true, for the product of two odd integers is odd.

II For all integers $k \geq 2$: if the product of k odd integers is odd, then the product of $k + 1$ odd integers is odd.

Proof of II Let $a_1, a_2, a_3, \ldots, a_k, a_{k+1}$ be odd integers. By hypothesis, $a_1 \times a_2 \times \cdots \times a_k$ is odd. Moreover, $a_1 \times \cdots \times a_k \times a_{k+1} = (a_1 \times \cdots \times a_k) \times a_{k+1}$ by definition. The expression in parentheses is odd by hypothesis, and so this is the product of two odd integers. Since such a product is odd, we conclude that $a_1 \times \cdots \times a_k \times a_{k+1}$ is odd.

From the Principle of Mathematical Induction we now conclude that the given proposition is true.

Illustration 4 (See Illustration 2.) Prove: For all integers $n \geq 2$ and all real numbers $a, b_1, \ldots, b_n$, it is true that

$$a(b_1 + \cdots + b_n) = ab_1 + \cdots + ab_n$$

We have: $\qquad a = 2 \qquad A = \{n \mid n \geq 2\}$

$\qquad p_n$: $a(b_1 + \cdots + b_n) = ab_1 + \cdots + ab_n$

I p_2 is true, for $a(b_1 + b_2) = ab_1 + ab_2$ by the distributive law.

II For all integers $k \geq 2$: if $a(b_1 + \cdots + b_k) = ab_1 + \cdots + ab_k$, then $a(b_1 + \cdots + b_{k+1}) = ab_1 + \cdots + ab_{k+1}$.

Proof of II By definition, $b_1 + \cdots + b_{k+1} = (b_1 + \cdots + b_k) + b_{k+1}$. Hence

$$a(b_1 + \cdots + b_{k+1}) = a[(b_1 + \cdots + b_k) + b_{k+1}]$$
$$= a(b_1 + \cdots + b_k) + ab_{k+1}$$

by the distributive law. By hypothesis, $a(b_1 + \cdots + b_k) = ab_1 + \cdots + ab_k$. Hence

$$a(b_1 + \cdots + b_{k+1}) = (ab_1 + \cdots + ab_k) + ab_{k+1}$$
$$= ab_1 + \cdots + ab_k + ab_{k+1}$$

The stated theorem is now proved.

Illustration 5   Prove: For all integers ≥ 1, $1 + 2 + \cdots + n = \dfrac{n(n + 1)}{2}$.

We have: $\qquad\qquad a = 1 \qquad A = \{n \mid n \geq 1\}$

$$p_n: \quad 1 + 2 + \cdots + n = \frac{n(n + 1)}{2}$$

I p_1 is true, for $1 = \dfrac{1(2)}{2}$.

II For all integers $k \geq 1$, if $1 + 2 + \cdots + k = \dfrac{k(k + 1)}{2}$, then

$$1 + 2 + \cdots + k + k + 1 = \frac{(k + 1)(k + 2)}{2}$$

Proof of II By hypothesis, $1 + 2 + \cdots + k = \dfrac{k(k + 1)}{2}$.   Adding $k + 1$ to each side of the equality, we have

$$1 + 2 + \cdots + k + k + 1 = \frac{k(k + 1)}{2} + k + 1$$
$$= \frac{k^2 + k + 2k + 2}{2}$$
$$= \frac{k^2 + 3k + 2}{2}$$
$$= \frac{(k + 1)(k + 2)}{2}$$

Illustration 6   Prove: For all integers $n \geq 1$, $3^{2n} - 1$ is divisible by 8.

We have: $\qquad\qquad a = 1 \qquad A = \{n \mid n \geq 1\}$
$$p_n: \quad 3^{2n} - 1 \text{ is divisible by 8.}$$

I p_1 is true, for $3^2 - 1 = 8$ is divisible by 8.
II For all integers $k \geq 1$, if $3^{2k} - 1$ is divisible by 8, then $3^{2k+2} - 1$ is divisible by 8.

Proof of II By hypothesis, $3^{2k} - 1$ is divisible by 8, so that there is an integer x such that $3^{2k} - 1 = 8x$, or $3^{2k} = 1 + 8x$.   Multiplying both sides of this equality by $3^2 = 9$, we get

$$3^{2k+2} = 9 + 8(9x)$$

or
$$3^{2k+2} = 1 + [8 + 8(9x)]$$
$$= 1 + 8(1 + 9x)$$

So $3^{2k+2} - 1 = 8(1 + 9x)$, or $3^{2k+2} - 1$ is divisible by 8.

Problems 2.7

Mathematical Induction

In Probs. 1 to 28 use mathematical induction to show the truth of the stated proposition for all integers $n \geq 1$.

1 $1 + 3 + 5 + \cdots + (2n - 1) = n^2$

2 $1 + 4 + 7 + \cdots + (3n - 2) = \dfrac{n(3n - 1)}{2}$

3 $2 + 7 + 12 + \cdots + (5n - 3) = \dfrac{n}{2}(5n - 1)$

4 $2 + 6 + 10 + \cdots + (4n - 2) = 2n^2$

5 $n^3 + 2n$ is divisible by 3.

6 $n^2 + n$ is divisible by 2.

7 $n(n + 1)(n + 2)$ is divisible by 3.

8 $n^3 + 5n$ is divisible by 3.

9 $1 + 2 + 4 + 8 + \cdots + 2^{n-1} = 2^n - 1$

10 $2 + 6 + 18 + \cdots + 2 \cdot 3^{n-1} = 3^n - 1$

11 $1 + 2 + 3 + \cdots + n = \dfrac{n(n + 1)}{2}$

12 $1^2 + 2^2 + 3^2 + \cdots + n^2 = \dfrac{n(n + 1)(2n + 1)}{6}$

13 $1^3 + 2^3 + 3^3 + \cdots + n^3 = \dfrac{n^2(n + 1)^2}{4}$

14 $2 + 5 + 13 + \cdots + (2^{n-1} + 3^{n-1}) = 2^n - 1 + \frac{1}{2}(3^n - 1)$

15 $a + (a + d) + (a + 2d) + \cdots + [a + (n - 1)d]$
$$= \dfrac{n}{2}[2a + (n - 1)d]$$

16 $a + ar + ar^2 + \cdots + ar^n = \dfrac{a(1 - r^{n+1})}{1 - r}, r \neq 1$

17 $4^n - 1$ is divisible by 3.

18 $5^n - 1$ is divisible by 4.

19 $3^{2n} - 1$ is divisible by 4.

20 $2^{2n} - 1$ is divisible by 3.

21 $x^n - 1$ is divisible by $x - 1$, where x is an integer $\neq 1$.

22 $x^{2n} - 1$ is divisible by $x + 1$, where x is an integer $\neq -1$.

23 $x^n - y^n$ is divisible by $x - y$, where x and y are integers and $x \neq y$.

24 $x^{2n} - y^{2n}$ is divisible by $x + y$, where x and y are integers and $x + y \neq 0$.

25 If a is a real number, we define $a^1 = a$ and $a^{n+1} = a^n \cdot a$. Prove that $1^n = 1$.

26 The product of $(2n + 1)$ negative real numbers $a_1 \times a_2 \times \cdots \times a_{2n+1}$ is negative.

27 The product of $2n$ negative real numbers $a_1 \times a_2 \times \cdots \times a_{2n}$ is positive.

28 $(ab)^n = a^n b^n$

29 For all integers $n \geq 2$, prove that the sum of n integers $a_1 + \cdots + a_n$ is an integer.

30 For all integers $n \geq 2$, prove that the product of n integers $a_1 \times a_2 \times \cdots \times a_n$ is an integer.

In Probs. 31 to 34 try to establish the indicated relation by mathematical induction. Point out why the method fails. (Each relation is false.)

31 $3 + 6 + 9 + \cdots + 3n = \dfrac{3n(n + 1)}{2} + 1$

32 $3 + 5 + 7 + \cdots + (2n + 1) = n^2 + 2$

33 $n^2 - 1$ is divisible by 3 for all $n \geq 2$.

34 $n^2 + n$ is divisible by 4 for all $n \geq 3$.

Integers

35 Verify that R1 to R8 and R10 to R11 are satisfied when $a = 2$, $b = -3, c = 4$.

36 Verify that R1 to R8 and R10 to R11 are satisfied when $a = 4, b = 3$, $c = -2$.

37 Show by a counterexample that the integers do not satisfy R9.

38 Are the integers closed under subtraction?

39 Are the integers closed under division?

40 Do the integers form a field?

2.8 Rational Numbers

A rational number is really nothing but a fraction whose numerator and denominator are both integers. Let us give a formal definition.

Definition A *rational number* is a real number which can be expressed in the form a/b where a and b are integers and $b \neq 0$.

As we know from the example $\frac{1}{2} = \frac{2}{4}$, there are many expressions of the form a/b which represent the same rational number. So that we can identify such cases easily, we need the following definition.

Definition The expressions a/b and c/d where a, b, c, and d are integers and $b \neq 0$ and $d \neq 0$ represent the *same rational number* if and only if $ad = bc$.

Hence we may write

$$a/b = c/d \qquad \text{when} \qquad ad = bc$$

We can now state the following theorem, which expresses the most important and useful property of rational numbers.

Theorem 9 Given any pair of integers a and b ($\neq 0$), there exists a rational number x such that $bx = a$. Moreover, any two rational numbers x_1 and x_2 with this property are equal.

Proof The existence of a solution is immediate, for $x = a/b$ has the required property. In order to establish the second part of the theorem, we suppose that x_1 and x_2 both satisfy $bx = a$. Then

$$bx_1 = a$$
$$bx_2 = a$$

Subtracting, we have

$$b(x_1 - x_2) = 0$$

or $\qquad\qquad x_1 - x_2 = 0 \qquad$ [Theorem 4, since $b \neq 0$]

Finally we must remind you of the rules for adding and multiplying rational numbers. These are given by the theorems below.

Theorem 10 $\dfrac{a}{b} + \dfrac{c}{d} = \dfrac{ad + bc}{bd}$

addition

Proof Let $x = a/b$; then $bx = a$.
Let $y = c/d$; then $dy = c$.

From these two equations we obtain

$$bdx = ad$$
$$bdy = bc$$

Adding and using the distributive law, we get

$$bd(x + y) = ad + bc$$

So $\qquad\qquad x + y = \dfrac{ad + bc}{bd}$

Theorem 11 $(a/b) \times (c/d) = ac/bd$

multiplication

Proof Using the notation in the proof of Theorem 10, we have again

$$bx = a$$
$$dy = c$$

Multiplying the left-hand sides and the right-hand sides separately, we have

$$(bd)(xy) = ac$$

Therefore
$$xy = \frac{ac}{bd}$$

With these concepts of addition and multiplication we can now check to see how many of R1 to R11 are satisfied by the rational numbers. As a matter of fact we find that all of these are satisfied. This means that the arithmetic of rational numbers is just like that of the real numbers. This might lead us to believe that there is no difference between the real numbers and their special case, the rational numbers. However, we shall see that real numbers such as $\sqrt{2}$ are not rational and hence that a distinction must be made.

2.9   Decimal Expansions

By carrying out the ordinary process of division, any rational number can be represented as a decimal. Some representations "terminate" after a finite number of steps; i.e., all later terms in the expansion are zero. For example,

$$\tfrac{1}{2} = 0.5000\cdots$$
$$\tfrac{1}{4} = 0.2500\cdots$$

But other expansions never terminate, such as

$$\tfrac{1}{3} = 0.3333\cdots$$
$$1\tfrac{1}{7} = 1.142857142857\cdots$$

By experimenting you may assure yourself that in each expansion the digits after a certain point repeat themselves in certain groups like (0), (3), and (142857) above. This is always true for rational numbers.

It is awkward to express numbers in this form since we cannot be sure what the $\cdots$ at the end really mean. To clear up this ambiguity, we place a bar over the set of numbers which is to be repeated indefinitely. In this notation we write

repeating
decimal

$$\tfrac{1}{2} = 0.5\overline{0}$$
$$\tfrac{1}{4} = 0.25\overline{0}$$
$$\tfrac{1}{3} = 0.\overline{3}$$
$$1\tfrac{1}{7} = 1.\overline{142857}$$

It is also true that any repeating decimal expansion of this type represents a rational number. We state this as Theorem 12.

Theorem 12   Every repeating decimal expansion is a rational number.

Before giving the general proof, we give several illustrations.

Illustration 1   Prove that $a = 3.\overline{3}$ is a rational number.

Solution   If we multiply by 10, we merely shift the decimal point; thus

$$10a = 33.\overline{3} = 30 + a$$

Hence

$$9a = 30$$

$$a = {}^{30}\!/_9 = 3\frac{1}{3}$$

Illustration 2   Now consider the harder case where $b = 25.\overline{12}$.

Solution

$$100b = 2512.\overline{12}$$

$$b = 25.\overline{12}$$

Subtracting, we find

$$99b = 2,487$$

$$b = \frac{2,487}{99} = \frac{829}{33} = 25\frac{4}{33}$$

Illustration 3   Finally consider $c = 2.3\overline{12}$.

Solution   The change here is that the repeating part begins one place to the right of the decimal point. We can correct this easily by writing

$$10c = 23.\overline{12}$$

and then proceeding as in Illustration 2.

To prove the general theorem, suppose that

$$c = a_0 . a_1 \cdots a_k \overline{b_1 \cdots b_p}$$

where the a's and b's represent digits in the expansion of c. Using the idea of Illustration 3, we write

$$d = 10^k c = a_0 a_1 \cdots a_k . \overline{b_1 \cdots b_p}$$

Then

$$10^p d = a_0 a_1 \cdots a_k b_1 \cdots b_p . \overline{b_1 \cdots b_p}$$

$$(10^p d) - d = (a_0 a_1 \cdots a_k b_1 \cdots b_p - a_0 a_1 \cdots a_k)$$

Solving, we find that d is rational and that therefore c is rational.

2.10   Some Irrational Numbers

We have seen that the set of rational numbers is identical with the set of repeating decimals. However, we may perfectly well conceive of a nonrepeating decimal. This might be constructed in infinite time by a man throwing a 10-sided die who records the results of his throws in sequence. Thus he might obtain the number (nonrepeating)

$$0.352917843926025\cdots$$

Such a number is not rational but is included among the reals; and thus the reals include irrationals as well as rationals.

Perhaps this example is farfetched, and therefore we consider the very practical question of solving the equation $x^2 = 2$. The value of x is equal to the length of the hypotenuse of a right triangle whose legs are each 1. We now wish to show that $x = \sqrt{2}$ is not rational. We first prove a preliminary theorem, in which a is assumed to be an integer.

Theorem 13 If a^2 is divisible by 2, then a is divisible by 2.

We have met this theorem before in Sec. 1.10, Illustration 4, but let us give another proof here.

Proof Every integer a can be written in one of the two forms

$$a = \begin{cases} 2n \\ 2n + 1 \end{cases} \quad \text{where } n \text{ is an integer}$$

Hence
$$a^2 = \begin{cases} 4n^2 \\ 4n^2 + 4n + 1 \end{cases}$$

Since a^2 is divisible by 2, according to the hypothesis, a^2 must equal $4n^2$. Hence $a = 2n$, and a is divisible by 2.

Theorem 14 $\sqrt{2}$ is not a rational number.

$\sqrt{2}$ is
irrational

Proof (Indirect) Suppose p/q is a rational number in lowest terms; that is, p and q have no common factor. Suppose also that $p^2/q^2 = 2$, or that $p^2 = 2q^2$.

Then p^2 is divisible by 2, and thus p is divisible by 2 (Theorem 13). Write $p = 2r$, where r is an integer. Then $4r^2 = 2q^2$, or $2r^2 = q^2$. Hence q^2 is divisible by 2, and thus q is divisible by 2 (Theorem 13). Hence p and q have a common factor contrary to our assumption. This proves the theorem.

Problems 2.10

Rational Numbers

1 Verify that R1 to R11 are satisfied when $a = \frac{1}{2}, b = 3, c = -4$.
2 Verify that R1 to R11 are satisfied when $a = 2, b = \frac{1}{3}, c = 4$.
3 Show that the natural numbers are special cases of the rationals.
4 Show that the integers are special cases of the rationals.
5 Prove: For any pair of rational numbers a and b, there exists a rational number x such that $a + x = b$. Moreover, any two rationals x_1 and x_2 with this property are equal.

6 Prove: For any three rational numbers a, b, and c, where $a \neq 0$, there exists a rational number x such that $ax + b = c$. Moreover, any two rationals x_1 and x_2 with this property are equal.

7 Prove: For any two rational numbers a/b and c/d, the quotient

$$\frac{a/b}{c/d} = \frac{ad}{bc}.$$ HINT: Consider $(c/d)x = a/b$.

8 Prove: The two expressions $(-a)/b$ and $-(a/b)$ represent the same rational number. HINT: Show that each represents the additive inverse of (a/b).

9 Prove: There is no rational number which is nearer to 0 than every other rational number.

10 Prove: Given any two rational numbers a and b, with $a < b$, there is at least one rational number greater than a and less than b. How many rational numbers are there between a and b?

In Probs. 11 to 16 find decimal expansions for the given rational numbers.

11 $\frac{4}{7}$ **12** $\frac{5}{6}$

13 $\frac{4}{9}$ **14** $\frac{3}{8}$

15 $3\frac{1}{9}$ **16** $4\frac{2}{11}$

In Probs. 17 to 22 find expressions of the form a/b for the given decimal expansions.

17 $0.\overline{7}$ **18** $4.2\overline{3}$

19 $12.\overline{26}$ **20** $5.4\overline{72}$

21 $3.7\overline{621}$ **22** $2.\overline{9}$ (comment)

23 Prove that the decimal expansion of any rational number is repeating. HINT: Try dividing, and see what happens.

24 When a/b is expressed as a repeating decimal, what is the maximum length of the period? HINT: Try dividing, and see what happens.

25 Prove that $\sqrt{3}$ is irrational. HINT: First prove the analogue of Theorem 13: "If a^2 is divisible by 3, then a is divisible by 3." To do so, note that every integer can be written in one of the forms:

$$a = \begin{cases} 3n \\ 3n + 1 \\ 3n + 2 \end{cases} \quad \text{where } n \text{ is an integer}$$

Hence

$$a^2 = \begin{cases} 9n^2 \\ 9n^2 + 6n + 1 \\ 9n^2 + 12n + 4 \end{cases}$$

Since a^2 is divisible by 3, according to the hypothesis, a^2 must equal $9n^2$, etc.

26 Prove that $\sqrt{5}$ is irrational.

27 Where does the method of Probs. 25 and 26 fail when we try to prove $\sqrt{4}$ to be irrational?

28 Prove that $1 + \sqrt{2}$ is irrational. HINT: Suppose that $1 + \sqrt{2} = a/b$. Then $\sqrt{2} = a/b - 1 = (a - b)/b$. Why is this impossible?

29 Prove that $2 - \sqrt{3}$ is irrational.

*30 Prove that $\sqrt{2} + \sqrt{3}$ is irrational.

2.11 Geometric Representation of Real Numbers

In the connection between arithmetic and geometry, the representation of real numbers as points on a line is most important. You are probably familiar with this idea, which is illustrated in Fig. 2.1. In order to obtain

Linear scale

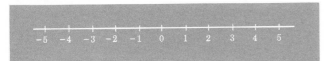

Figure 2.1

this representation, we start with the points 0 and 1 chosen at random, except that 0 is to the left of 1. The segment [0,1] is said to have length 1 by definition of "length". It is now assumed that this segment can be slid along the line without altering its length. Doing this step by step, we locate the other integers, so that the length of the segment between any two successive integers is still equal to 1. The location of $1/b$ (where b is a natural number) is found by dividing [0,1] into b equal parts by the usual geometric construction. Then by sliding the segment [0,1/b] along the line we locate the points a/b. The location of the irrational numbers is more complicated, and we pass over this point. Their approximate positions, however, can be obtained from the first few decimals in their decimal expansions.

The most important fact about this representation is *that every point corresponds to one and only one real number and that every real number corresponds to one and only one point.* This property is often taken as one of the axioms of plane geometry.

This representation has another important property, namely, it preserves *order*. Before we can state this precisely, we must define a notion of order for real numbers and a similar notion for points on a line. Let us start with real numbers.

Definition We say that a is greater than b (written $a > b$) if $a - b$ is *positive*. Similarly a is less than b (written $a < b$) if $a - b$ is *negative*.

inequality

The symbols $\geq$ and $\leq$ mean, respectively, "greater than or equal to" and "less than or equal to". It is easy to see that, for any pair of real numbers, one and only one of the following relations is true:

$$a < b \qquad a = b \qquad a > b$$

We shall study the properties of these inequalities in some detail in Chap. 8.

For a line, we introduce order by means of the notion "beyond". First of all we place an arrow on one end of the line and thus define a "positive direction" on the line as the direction toward the arrow. We now call this line a "directed line". It is customary to direct horizontal lines to the *right* and vertical lines *upward*. Then we define "beyond" as follows.

Definition A point P is *beyond* a point Q on a directed line if the segment (or vector) from Q to P points in the given positive direction. If P is beyond Q, we write $P > Q$.

Let us now return to our assumption about real numbers and points on the line and describe how this preserves order. Let P_a be the point corresponding to the real number a and P_b to b. Then our correspondence is such that

$$P_a > P_b \text{ if and only if } a > b$$

In summary, we have defined a correspondence between real numbers and points on a line which is 1 to 1 and which preserves *order*. The number associated with a point is called its "coordinate", and we can use coordinates to identify points. Thus, the point whose coordinate is the number a will henceforth be written as the point a and not as P_a, as was done above. This use of coordinates is the foundation of the application of real numbers to geometry and to geometrical representations of nature.

length

By means of coordinates we can now define the length of an arbitrary segment whose end points are a and b. The notation for such a segment is $[a,b]$.

Definition The *length* of the segment $[a,b]$ is the real number $|b - a|$.

2.12 The Use of Real Numbers in the Plane

We shall now use the correspondence of the last section to set up a relationship between ordered pairs of real numbers and points in the plane. This is based upon an idea of René Descartes (1596–1650). Since ordered pairs will turn up in several other places in this book, let us say what they are.

Definition An *ordered pair* (x,y) of real numbers is a pair in which x is the
ordered pair first element and in which y is the second element. Because of the order-
ing (x,y) is to be distinguished from (y,x) if $x \neq y$.

First, we construct two perpendicular lines in the plane (Fig. 2.2) which
we call the X-axis and the Y-axis. Their point of intersection is called the
origin O. We put the X-axis into an exact correspondence with the real
numbers by placing zero at O, the positive reals to the right of O and the
negative reals to its left. We do the same for the Y-axis, putting the positive reals above O and the negative reals below O. We remind ourselves
of these conventions by putting arrows on the right end of the X-axis and
the upper end of the Y-axis. These lines divide the plane into four regions
called "quadrants" which are numbered I, II, III, and IV in Fig. 2.2.

Rectangular coordinate system

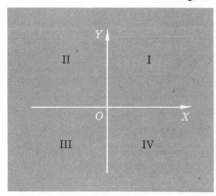

Figure 2.2 Roman numerals represent quadrants.

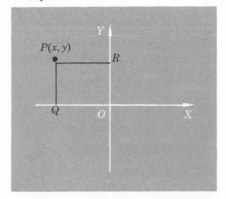

Figure 2.3 An ordered pair of numbers (x,y)
determines a point in the plane.

Using this scheme, we can now associate an ordered pair of real numbers (x,y) with each point P of the plane. Let P be a point on the X-axis.
It corresponds to a real number x on this axis. We associate with P the
ordered pair $(x,0)$. Now let P be a point on the Y-axis. Similarly we associate the ordered pair $(0,y)$ with it. When P is not on either axis, draw
PQ perpendicular to the X-axis and PR perpendicular to the Y-axis
(Fig. 2.3). Suppose that Q corresponds to the real number x on the X-axis
and that R corresponds to the real number y on the Y-axis. Then we
associate the ordered pair (x,y) with P.

By this process we find an ordered pair (x,y) which corresponds to each
P in the plane. It is also evident that every pair (x,y) determines a point
in the plane, for suppose (x,y) is given. These locate points Q and R
(Fig. 2.3). Draw PQ and PR as perpendiculars to the X-axis and the Y-axis
at Q and R, respectively. These lines intersect at P, which is the desired
point.

*Thus we have established a 1 to 1 correspondence between the points of
the plane and the ordered pairs* (x,y).

Definition   The real numbers x and y in the ordered pair (x,y) are called the *coordinates* of the point P.   Sometimes x is called the *x-coordinate*, or the *abscissa;* and y is called the *y-coordinate*, or the *ordinate*.

We often identify the point P with its pair of coordinates and speak of the "point (x,y)".   By using this identification, we can convert geometric statements about points into algebraic statements about numbers and can convert geometric reasoning into algebraic manipulation.   The methods of algebra are usually simpler than those of geometry, and therefore the algebraic approach is now the common one.   The detailed elaboration of this method is called "analytic geometry".

2.13   Lengths of Segments; Units on the Axes

Suppose that P_1 and P_2 lie on a line parallel to the X-axis.   Then we may write their coordinates as $P_1(x_1,a)$ and $P_2(x_2,a)$ (Fig. 2.4).   We wish to have an expression for the length of P_1P_2.   Draw P_1R and P_2S perpendicular to the X-axis.   Then R has coordinates $(x_1,0)$ and S has coordinates $(x_2,0)$.   Moreover, the lengths P_1P_2 and RS are equal, since opposite sides of a rectangle are equal.

Length of a horizontal segment

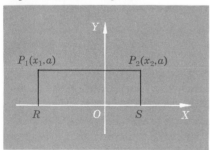

Figure 2.4
For a horizontal segment, the length of P_1P_2 is $|x_2 - x_1|$.

From Sec. 2.11 we know that the length $RS = |x_2 - x_1|$.   Hence $P_1P_2 = |x_2 - x_1|$.   This gives us Theorem 15.

Theorem 15   The length of the segment between $P_1(x_1,a)$ and $P_2(x_2,a)$ is given by $P_1P_2 = |x_2 - x_1|$.

A similar proof gives Theorem 16.

Theorem 16   The length of the segment between $Q_1(a,y_1)$ and $Q_2(a,y_2)$ is given by $Q_1Q_2 = |y_2 - y_1|$.

We have said nothing about the relation of distance on the X-axis to that on the Y-axis, and we prefer not to make any rigid requirements about this at present. Indeed, it is often useful to use different scales of measurement on the two axes. Unequal scales are used for a variety of reasons of which the following are the most common:

1 The range of values to be plotted on the Y-axis is much greater (or smaller) than the range to be plotted on the X-axis. In this case we must contract (or expand) the scale on the Y-axis in order to get a graph on a reasonably shaped piece of paper.

Illustration 1 Suppose that we are plotting $y = x^{10}$ for x in the range 0 to 2. Then y lies in the range 0 to 1,024. In this case it would be extremely awkward to use equal scales on the two axes.

2 In applications to science the physical significance of the numbers on the two axes may be very different. In such cases the physical units of measurement (such as time, distance, velocity, etc.) are not comparable; and suitable scales on the two axes should be chosen independently.

Illustration 2 In order to illustrate the motion of a particle, it is customary to plot the distance traveled on the vertical axis and the corresponding time on the horizontal axis. The units of measurement are *feet* and *seconds*, respectively, and it would be absurd to equate feet and seconds. Hence separate, convenient scales are used on the two axes.

In geometry, however, it is necessary to plot distance on each of the axes and it is also necessary to use the same scale on each axis. When we do these things, it is meaningful and helpful to speak of the lengths of segments on slanting lines. The notion of slant distance, however, is quite meaningless in situations like that described in Illustration 2.

Problems 2.13

1 Use the symbol $>$ to represent the correct inequality between each of the following pairs of numbers: 2 and 5; -1 and 7; 4 and -4; -2 and -6; -3 and 0.

2 Use the symbol $<$ to represent the correct inequality between each of the following pairs of numbers: 8 and -7; -4 and -5; 6 and 3; -4 and 9; 0 and 4.

3 Write a set of inequalities expressing the fact that c lies inside the segment $[a,b]$ where $a < b$.

4 Write a set of inequalities expressing the fact that c lies outside the segment $[a,b]$ where $a < b$.

5 Find the lengths of the following segments: $[16,2]$, $[-7,5]$, $[4,-2]$, $[-5,-3]$, $[21,-13]$.

6 Find the lengths of the following segments: $[12, -6]$, $[-9, 5]$, $[6, 9]$, $[-20, 0]$, $[-12, -25]$.

7 Plot the points whose coordinates are $(2, 6)$, $(-3, 2)$, $(-4, -6)$, $(5, -1)$, $(0, 0)$.

8 Plot the points whose coordinates are $(1, 5)$, $(-2, 4)$, $(-3, -2)$, $(2, -3)$, $(0, 0)$.

9 What signs do the coordinates of points in quadrant I have; in quadrant III?

10 What signs do the coordinates of points in quadrant II have; in quadrant IV?

11 State the quadrant in which each of the following points lies: $(2, -4)$, $(3, 7)$, $(-5, -6)$, $(-9, 11)$, $(-4, 7)$.

12 State the quadrant in which each of the following points lies: $(-8, -6)$, $(19, -4)$, $(-6, 9)$, $(2, 2)$, $(8, -4)$.

13 Find the lengths of the segments joining the following pairs of points: $(2, 3)$ and $(5, 3)$; $(-3, -5)$ and $(-3, -7)$; $(-4, 2)$ and $(7, 2)$; $(5, 5)$ and $(5, -6)$; $(0, 0)$ and $(0, 3)$.

14 Find the lengths of the segments joining the following pairs of points: $(4, 7)$ and $(4, -10)$; $(3, -5)$ and $(7, -5)$; $(6, 6)$ and $(13, 6)$; $(4, 0)$ and $(0, 0)$; $(-6, -9)$ and $(-6, -11)$.

15 If P_1P_2 is parallel to the X-axis, show that the length of P_1P_2 is equal to the length of P_2P_1.

16 If Q_1Q_2 is parallel to the Y-axis, show that the length of Q_1Q_2 is equal to the length of Q_2Q_1.

17 If P_1P_2 is parallel to the X-axis, show that the square of its length is $(x_2 - x_1)^2 = (x_1 - x_2)^2$.

18 If Q_1Q_2 is parallel to the Y-axis, show that the square of its length is $(y_2 - y_1)^2 = (y_1 - y_2)^2$.

Transformation of Coordinates   If we are given a coordinate system on a line in terms of numbers x, we can define a new coordinate system x' by giving a relationship between x and x'. This relabels the points with new numbers and is called a *transformation of coordinates*. The following problems give some important illustrations of these.

In Probs. 19 to 22 we take $x' = a + x$. This transformation is called a *translation*.

19 Prove that a translation leaves the lengths of segments unchanged. HINT: Prove that $|x_2' - x_1'| = |x_2 - x_1|$.

20 Prove: If the coordinate of any one point is left unchanged by a translation, then the coordinates of all points are unchanged. HINT: Prove that $a = 0$.

21 Express as a translation the relationship between absolute temperature K (degrees Kelvin) and centigrade temperature C (degrees centigrade).

22 Express as a translation the relationship between the distance s of a rocket from the center of the earth and its height h above the surface of the earth.

In Probs. 23 to 26 we take $x' = ax$, where $a \neq 0$. This transformation is called a *dilatation*.

23 Prove that a dilatation multiplies the lengths of segments by $|a|$.
24 Prove that, if the coordinate of any *one* point (other than $x = x' = 0$) is left unchanged by a dilatation, then the coordinates of all points are unchanged.
25 Express the relationship between feet F and inches I as a dilatation.
26 Express the relationship between seconds S and hours H as a dilatation.

In Probs. 27 to 30 we take $x' = ax + b$, where $a \neq 0$. This transformation is called a *linear transformation*.

27 Prove that a linear transformation multiplies the lengths of segments by $|a|$.
28 Prove that a linear transformation with $a \neq 1$ leaves the coordinate of just one point unchanged. Find this point.
29 Express the relationship between degrees Fahrenheit F and degrees centigrade C as a linear transformation. What temperature is the same in both systems?
30 Express the relationship between degrees Fahrenheit F and degrees Kelvin K as a linear transformation. What temperature is the same in both systems?

2.14 Complex Numbers

There are, unfortunately, many problems that cannot be solved by the use of real numbers alone. For instance, we are unable to solve $x^2 = -1$.

In order to handle such situations the new symbol i is introduced, which, by definition, is to have the property that $i^2 = -1$. We then write expressions like $a + bi$ where a and b are real numbers and call these expressions *complex numbers*. The number a is the *real part of $a + bi$*, and bi is its *imaginary part*. So that we can treat these like numbers, we must define the usual arithmetic operations on them.

Definitions The arithmetic operations on complex numbers are defined as follows:

Equality: $a + bi = c + di$ if and only if $a = c$ and $b = d$.
Addition: $(a + bi) + (c + di) = (a + c) + (b + d)i$
Multiplication: $(a + bi) \times (c + di) = (ac - bd) + (bc + ad)i$

Note that the definition of multiplication is consistent with the property that $i^2 = -1$. For we can multiply $(a + bi)(c + di)$ by ordinary algebra and obtain $ac + i(bc + ad) + i^2(ad)$. When we replace i^2 with -1 and rearrange, we obtain the formula in the definition.

Illustration 1

a $(3 + 6i) + (2 - 3i) = 5 + 3i$
b $(7 + 5i) - (1 + 2i) = 6 + 3i$
c $(5 + 7i)(3 + 4i) = 15 + 41i + 28i^2$
$$= (15 - 28) + 41i = -13 + 41i$$
d $(2 - 3i)(-1 + 4i) = -2 + 11i - 12i^2$
$$= (-2 + 12) + 11i = 10 + 11i$$

conjugate

We also must consider division. We wish to express $1/(a + bi)$ as a complex number. This is best approached through the use of the conjugate complex number $a - bi$.

Definition   The complex numbers $a + bi$ and $a - bi$ are called *conjugates*.

We write

$$\frac{1}{a + bi} = \left(\frac{1}{a + bi}\right)\left(\frac{a - bi}{a - bi}\right) = \frac{a - bi}{a^2 + b^2} = \frac{a}{a^2 + b^2} + \frac{-b}{a^2 + b^2}i$$

which is the required complex number equal to $1/(a + bi)$. By extension of this method we can evaluate general quotients $(a + bi)/(c + di)$:

$$\frac{a + bi}{c + di} = \left(\frac{a + bi}{c + di}\right)\left(\frac{c - di}{c - di}\right) = \frac{(ac + bd) + (bc - ad)i}{c^2 + d^2}$$

Hence we have the rule for division.

division

Division   In order to form the quotient $(a + bi)/(c + di)$, multiply numerator and denominator by the conjugate complex number $c - di$ and simplify the result.

Illustration 2

$$\frac{4 + i}{2 - 3i} = \frac{4 + i}{2 - 3i} \times \frac{2 + 3i}{2 + 3i} = \frac{(4 + i)(2 + 3i)}{(2 - 3i)(2 + 3i)} = \frac{5 + 14i}{13}$$

We could write this answer as $5\!/\!13 + {}^{14}\!/\!13\,i$, but this is an unnecessary refinement.

Finally, let us solve some equations involving complex numbers. The general method is suggested by the illustration below.

Illustration 3   Solve: $(x + yi)(2 - 3i) = 4 + i$. We could do this by writing $x + yi = (4 + i)/(2 - 3i)$ and evaluating the quotient on the right. But let us use another method. If we multiply out the left-hand side, we get

$$(2x + 3y) + (-3x + 2y)i = 4 + i$$

From our definition of the equality of two complex numbers, the *real parts* of both sides must be equal, and similarly the *imaginary* parts must be equal. Hence:

$$2x + 3y = 4$$
$$-3x + 2y = 1$$

We can solve these simultaneous equations and obtain $x = \frac{5}{13}$ and $y = \frac{14}{13}$.

This method of equating real and imaginary parts is of great importance in the application of complex numbers to engineering, and you should be certain that you understand it.

There are a number of other important properties of complex numbers that need to be discussed. Since these depend upon a knowledge of trigonometry, we defer their treatment to Sec. 10.11.

The above definition of a complex number is somewhat lacking in intuitive appeal. We have said that there is no real number x such that $x^2 = -1$, but immediately we introduce i with this property. What, then, is i? Mathematicians were sufficiently disturbed about this to call i an *imaginary* number and $a + bi$ *complex* numbers; by contrast, other numbers in our system are *real*. Our purpose now is to give an alternative

another
approach

development of the complex numbers in a logical and nonimaginary fashion.

Definitions

Complex number: A *complex number* is an ordered pair of real numbers (a,b).

Real part of a complex number: The complex number $(a,0)$ is called the *real part* of the complex number (a,b). We shall see that the pairs $(a,0)$ can be identified with the real numbers a in a natural fashion.

Imaginary part of a complex number: The complex number $(0,b)$ is called the *imaginary part* of the complex number (a,b). We shall also call complex numbers of this form *pure imaginary numbers*.

The arithmetic of complex numbers is given by the following basic definitions:

Definitions

Equality: Two complex numbers (a,b) and (c,d) are said to be *equal* if and only if $a = c$ and $b = d$.
Addition: $(a,b) + (c,d) = (a + c, b + d)$
Multiplication: $(a,b) \times (c,d) = (ac - bd, bc + ad)$

It is evident that there is a 1 to 1 correspondence between the complex numbers $(a,0)$ and the real numbers a which is defined by $(a,0) \leftrightarrow a$. This is a particularly useful correspondence, for under it sums correspond to sums and products to products. That is:

$$(a,0) + (c,0) = (a + c, 0) \qquad (a,0) \times (c,0) = (ac,0)$$

$$\updownarrow \qquad \updownarrow \qquad \updownarrow \qquad\qquad \updownarrow \qquad \updownarrow \qquad \updownarrow$$

$$a \;+\; c \;=\; a + c \qquad\quad a \;\times\; c \;=\; ac$$

Such a correspondence is called an *isomorphism,* and we say that the set of complex numbers $(a,0)$ is isomorphic to the set of real numbers a relative to addition and multiplication. We are therefore justified in identifying these two symbols and in calling $(a,0)$ a real number when there is no source of confusion.

Although the complex numbers $(a,0)$ are really nothing new, the pure imaginaries $(0,b)$ *are* something new. Their arithmetic, as derived from the definitions, is given by the following rules.

Addition $(0,b) + (0,d) = (0, b + d)$

Multiplication $(0,b) \times (0,d) = (-bd, 0)$

It is important to note that the product of two pure imaginaries is a real number. In particular,

$$(0,1) \times (0,1) = (-1,0)$$

We now recall that our motivation for introducing the complex numbers was our inability to solve the equation $x^2 = -1$ in terms of real numbers. Let us see how the introduction of complex numbers enables us to provide such a solution. By means of the isomorphism above, we see that the equation $x^2 = -1$ corresponds to the equation

$$(x,y)^2 = (x,y) \times (x,y) = (-1,0)$$

As we have noted, $(x,y) = (0,1)$ is a solution of this equation, and we also see that $(x,y) = (0,-1)$ is another solution. Therefore our introduction of complex numbers permits us to solve equations of this type, which had no solution in terms of real numbers.

In order to complete our discussion we need to show the correspondence between our two definitions of complex numbers. In preparation for this we note the following identities:

$$(0,b) = (b,0) \times (0,1)$$
$$(a,b) = (a,0) + [(b,0) \times (0,1)]$$

Exercise

A   Verify the above identities.

We then set up the following relationship between the two notations:

	(a,b) notation	a + bi notation
Real numbers	$(a,0)$	a
Unit imaginary	$(0,1)$	i

Using the identities above, we then derive the correspondences:

	(a,b) notation	a + bi notation
Pure imaginaries	(0,b)	bi
Complex numbers	(a,b)	a + bi

From these we show that the rules for the equality, addition, and multiplication of complex numbers in the $a + bi$ notation, which were stated as definitions at the beginning of this section, are in agreement with the corresponding definitions in the (a,b) notation.

Finally, we observe that with these definitions the complex numbers form a field. The details of the proof of this are included in the problems.

Problems 2.14

In Probs. 1 to 14 find the sum or difference of the given complex numbers.

1 $(9 + 8i) + (4 - 7i)$ **2** $(-19 + 5i) + (7 + 2i)$
3 $(15 - 7i) - (20 + 9i)$ **4** $(-13 + 9i) - (21 + 5i)$
5 $(10 + 6i) + (-13 - 18i)$ **6** $(6 - 5i) + (-4 + 2i)$
7 $-(5 - 3i) + (7 + 4i)$ **8** $-(33 + 5i) + (13 - 7i)$
9 $-(3 + i) - (-8 + 6i)$ **10** $-(-8 + 6i) - (6 - 4i)$
11 $6 - (4 - 7i)$ **12** $(8 - 9i) + 6$
13 $(5 + 13i) - 11i$ **14** $(21 - 16i) + 29i$

In Probs. 15 to 30 find the product of the given complex numbers.

15 $(3 + 7i)(-5 + 2i)$ **16** $(5 - 9i)(4 + 3i)$
17 $(7 + 3i)(5 - 2i)$ **18** $(6 + 2i)(5 + 9i)$
19 $(\sqrt{3} + i)(\sqrt{3} - i)$ **20** $(\sqrt{5} - 3i)(\sqrt{5} + 3i)$
21 $(8 + 5i)(8 - 5i)$ **22** $(3 - 7i)(3 + 7i)$
23 $3(7 - 5i)$ **24** $8(-3 + 6i)$
25 $(9i)(1 + 5i)$ **26** $(-4i)(-13 + 7i)$
27 $(3i)(7i)$ **28** $(4i)(-6i)$
29 i^5 **30** i^6

In Probs. 31 to 40 find the quotient of the given complex numbers.

31 $(5 + 4i)/(5 + 6i)$ **32** $(1 + 2i)/(3 + 5i)$
33 $(8 + 7i)/(4 - 3i)$ **34** $(5 + 9i)/(3 - i)$
35 $(-7 + 4i)/(-2 + 3i)$ **36** $(-8 + 4i)/(-5 - 6i)$
37 $5i/(6 + 9i)$ **38** $7i/(6 - 3i)$
39 $(3 + 8i)/(6i)$ **40** $(4 - i)/(3i)$

In Probs. 41 to 46 solve for x and y by equating real and imaginary parts.

41 $(x + iy)(1 - 5i) = 2 + 5i$ **42** $(x + iy)(3 - 7i) = 2 + 4i$
43 $(x + iy)(-1 + 4i) = -9 + 5i$ **44** $(x + iy)(2 + i) = 1 + 3i$
45 $(x + iy)(6i) = 9$ **46** $(x + iy)(-3i) = 13i$

In Probs. 47 to 52 show that the given complex number satisfies the given equation.

47 $3 + 3i; x^2 - 6x + 18 = 0$
48 $3 - 3i; x^2 - 6x + 18 = 0$
49 $3 - i; x^2 - (1 + 4i)x + (-1 + 17i) = 0$
50 $-2 + 5i; x^2 - (1 + 4i)x + (-1 + 17i) = 0$
51 $3 - 5i; x^2 + (-8 + 3i)x + (25 - 19i) = 0$
52 $5 + 2i; x^2 + (-8 + 3i)x + (25 - 19i) = 0$
53   Verify that R1 to R11 are satisfied when $a = 1 + 2i$, $b = 2 - i$, and $c = 2 + 3i$.
54   What is the additive inverse of $a + bi$? The multiplicative inverse?

In Probs. 55 to 60 carry out the indicated operations on complex numbers in the (a,b) notation.

55 $(3, -4) + (2,3)$ **56** $(1,2) - (-4,1)$
57 $(2,3) \times (1,4)$ **58** $(-1,2) \times (2,-3)$
59 $(4,1)/(3,-2)$ **60** $(-2,3)/(3,4)$

2.15   Graphical Representation of Complex Numbers

In contrast to the real numbers, the complex numbers cannot be represented in a useful way by the points on a line. By the same token we shall not define the notion of inequality between two complex numbers.

If we do wish to have a graphical representation of complex numbers, the most convenient procedure is to plot them as points in the plane (Fig. 2.5). To do so we measure the real part a of $a + bi$ along the hori-

Complex plane

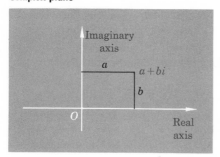

Figure 2.5
The real part, a, of $a + bi$ is measured horizontally, the imaginary part, b, vertically.

zontal (or *real*) axis and the imaginary part b along the vertical (or *imaginary*) axis. The details are the same as those for plotting the ordered pair (a,b) in Sec. 2.12. In this way we can establish a 1 to 1 correspondence between the complex numbers and the points of the plane.

addition

Addition of complex numbers has a convenient graphical interpretation within this framework (Fig. 2.6). Let P represent $a + bi$, and Q represent $c + di$. Then, as in the figure, complete the parallelogram $POQR$. The point R now represents $(a + bi) + (c + di)$ or $(a + c) + (b + d)i$.

Sum of two complex numbers

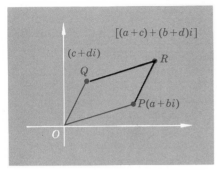

Figure 2.6 To add two complex numbers, complete the indicated parallelogram.

Figure 2.7 This diagram is used to prove that the construction used in Fig. 2.6 is correct.

To prove this is a simple exercise in geometry. In Fig. 2.7, $OT = a$ and $OS = c$. Since triangle PRU is congruent to triangle OQS, $PU = OS = c$. Hence $OV = OT + TV = OT + PU = a + c$. A similar argument shows that $RV = b + d$.

Exercises

A Add $2 + 3i$ and $4 + i$ algebraically and graphically.

B Devise a graphical construction for the difference $(a + bi) - (c + di)$. Check this by computing $(4 + 2i) - (3 + 5i)$ both algebraically and graphically.

Unfortunately the multiplication of complex numbers by graphical means is more complicated. We shall discuss this in Sec. 10.11.

2.16 Solutions of Other Algebraic Equations

Since we have extended our number system by considering progressively more and more complicated equations, it is reasonable to suppose that we may be led to "supercomplex" numbers in an effort to solve equations of the form

$$ax^n + bx^{n-1} + \cdots + cx + d = 0 \qquad a \neq 0$$

where $a, b, \ldots, c, d$ are complex numbers. As a matter of fact no new types of numbers need to be introduced for this purpose. This is a consequence of the so-called "Fundamental Theorem of Algebra":

Theorem 17 The equation above with complex coefficients always has a complex number $x = u + iv$ as a solution.

The proof of this theorem is beyond the scope of this book.

Since we do not have to invent any new numbers to solve this equation, we say that the set of complex numbers is *algebraically closed*. As a consequence we end our development of the number system at this point. It should be remarked, however, that for other purposes mathematicians have developed systems of "hypercomplex" numbers, two of which are called "quaternions" and "Cayley Numbers". We do not discuss these here.

2.17 Classification of Numbers

The following chart shows the various types of numbers and their relationships.

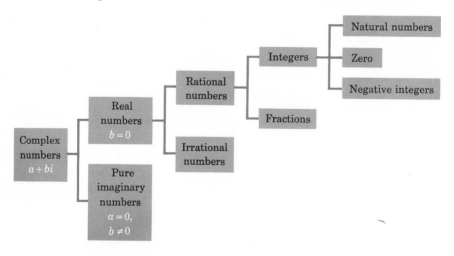

References

Birkhoff, Garrett, and Saunders MacLane: "A Survey of Modern Algebra", chaps. 1, 2, and 3, Macmillan, New York (1965).

Courant, Richard, and Herbert Robbins: "What Is Mathematics?" chaps. 1 and 2, Oxford, New York (1941).

Dantzig, Tobias: "Number, the Language of Science", Macmillan, New York (1930).

Henkin, Leon: "Mathematical Induction", Mathematical Association of America, Buffalo, N.Y. (1961).

Niven, Ivan, and Herbert S. Zuckerman: "An Introduction to the Theory of Numbers", Wiley, New York (1960).

Sorinskii, I. S.: "The Method of Mathematical Induction", Blaisdell, New York (1961).

Youse, B. K.: "Mathematical Induction", Prentice-Hall, Englewood Cliffs, N.J. (1964).

POLYNOMIALS

3.1 Algebraic Expressions

In this chapter and in the following two chapters we present a review of some of the most important notions in elementary algebra. These deal with the properties of *algebraic expressions*. In order to form such an expression, we first fix upon a particular field of numbers, which is generally the real or complex numbers. We then select a finite set of numbers belonging to this field and a finite collection of variables whose universal set is this field. Then we combine these in a finite number of steps through the processes of addition, subtraction, multiplication, division, and extraction of roots. The end result is an algebraic expression. The following are examples of such expressions:

$$2a + \frac{3}{b^2} \qquad \frac{6x + 5x^2}{2 - 3x} \qquad \sqrt{3r + 5s^2}$$

polynomials For the present we shall restrict ourselves to the operations of addition, subtraction, and multiplication; and we call the resulting algebraic expressions *polynomials*. A polynomial is the sum of a finite number of *terms*, each of which is the product of a finite collection of numbers and variables. A term may also include a negative sign, corresponding to a -1 contained in this product. In writing the expressions for these terms we write, as usual, a^2 for $a \times a$; b^3 for $b \times b \times b$; etc. You should note that the only exponents which occur on variables are *positive integers;* fractional and negative exponents are excluded. The numbers involved, however, may be any elements of the underlying field. Examples of polynomials are

$$4x^3 - 2x^2 + x - 1 \qquad \tfrac{1}{2}a + 3bc$$
$$5xy^2 - \tfrac{1}{3}yz + \sqrt{2} \qquad (3 + 2i)x^2 - 6i$$

Polynomials, such as $8x^2y^6$, which contain only one term are called *monomials;* those with two terms, such as $4a^2 - 7bc$, are called *binomials;* and those with three terms, such as $7a^2 + 4ab + 3b^2$, are called *trinomials.* No special names are given to polynomials with more than three terms.

In writing a term it is customary to multiply together all the numbers involved and to put this product in front. This number is called the *numerical coefficient* of the term. Similarly, all factors involving the same variable are brought together and are written as this variable with a suitable exponent. Thus we write

$$2 \times a \times 4 \times b \times a \times b \times b \qquad \text{as} \qquad 8a^2b^3$$

like terms

Two terms which differ only in their numerical coefficient are called *like terms.* For instance, $4x^2y^3$ and $6x^2y^3$ are like terms, but $3xy^2$ and $7x^2y$ are *not* like terms; similarly $4x^2$ and $3a^2$ are *not* like terms. In writing a polynomial it is customary to combine like terms by the use of the distributive law; for example:

$$4x^2y^3 + 6x^2y^3 = 10x^2y^3$$

As a result a polynomial will be expressed as a finite sum of unlike terms, each of which either is a number or consists of a numerical coefficient multiplied by the product of a finite number of distinct variables, each of which carries an exponent which is a positive integer.

Exercise

A Which of the following expressions are polynomials?

1 $3x^2 - x + \tfrac{1}{2}$

2 $x^3 + \dfrac{1}{x} + 3$

3 $\sqrt{2x} + 4x^2$

4 $2\sqrt{x} + 5x^2 - 2$

5 $\dfrac{3}{2x^2 - x + 6}$

3.2 Addition of Polynomials

The procedure for adding polynomials is a direct consequence of the commutative and associative laws for addition and of the distributive law (R2, R5, R11, Sec. 2.5). Let us illustrate by adding:

$$(5x^2y + x - 3xy^2 + 2) + (-2x + 3y + 7xy^2 - 5)$$

The first step is to use the commutative and associative laws to group like terms together. In the above example we obtain

$$(5x^2y) + (x - 2x) + (-3xy^2 + 7xy^2) + (3y) + (2 - 5)$$

Then we use the distributive law to combine the like terms. In our example we have the final result

$$5x^2y - x + 4xy^2 + 3y - 3$$

The process of addition can be conveniently carried out by arranging the work in columns, where each column contains like terms. In the above example we write

$$
\begin{array}{l}
5x^2y + x - 3xy^2 + 2 \\
\underline{ - 2x + 7xy^2 - 5 + 3y} \\
5x^2y - x + 4xy^2 - 3 + 3y
\end{array}
$$

This arrangement of the work is particularly helpful when three or more polynomials are to be added.

In order to subtract two polynomials, we convert the problem to addition (see the definition of subtraction, Sec. 2.2) and then proceed as above. For example:

$$(4x^2 - 3xy + 2) - (5x^2 + x - 3)$$

is written as the sum

$$(4x^2 - 3xy + 2) + (-5x^2 - x + 3)$$

and equals $\qquad$ $-x^2 - 3xy - x + 5$

Problems 3.2

Perform the indicated operations:

1   $(2a^2 + 3ab + c) + (2c - 5a^2 - ab)$
2   $(4x^2 - 2y^2 + 3x) + (4x + 6y^2 - 2x^2)$
3   $(-pq + q^3 + 2p^2) - (2p^2 - 2q^3 + 4pq)$
4   $(7r + 4s - 3r^2s^3) - (3s + 6r + 3r^2s^3)$
5   $(12x^2y + 4xy^2 - 2xy + 5) + (4xy + x^2 - 2y^2)$
6   $(7abc + 4ab^2 - 6ab + 4) + (7ab + 2ab^2 + 3b)$
7   $(8x^3 - 5y^3 + 2xy) - (5x^3 + 6y + 2xy)$
8   $(-2pq + 6p^2 - 4q) - (4q^2 + q - 5p^2)$
9   $(x^2 - 2y^2 + 3x) + (5a^2 + 3b^2 + 2y)$
10   $(6r^2 + 4rs + 2s^2) + (2x^2 + y^2 + s^3)$
11   $(2x^2y - 4xy^2 + 6xy) + (3x^3 + 7x^2y + 2xy^2) + (-xy^2 + 4xy + 2)$
12   $(-8a^3 + 14b^2 - 6a^2b) + (9a^3 + 9b^3 - a^2b) + (3a^2 + 8ab + b^2)$
13   $(2x^2 - 16y^2) + (3x^2 + 8y^2) + (-2r^2 + 4s^2)$
14   $(-pq + 3q^2) + (4pq - q^2) + (2a^2 + b^2)$
15   $(5x^4 - 2a^2 + 4xy) - (2x^4 + 5a^2 - xy) + (3x^4 + 2a^2 + 3xy)$
16   $(2a^2 - 2b^2 + 8c^3) + (2a^2 + 4b^2 - 5c^3) - (a^2 + b^2 + c^3)$
17   $(3xy + x^2 - 3y^4) + (x^2 + y^2 - 2xy) - (3x^2 - 2y^2 + 6xy)$
18   $(5r^2s + r^3 + 3s^5) - (2r^3 - 4s^5 + 5rs^2) + (10r^2s + 15rs^2)$
19   $[(3a^2 - 2b^2) - (6ab + b^2)] - (4a^2 + 6ab + 10b^2)$
20   $(25x^2 - 22xy + 15y^2) - [(4x^2 + 2y^2) - (15xy - 10y^2)]$

3.3 Multiplication of Polynomials

The procedure for the multiplication of polynomials is based upon the method of multiplying monomials, together with the repeated use of the distributive law (R11). We recall that a monomial is either a number or the product of a numerical coefficient and a collection of variables carrying positive integral exponents. The product of two monomials is, therefore, the product of all the factors of the two given monomials taken together.

Illustration 1

a $(2x^2y)(5x^3y^4) = 10x^5y^5$
b $(-5a^3bc)(4ab^2g^3) = -20a^4b^3cg^3$

In carrying out this product we recall that

$$a^p \cdot a^q = a^{p+q}$$

for a^p is the product of p a's and a^q is the product of q a's. There are, therefore, $(p + q)$ a's in the combined product.

We now use the distributive law to reduce the problem of multiplying polynomials to that of multiplying monomials.

Illustration 2 Multiply $(a^2 + 2b)(3a^2 + 4b + 1)$. One use of the distributive law permits us to write this as

$$a^2(3a^2 + 4b + 1) + 2b(3a^2 + 4b + 1)$$

A second application of the distributive law gives

$(3a^4 + 4a^2b + a^2) + (6a^2b + 8b^2 + 2b)$
$$= 3a^4 + 10a^2b + a^2 + 8b^2 + 2b$$

The work can be conveniently arranged as shown below:

$$
\begin{array}{l}
3a^2 + 4b + 1 \\
a^2 + 2b \\
\hline
3a^4 + 4a^2b + a^2 \\
 + 6a^2b + 8b^2 + 2b \\
\hline
3a^4 + 10a^2b + a^2 + 8b^2 + 2b
\end{array}
$$

A particularly important special case is that of the product of two polynomials which involve powers of a single variable only. In this case it is convenient to arrange the order of the terms so that the exponents decrease from term to term, i.e., "according to decreasing powers". Thus we would rearrange

$$7x^2 + 21x^5 - x^3 + 2x - 1 + 5x^4$$

to read $\qquad 21x^5 + 5x^4 - x^3 + 7x^2 + 2x - 1$

This will help us to keep things straight in our multiplications and later in our divisions.

Illustration 3   Multiply:

$$
\begin{array}{r}
5x^4 - 8x^3 + x^2 + 5x - 3 \\
x^2 + 2x - 1 \\
\hline
5x^6 - 8x^5 + x^4 + 5x^3 - 3x^2 \\
+ 10x^5 - 16x^4 + 2x^3 + 10x^2 - 6x \\
- 5x^4 + 8x^3 - x^2 - 5x + 3 \\
\hline
5x^6 + 2x^5 - 20x^4 + 15x^3 + 6x^2 - 11x + 3
\end{array}
$$

Problems 3.3

Perform the indicated operations:

1 $(2a^4b^2)(-3a^3b^4)$ **2** $(-5p^3q^5)(2p^4q^8)$

3 $(10x^2yz^4)(5x^3wy^4)$ **4** $(-6r^3s^4t)(9s^3t^7u^3)$

5 $(4a + 7b)(2a - 4b)$ **6** $(5x - 7y)(3x - 6y)$

7 $(4x + 2y)^2$ **8** $(-2a + 7b)^2$

9 $(3p + q)^3$ **10** $(2r - 3s)^3$

11 $(3x + 4y)(3x - 4y)$ **12** $(4a - 9b)(4a + 9b)$

13 $(4x^4 - 5x^3 + x^2 - 2x + 2)(x^2 + 4x + 3)$

14 $(3x^3 + 7x^2 - 2x + 5)(2x^3 - 4x + 2)$

15 $(2x^2 - x^3 - 5x^4 + 3)(x + 8x^2)$

16 $(4x + x^4 - 5x^3 + 2)(2x - 3 + x^2)$

17 $(3a^4 - 7ab^2 + b^3)(5a^3 - b^2 + 4)$

18 $(r^3 + 5r^2s^2 + s^4)(s^2 - 5s + 7)$

19 $(8x^4 - 5x^3y + 6x^2y^2 + 2xy^3 - y^4)(3x^2 + xy - y^2)$

20 $(5a^4 - 4a^3b + 6a^2b^2 - 2ab^3 + b^4)(a^2 - 2ab - b^2)$

3.4   Binomial Theorem

By direct multiplication, as in the last section, we can easily establish the following formulas:

$$(a + b)^2 = a^2 + 2ab + b^2$$
$$(a + b)^3 = a^3 + 3a^2b + 3ab^2 + b^3$$
$$(a + b)^4 = a^4 + 4a^3b + 6a^2b^2 + 4ab^3 + b^4$$
$$(a + b)^5 = a^5 + 5a^4b + 10a^3b^2 + 10a^2b^3 + 5ab^4 + b^5$$

Pascal's Triangle

The coefficients in these products have a pattern which is illustrated by the following scheme, known as *Pascal's Triangle*.

$$
\begin{array}{cccccccccccc}
(a + b)^0 & & & & & & 1 & & & & & \\
(a + b)^1 & & & & & 1 & & 1 & & & & \\
(a + b)^2 & & & & 1 & & 2 & & 1 & & & \\
(a + b)^3 & & & 1 & & 3 & & 3 & & 1 & & \\
(a + b)^4 & & 1 & & 4 & & 6 & & 4 & & 1 & \\
(a + b)^5 & 1 & & 5 & & 10 & & 10 & & 5 & & 1 \\
& \cdots & & & & & & & \cdots & & &
\end{array}
$$

In this array each horizontal line begins and ends with a 1, and each other entry is the sum of the two numbers to its left and right in the horizontal row above.

Exercises

A By direct multiplication verify the above formulas for $(a + b)^2$, $(a + b)^3$, $(a + b)^4$, $(a + b)^5$.

B From Pascal's Triangle determine the coefficients of the terms in the expansion of $(a + b)^6$, and verify your result by direct multiplication. (*Ans.:* 1, 6, 15, 20, 15, 6, 1.)

Whenever we discover a pattern like this, we suspect that there must be some general way of describing it and so we are led to ask whether there is some general formula for $(a + b)^n$, where n is any positive integer. There is indeed such a formula, known as the *Binomial Formula*, and we shall now proceed to develop it. By way of preparation we must introduce some new notations.

Definition The symbol $n!$ (for n a positive integer), read "*n factorial*", stands

factorial for the product

$$n! = 1 \times 2 \times 3 \times \cdots \times n$$

Further, $0!$ is defined to be 1. Factorials are not defined for negative integers or for other real numbers.

Exercises

C Compute $2!$, $3!$, $4!$, $5!$, $6!$.

D Show that $n!/n = (n - 1)!$.

E Compute $7!/3!$; $n!/(n - 1)!$; $n!(n - 2)$; $n!/r!$, where $r < n$.

Definition The symbol $\binom{n}{r}$, where n and r are integers ≥ 0 and $n \geq r$, is

defined to be

$$\binom{n}{r} = \frac{n!}{(n - r)!r!}$$

These symbols are called *binomial coefficients*.

Exercises

F Show that

$$\binom{4}{2} = 6 \qquad \binom{6}{4} = 15 \qquad \binom{5}{3} = 10 \qquad \binom{4}{4} = 1 \qquad \binom{5}{1} = 5$$

G From the definition show that

$$\binom{n}{r} = \binom{n}{n - r}$$

The connection between these symbols and the expression for $(a + b)^n$ is easily seen from the fact that Pascal's Triangle now can be written in the form

$$(a + b)^0 \qquad\qquad \binom{0}{0}$$

$$(a + b)^1 \qquad\qquad \binom{1}{0} \quad \binom{1}{1}$$

$$(a + b)^2 \qquad\qquad \binom{2}{0} \quad \binom{2}{1} \quad \binom{2}{2}$$

$$(a + b)^3 \qquad\qquad \binom{3}{0} \quad \binom{3}{1} \quad \binom{3}{2} \quad \binom{3}{3}$$

$$(a + b)^4 \qquad\qquad \binom{4}{0} \quad \binom{4}{1} \quad \binom{4}{2} \quad \binom{4}{3} \quad \binom{4}{4}$$

$$(a + b)^5 \qquad \binom{5}{0} \quad \binom{5}{1} \quad \binom{5}{2} \quad \binom{5}{3} \quad \binom{5}{4} \quad \binom{5}{5}$$

$$\cdots \qquad\qquad\qquad \cdots$$

Exercise

H Verify that this representation of Pascal's Triangle agrees with the one given earlier in this section.

We have now given you a broad hint regarding the nature of the Binomial Formula. Can you guess the correct expression for it? Cover up the next few lines of this page, and write down your guess. Please do not peek until you have written it down, for intelligent guessing is a most important part of the process of mathematical discovery, and you need practice in doing it. Now you can look, and we hope that you have written something like the following.

Theorem 1 Binomial Theorem Let n be a positive integer. Then

$$(a + b)^n = a^n + \binom{n}{1} a^{n-1}b + \binom{n}{2} a^{n-2}b^2 + \cdots$$

$$+ \binom{n}{n-2} a^2 b^{n-2} + \binom{n}{n-1} ab^{n-1} + b^n$$

Or, in the expansion of $(a + b)^n$, the coefficient of $a^{n-r}b^r$ is $\binom{n}{r}$.

Proof Since this theorem is to be proved for all values of n, a reasonable approach is to try mathematical induction. The formula is trivially verified for $n = 1$, and indeed we have verified it for $n = 1, 2, 3, 4, 5$. We must now establish the truth of step II of the Principle of Mathematical Induction (Sec. 2.7). In this case we must show that:

For all $k \geq 1$: If the coefficient of $a^{k-r}b^r$ in the expansion of $(a + b)^k$ is

$\binom{k}{r}$, then the coefficient of $a^{k+1-r}b^r$ in the expansion of $(a + b)^{k+1}$ is $\binom{k + 1}{r}$. We write

$$(a + b)^k = a^k + \cdots + \binom{k}{r - 1} a^{k+1-r}b^{r-1} + \binom{k}{r} a^{k-r}b^r + \cdots + b^k$$

Then $(a + b)^{k+1}$ is given by the product

$$a^k + \cdots + \binom{k}{r - 1} a^{k+1-r}b^{r-1} + \binom{k}{r} a^{k-r}b^r + \cdots + b^k$$

$$\underline{a + b}$$

$$a^{k+1} + \cdots + \binom{k}{r - 1} a^{k+2-r}b^{r-1} + \binom{k}{r} a^{k+1-r}b^r + \cdots + ab^k$$

$$a^k b + \cdots \qquad\qquad + \binom{k}{r - 1} a^{k+1-r}b^r + \binom{k}{r} a^{k-r}b^{r+1}$$

$$\underline{\qquad\qquad\qquad\qquad\qquad\qquad + \cdots + b^{k+1}}$$

$$a^{k+1} + \cdots \qquad\qquad + \left[\binom{k}{r} + \binom{k}{r - 1}\right] a^{k+1-r}b^r + \cdots + b^{k+1}$$

Therefore the coefficient of $a^{k+1-r}b^r$ is $\binom{k}{r} + \binom{k}{r - 1}$. We must now simplify this.

$$\binom{k}{r} + \binom{k}{r - 1} = \frac{k!}{r!(k - r)!} + \frac{k!}{(r - 1)!(k - r + 1)!}$$

$$= \frac{k!(k - r + 1) + k!r}{r!(k - r + 1)!}$$

$$= \frac{k!(k + 1)}{r!(k + 1 - r)!} = \frac{(k + 1)!}{r!(k + 1 - r)!} = \binom{k + 1}{r}$$

Exercise

I   Relate the last computation in the proof above to the method of constructing Pascal's Triangle.

The Binomial Theorem permits us to write down rather quickly the expansions of powers of binomials which are tedious to compute by repeated multiplication.

Illustration 1

a   Expand $(2x + 5y)^3$.

$$(2x + 5y)^3 = (2x)^3 + 3(2x)^2(5y) + 3(2x)(5y)^2 + (5y)^3$$
$$= 8x^3 + 60x^2y + 150xy^2 + 125y^3$$

b Expand $(3x - 2y)^4$.

$$(3x - 2y)^4 = (3x)^4 + 4(3x)^3(-2y) + 6(3x)^2(-2y)^2$$
$$+ 4(3x)(-2y)^3 + (-2y)^4$$
$$= 81x^4 - 216x^3y + 216x^2y^2 - 96xy^3 + 16y^4$$

c Compute the term involving x^4y^3 in the expansion of $(3x - 5y)^7$. This term will involve $(3x)^4(-5y)^3$ with an appropriate coefficient.

The theorem tells us that this coefficient is $\binom{7}{3} = 35$. So the term is

$$35(3x)^4(-5y)^3 = -354,375x^4y^3$$

Problems 3.4

1 Compute: $\binom{5}{3}$; $\binom{7}{2}$; $\binom{6}{4}$; $\binom{5}{0}$; $\binom{4}{1}$.

2 Compute: $\binom{6}{3}$; $\binom{4}{0}$; $\binom{7}{3}$; $\binom{8}{3}$; $\binom{7}{5}$.

In Probs. 3 to 6 verify the given formulas by direct computation.

3 $\binom{5}{3} + \binom{5}{4} = \binom{6}{4}$

4 $\binom{7}{4} + \binom{7}{5} = \binom{8}{5}$

5 $1 + \binom{4}{1} + \binom{4}{2} + \binom{4}{3} + \binom{4}{4} = 2^4$

6 $1 + \binom{5}{1} + \binom{5}{2} + \binom{5}{3} + \binom{5}{4} + \binom{5}{5} = 2^5$

In Probs. 7 to 16 expand by the Binomial Theorem.

7 $(2x + y)^5$ **8** $(a + 3b)^4$

9 $(r - 2s)^6$ **10** $(4p - 3q)^4$

11 $(x + \frac{1}{2}y)^6$ **12** $(\frac{1}{3}x + y)^6$

13 $(2/x + x^2)^3$ **14** $(x^2 - 3/x)^4$

15 $(1.01)^5 = (1 + 0.01)^5$ **16** $(1.98)^4 = (2 - 0.02)^4$

In Probs. 17 to 22 find the coefficient of the given term in the given expansion.

17 a^8b^5 in $(a + b)^{13}$ **18** r^3s^6 in $(2r + 3s)^9$

19 x^3y^4 in $(2x - 3y)^7$ **20** p^3q^3 in $(4p - 5q)^6$

21 x^4y^6 in $(x^2 + y^3)^4$ **22** x^4 in $(x^2 + 1/x)^5$

23 Write $(x + y + z)^3 = [x + (y + z)]^3$, and expand through repeated use of the Binomial Theorem.

24 Write $(x + y - z)^3 = [(x + y) - z]^3$, and expand through repeated use of the Binomial Theorem.

25 In the Binomial Formula put $a = 1$ and $b = 1$, and then prove that

$$2^n = 1 + \binom{n}{1} + \binom{n}{2} + \cdots + \binom{n}{n-2} + \binom{n}{n-1} + \binom{n}{n}$$

3.5   Division of Polynomials

In this section we shall restrict ourselves to polynomials which involve only a single variable, and we shall assume them to be arranged in descending powers of that variable. Examples of such polynomials are

$$15x^5 + 8x^3 - 4x^2 + x + 7 \qquad 8a^4 + \tfrac{1}{2}a^3 - a^2 + 2a + 5$$

The highest exponent which appears is called the *degree* of the polynomial. The degrees of the two examples above are 5 and 4, respectively. We shall denote such polynomials by the symbols $P(x)$, $Q(x)$, $D(x)$, etc., where the variable in the parentheses indicates the variable of the polynomial and P, Q, R, etc., are names to represent different polynomials of this type.

Suppose that we have two polynomials $P(x)$ of degree n and $D(x)$ of degree r, where $n \geq r$. We wish to consider what happens when we divide $P(x)$ by $D(x)$. This process of division needs a definition, but we shall postpone giving this until the method is clear. We illustrate by an example.

Illustration 1   Divide $6x^4 + 7x^3 + 12x^2 + 10x + 1$ by $2x^2 + x + 4$.

$$
\begin{array}{r}
3x^2 + 2x - 1 \\
2x^2 + x + 4\,\overline{\big)\,6x^4 + 7x^3 + 12x^2 + 10x + 1} \\
6x^4 + 3x^3 + 12x^2 \\
\hline
4x^3 + 0x^2 + 10x + 1 \\
4x^3 + 2x^2 + 8x \\
\hline
-\,2x^2 + 2x + 1 \\
-\,2x^2 - x - 4 \\
\hline
3x + 5
\end{array}
$$

We have a quotient of $3x^2 + 2x - 1$ and a remainder of $3x + 5$.

The process of division as illustrated above is straightforward and should already be familiar to you. Let us suppose that we have carried through the division

$$D(x)\,\overline{\big)\,P(x)}$$

and obtained a *quotient* $Q(x)$ and a *remainder* $R(x)$. $D(x)$ is called the *divisor*, and $P(x)$ is called the *dividend*. There are several points worthy of mention:

1   The division continues step by step until a remainder is reached whose *degree is less than the degree of the divisor.*

2   When the remainder is zero, the division is said to be *exact.*

3   By reversing the steps of the computation we can show that

$$P(x) = D(x) \cdot Q(x) + R(x)$$

for all values of x.

This leads us to the following statement.

Theorem 2   The Division Algorithm Let $P(x)$ and $D(x)$ be polynomials of degrees n and r, respectively, where $n \geq r$. Then there exist polynomials $Q(x)$, called the quotient, and $R(x)$, called the remainder, such that:

division algorithm

1   $P(x) = D(x) \cdot Q(x) + R(x)$ for all x.
2   The degree of $R(x)$ is less than the degree of $D(x)$.

The proof of this algorithm can be obtained by a generalization of the process shown in Illustration 1 or by more sophisticated means. Since either of these is too complicated to explain in detail here, we shall omit the proof.

In the problems below, follow the method of the illustration. You may run into trouble if a term is missing from the dividend or the divisor as in

$$4x^3 + 2x^2 - 3$$

which has no term in x. It will help you to keep matters straight if you will supply the missing term (or terms) with zero coefficients and write the above as

$$4x^3 + 2x^2 + 0x - 3$$

Problems 3.5

In Probs. 1 to 20 obtain the quotient and remainder, and check your result by substituting back into the equation $P(x) = D(x) \cdot Q(x) + R(x)$.

	Dividend	*Divisor*
1	$2x^3 + 7x^2 + 10x + 6$	$2x + 3$
2	$6x^3 - 7x^2 + 14x - 8$	$3x - 2$
3	$3x^3 + 10x^2 - 2x + 20$	$x + 4$
4	$-12x^3 + 7x^2 + x + 10$	$3x + 2$
5	$2x^4 + 5x^3 + 13x^2 + 15x + 10$	$x^2 + 2x + 3$
6	$6x^4 + 8x^3 + 5x^2 + 15x + 8$	$3x^2 + 4x - 2$
7	$12x^5 - 3x^4 + 11x^3 - 17x^2 + 2$	$3x^2 + 2$
8	$3x^5 - 5x^4 - 2x^3 + 28x^2 + 5x + 6$	$3x^2 + x - 3$
9	$4x^5 - x^4 + 8x^3 + x^2 + 7$	$4x^3 - x^2 + 1$
10	$5x^5 - 4x^4 + x^3 - 3x^2 + 6x + 3$	$x^3 - 1$
11	$25x^2 + 6x^4 - 11x^3 - 12x^5 + 7 + 3x$	$2 + 3x^2$
12	$30x^3 - 18x^5 - 41x^2 + 21x^4 + 5$	$5 - 3x^2$

	Dividend	*Divisor*
13	$x^4 - 1$	$x^2 - 1$
14	$x^6 - 1$	$x^3 - 1$
15	$x^6 - 1$	$x^2 + 1$
16	$x^5 - 1$	$x^2 + 1$
17	$x^6 - y^6$	$x - y$
18	$x^8 - y^8$	$x + y$
19	$\frac{2}{3}x^3 - 3x^2 - 5x + 1$	$2x + 3$
20	$(2 - i)x^3 + (6 - 4i)x^2 + (4 - i)x - (3 + 9i)$	$x + (3 - i)$

21 If $P(x)$ is of degree n and $D(x)$ of degree r and if $r > n$, show that the division algorithm is satisfied trivially with $Q(x) = 0$ and $R(x) = P(x)$.

3.6 Factoring

In factoring we seek to undo the process of multiplication. We are given a polynomial and are asked to discover how it can be expressed as a product of other polynomials, called its factors. You will find factoring important when you are asked to simplify algebraic fractions or to solve certain types of equations.

Definition   If a polynomial $P(x)$ is the product of r polynomials $Q_1(x), \ldots,$ $Q_r(x)$, that is, if

$$P(x) = Q_1(x) \cdot Q_2(x) \cdots Q_r(x)$$

then $Q_1(x), \ldots, Q_r(x)$ are called *factors* of $P(x)$.

Exercise

A   If $Q(x)$ is a factor of $P(x)$, show that the quotient of $P(x)$ by $Q(x)$ involves no remainder, i.e., that the division is exact.

In general, the factors $Q_1(x) \cdots Q_r(x)$ may have any coefficients in the field over which $P(x)$ is defined. In special cases, however, we may specify that the coefficients of the factors are to be integers and hence ask for the "integral factors" of $P(x)$. Similarly we may require the coefficients of the factors to be rational numbers or real numbers, etc. Unless some such restriction is stated, it is to be understood that the coefficients of the factors are unrestricted, i.e., that they may be of the general form $a + bi$.

The problem of factoring a general polynomial can be a quite complicated affair. As we shall see in Chap. 10, it is equivalent to the problem of finding all the roots of a polynomial equation, and this is by no means an elementary topic. We have a more limited objective here, namely, to factor certain simple polynomials by methods which are elementary and straightforward. Although these methods are not adequate to factor a general polynomial, they are very useful when they do apply. Hence they are worth mastering.

Removal of a Common Factor   When all the terms of the given polynomial have a factor in common, this expression may be factored out by
use of the distributive law in reverse.   These common factors should be
removed before other methods of factoring are employed.

Illustration 1

a $2x^3 + 6x^2 - 10 = 2(x^3 + 3x^2 - 5)$
b $x^4 - x^2 = x^2(x^2 - 1)$
c $ac + bc + ad + bd = (ac + bc) + (ad + bd)$
$= c(a + b) + d(a + b) = (c + d)(a + b)$

Trinomials with Integral Coefficients   We consider here trinomials of the
form

$$ax^2 + bx + c$$

where a, b, and c are *integers*.   We seek to write this as the product of two
linear polynomials with *integral* coefficients, i.e.,

$$ax^2 + bx + c = (px + q)(rx + s)$$

where p, q, r, and s are integers.   This factorization is not always possible,
but our method produces the required factors whenever they exist.

Let us first consider the simpler situation where $a = 1$.   Then p and r
must also equal 1, and we write

$$x^2 + bx + c = (x + q)(x + s)$$

Multiplying out the right-hand side, we find

$$x^2 + bx + c = x^2 + (q + s)x + qs$$

Thus, we are looking for two integers, q and s, such that

$$q + s = b \qquad qs = c$$

To find these, we factor c into all possible pairs (q,s) such that $qs = c$.
Then we examine these pairs (q,s) to determine whether or not in any of
them $q + s = b$.   If we find such a pair, we have solved the problem;
otherwise there are no factors of the prescribed form.   The details of the
method are best shown by illustrations.

Illustration 2

a Factor $x^2 + 5x + 6$.
The integral pairs of factors of 6 are

$$(1,6) \qquad (2,3) \qquad (-1,-6) \qquad (-2,-3)$$

We exclude the last two immediately since the sum of two negative
integers cannot be $+5$.   Examining the other two in turn, we find that
$(2,3)$ is a solution since $2 + 3 = 5$.   Therefore,

$$x^2 + 5x + 6 = (x + 2)(x + 3)$$

b Factor $x^2 - 6x + 8$.
The integral pairs of factors of 8 are

$$(1,8) \quad (2,4) \quad (-1,-8) \quad (-2,-4)$$

We exclude the first two immediately since the middle coefficient, -6, is negative. We find that $(-2,-4)$ is a solution since $(-2) + (-4) = -6$. Therefore

$$x^2 - 6x + 8 = (x - 2)(x - 4)$$

c Factor $x^2 + 3x - 10$.
The integral pairs of factors of -10 are

$$(1,-10) \quad (-1,10) \quad (2,-5) \quad (5,-2)$$

Examining these in turn, we find that $(5,-2)$ is a solution since $5 + (-2) = 3$. Therefore

$$x^2 + 3x - 10 = (x + 5)(x - 2)$$

d Factor $x^2 + 3x + 4$.
The integral pairs of factors of 4 are

$$(1,4) \quad (2,2) \quad (-1,-4) \quad (-2,-2)$$

None of these is a solution, and hence there are no factors of the prescribed form.

In the case of the general trinomial, we write

$$ax^2 + bx + c = (px + q)(rx + s)$$
$$= prx^2 + (ps + qr)x + qs$$

Therefore, $a = pr$, $b = ps + qr$, $c = qs$. The method is similar to the special case above, but here we have more possibilities. We find the pairs (p,r) which factor a and the pairs (q,s) which factor c. Then we examine each pair (p,r) in connection with each pair (q,s) to see whether or not for any of these combinations $ps + qr = b$. If so, we have a solution; otherwise, there are no factors of this form.

The number of possibilities will be reduced if we always take $a > 0$. This can always be arranged by removing the common factor, -1, if a is initially negative. Moreover, we can assume that p and r are both positive without losing any possible solutions; we must, however, allow q and s to take all appropriate positive and negative signs.

Illustration 3 Factor $8x^2 + 2x - 15$.
The pairs of factors of 8 are $(1,8)$ and $(2,4)$. The pairs of factors of -15 are $(1,-15)$, $(-1,15)$, $(3,-5)$, $(5,-3)$, $(-15,1)$, $(15,-1)$, $(-5,3)$, $(-3,5)$. In this case it is necessary to write the numbers in each pair in the two possible orders in order to cover all cases.

Now write one of the first pairs and one of the second pairs as shown below:

$$(1,8), (1,-15)$$

Multiply the two outside numbers, 1 and -15, and add to this the product of the two inside numbers, 8 and 1. This gives $(1)(-15) + (8)(1) = -7$. This should equal the coefficient of x, namely 2. Since it does not, these two pairs do not give a solution. Try each combination of a first pair and a second pair. Among these we find that the solution is

$$(2,4), \ (3,-5)$$

for $(2)(-5) + (4)(3) = 2$. Hence

$$8x^2 + 2x - 15 = (2x + 3)(4x - 5)$$

Difference of Two Squares We consider expressions of the form

difference
of two
squares

$$x^2 - a^2$$

An elementary calculation shows that

$$x^2 - a^2 = (x + a)(x - a)$$

Hence, the factors may be written down at sight.

Illustration 4

a $x^2 - 9 = (x + 3)(x - 3)$
b $16x^2 - 25 = (4x + 5)(4x - 5)$
c $25x^4 - 36y^6 = (5x^2 + 6y^3)(5x^2 - 6y^3)$
d $(3x + 5)^2 - (2x - 1)^2 = [(3x + 5) + (2x - 1)]$
$\qquad\qquad\qquad \times [(3x + 5) - (2x - 1)] = (5x + 4)(x + 6)$
e $x^2 - 2 = (x + \sqrt{2})(x - \sqrt{2})$

This method may be extended to cover expressions of the form

$$x^2 + a^2$$

For
$$\begin{aligned} x^2 + a^2 &= x^2 - (ia)^2 \\ &= (x + ia)(x - ia) \end{aligned}$$

Illustration 5

a $4x^2 + 9 = (2x + 3i)(2x - 3i)$
b $x^2 + 3 = (x + i\sqrt{3})(x - i\sqrt{3})$

Sum and Difference of Two Cubes We rely upon the two formulas

$$\begin{aligned} x^3 + a^3 &= (x + a)(x^2 - ax + a^2) \\ x^3 - a^3 &= (x - a)(x^2 + ax + a^2) \end{aligned}$$

which require no further explanation.

Combinations of the Above Methods It is often possible to factor complicated looking expressions by using two or more of these methods in turn.

Illustration 6 Factor:

$$\begin{aligned} &4x^2 + 24x + 32 - 16y^2 + 16y \\ &= 4x^2 + 24x + 36 - 16y^2 + 16y - 4 \end{aligned}$$

$$= 4[(x^2 + 6x + 9) - (4y^2 - 4y + 1)]$$
$$= 4[(x + 3)^2 - (2y - 1)^2]$$
$$= 4[(x + 3 + 2y - 1)(x + 3 - 2y + 1)]$$
$$= 4(x + 2y + 2)(x - 2y + 4)$$

Problems 3.6

In Probs. 1 to 6 factor the given expression.

1 $xy + 3x + 3y + 9$ 2 $xy - 20 + 4x - 5y$

3 $x^3 + x^2 - x - 1$ 4 $cy + dy - 5xy$

5 $sx - sy + vx - vy$ 6 $x^2y - 4y + x^2 - 4$

In Probs. 7 to 32 find the factors with integral coefficients, if any.

7 $x^2 - 2x - 8$ 8 $x^2 + 3x - 4$

9 $x^2 - 20x + 36$ 10 $x^2 + 12x + 27$

11 $32 - 12x + x^2$ 12 $15 - 8x + x^2$

13 $y^2 - 18xy + 45x^2$ 14 $a^2 - 11ab + 30b^2$

15 $x^2 + 6x + 10$ 16 $x^2 - 3x - 6$

17 $3x^2 + 13x + 4$ 18 $2x^2 + x - 15$

19 $3x^2 - 11x - 4$ 20 $2x^2 - 9x + 10$

21 $4x^2 + 5x - 21$ 22 $7x^2 + 32x - 15$

23 $5x^2 - 9xy - 18y^2$ 24 $2x^2 - xy - y^2$

25 $10x^2 + 7x - 12$ 26 $6x^2 + x - 35$

27 $12x^2 + 7x - 45$ 28 $24x^2 - 26x - 15$

29 $12x^2 + 67x - 275$ 30 $45x^2 + 12x - 105$

31 $3x^2 + 11x + 5$ 32 $4x^2 + 7x - 3$

In Probs. 33 to 56 factor the given expression.

33 $9x^2 - 16$ 34 $25x^2 - 144$

35 $x^2 - 7$ 36 $x^2 - 6$

37 $9x^2 + 4$ 38 $16x^2 + 25$

39 $(x + 3)^2 - (x - 2)^2$ 40 $(x + 7)^2 - (x + 3)^2$

41 $x^4 - y^4$ 42 $x^8 - y^8$

43 $x^2 + 4x - 5 - y^2 - 6y = (x^2 + 4x + 4) - (y^2 + 6y + 9)$

44 $x^2 - 6x - 7 - y^2 - 8y = (x^2 - 6x + 9) - (y^2 + 8y + 16)$

45 $27x^3 - y^3$ 46 $64x^3 - 1$

47 $x^3 + 125$ 48 $x^3 + 216$

49 $(a^2 - 9b^2) + 2a + 1$ 50 $(h^2 + 4) + 4h - x^2$

51 $x^4 - 10x^2 + 9$ 52 $x^4 - 20x^2 + 64$

53 $2x^3 + 7x^2y + 6xy^2$ 54 $3x^2y - 20xy^2 - 7y^3$

55 $2x^2 + 5xy - 2x + 2y^2 + 5y - 12$

56 $3x^2 - 2xy + 2x - y^2 + 6y - 8$

57 By mathematical induction show that for all $n \geq 1$, $x - y$ is a factor of $x^n - y^n$.

58 By mathematical induction show that for all $n \geq 1$, $x + y$ is a factor of $x^{2n} - y^{2n}$.

ALGEBRAIC FRACTIONS

Chapter

4.1 Introduction

An algebraic fraction is the quotient of two algebraic expressions. Examples of these are

$$(1) \quad \frac{3x^2 + 4a}{x + 1} \qquad (2) \quad \frac{\sqrt{x^3 - 1} + 4x}{\sqrt[3]{x + 7}} \qquad (3) \quad \frac{1/x^2 + \sqrt{3}\,x}{2/(x + 5)}$$

We recall that the variables in these expressions stand for arbitrary numbers. In the most general circumstances these numbers are complex, but in particular situations it may be specified that they are real or rational. This leads to two important remarks:

1 The algebra of fractions can be derived from properties R1 to R11 of Sec. 2.5.

2 It is understood that we cannot assign values to any variable which makes any denominator equal to zero. Thus in example (1) above we exclude $x = -1$; in example (2) we exclude $x = -7$; in example (3) we exclude $x = 0$ and $x = -5$. It is tedious to state these exclusions each time that we write a fraction, and so you will have to supply this information yourself and to take necessary precautions.

Throughout this chapter we shall restrict ourselves to fractions whose numerators and denominators are either polynomials or quotients of polynomials. We do this for simplicity of exposition, and not because the theory is restricted to such cases. Fractions containing radicals will be treated in Chap. 5.

4.2 Simplification of Fractions

Since fractions are troublesome enough in any case, we wish to be able to simplify any fraction that turns up as much as possible before putting it back into some further calculation. The most important way of doing this is nothing more than the familiar "reducing to lowest terms". In spite of the simplicity of this method, its misuse is a source of frequent errors on the part of careless students—so read this section carefully.

Basic Principle The method depends upon the familiar relation

$$\frac{ka}{kb} = \frac{a}{b} \qquad \text{for } k \neq 0$$

In other words: If we divide the numerator and denominator of a given fraction by the same quantity (not zero) the result is a fraction equal to the given one.

In order to apply this principle to algebraic fractions, we factor the numerator and denominator, look for common factors, and divide top and bottom by any factor which is common to both.

Illustration 1

a $$\frac{x^2 - 5x + 6}{x^2 - 4x + 3} = \frac{(x - 3)(x - 2)}{(x - 3)(x - 1)} = \frac{x - 2}{x - 1}$$

b $$\frac{4x^2 + 7x}{x^2} = \frac{x(4x + 7)}{x(x)} = \frac{4x + 7}{x}$$

c $$\frac{x^2 + 4x + 4}{x^2 + 4x + 3} = \frac{(x + 2)(x + 2)}{(x + 1)(x + 3)}$$

This does not simplify since the numerator and denominator have no common factor.

The matter of excluded values of x raises the question of what we mean by the equality of two fractions. In Illustration 1a, the fraction on the left is defined for all values of x except $x = 1$ and $x = 3$; on the other hand, the fraction on the right is defined for all values of x except $x = 1$. A strict use of equality between these two would lead us to the following non-sensical relation when we put $x = 3$:

$$\frac{0}{0} = \text{nonsense} = \frac{3 - 2}{3 - 1} = \frac{1}{2}$$

In order to avoid such difficulties, let us define equality of algebraic fractions as follows.

Definition *Two algebraic fractions* involving the variable x are *equal* if and only if they have the same numerical values when x is put equal to any number for which both fractions are defined.

equality

Common Errors   Some of the mistakes which students make are based upon the following erroneous relation:

$$\frac{k + a}{k + b} = \frac{a}{b} \qquad\qquad FALSE$$

In other words, it is incorrect to simplify a fraction by subtracting the same quantity from numerator and denominator!

Illustration 2

a $\dfrac{x^2 + 4x + 4}{x^2 + 4x + 3}$ does not equal $\dfrac{4}{3}$.   See Illustration 1c.

b $\dfrac{2x + 3}{2x + 1}$ does not equal $\dfrac{3}{1} = 3$.

Exercise

A   Disprove the statements $\dfrac{x^2 + 4x + 4}{x^2 + 4x + 3} = \dfrac{4}{3}$ and $\dfrac{2x + 3}{2x + 1} = 3$ by find-

ing suitable counterexamples.

Other errors are caused by failure to remember the distributive law. These are based upon the following erroneous relation:

$$\frac{k + a}{k} = a \qquad\qquad FALSE$$

The trouble here is that k has been treated as a factor of the numerator. Proceeding correctly, we can write, however,

$$\frac{k + a}{k} = \frac{k\left(1 + \dfrac{a}{k}\right)}{k} = 1 + \frac{a}{k} \qquad\qquad TRUE$$

Another approach to this is the following:

$$\frac{k + a}{k} = \frac{1}{k}(k + a)$$

Now apply the distributive law, which gives

$$\frac{1}{k}(k + a) = \frac{1}{k}(k) + \frac{1}{k}(a) = 1 + \frac{a}{k} \qquad\qquad TRUE$$

Illustration 3

a $\dfrac{5x + 7}{5x}$ is not equal to 7 but does equal $1 + \dfrac{7}{5x}$.

b $\dfrac{(x + 3)^2 + x - 2}{x + 3}$ is not equal to $(x + 3) + x - 2$ but does equal

$x + 3 + \dfrac{x - 2}{x + 3}$.

Exercise

B   Disprove the statements $\dfrac{5x + 7}{5x} = 7$ and $\dfrac{(x + 3)^2 + x - 2}{x + 3} =$

$(x + 3) + x - 2$ by finding suitable counterexamples.

In the problems which follow, some fractions will simplify, and others will not. In your zeal to effect a simplification do not commit either of the common errors illustrated above.

Problems 4.2

In Probs. 1 to 20 simplify where possible.

1 $\dfrac{x^2 - 9}{x^2 + 6x + 9}$

2 $\dfrac{x^2 + 4x + 4}{x^2 - 4}$

3 $\dfrac{x^2 + 7x + 12}{x^2 + 8x + 15}$

4 $\dfrac{x^2 - 6x + 8}{x^2 - 3x + 2}$

5 $\dfrac{x^2 + 5x}{x + 3}$

6 $\dfrac{x^2 + 7x}{7x}$

7 $\dfrac{2x^2 - 5x - 3}{x^2 + x - 12}$

8 $\dfrac{x^2 + 7x + 12}{x + 7}$

9 $\dfrac{9x^2 - 4}{6x^2 + x - 2}$

10 $\dfrac{6x^2 + 7x - 20}{2x^2 + 9x + 10}$

11 $\dfrac{ac + bc + ad + bd}{4a + 4b - ay - by}$

12 $\dfrac{y^2 + 4zy + 3z^2}{ay + az + by + bz}$

13 $\dfrac{x^3 - y^3}{x^2 - y^2}$

14 $\dfrac{x^3 + y^3}{x^2 + y^2}$

15 $\dfrac{a^4 + b^4}{a^2 - b^2}$

16 $\dfrac{4a^2x^2 - 25x^2}{2ax + 5x}$

17 $\dfrac{a^2x^3 + 7ax^3 + 12x^3}{a^2x - 16x}$

18 $\dfrac{a^4 - b^4}{a^2 - b^2}$

19 $\dfrac{x^3 + 9x^2 + 20x}{x + 9x^2 + 20}$

20 $\dfrac{ax^2 + 9ax + 18a}{y + 6}$

In Probs. 21 to 26 find counterexamples which disprove the given statements.

21 $\dfrac{3x + 9}{3x} = 9$

22 $\dfrac{5x - 3}{5x} = -3$

23 $\dfrac{11x + x^3}{11x + x} = x^2$

24 $\dfrac{15x + 8}{15x - 4} = -2$

25 $\dfrac{5x}{3} + \dfrac{7x}{5} = \dfrac{12x}{8}$

26 $\dfrac{3x}{7} - \dfrac{8x}{3} = -\dfrac{5x}{4}$

4.3   Addition

The addition of fractions is a straightforward application of the formula below, which we derived in Sec. 2.8, Theorem 10:

$$\frac{a}{b} + \frac{c}{d} = \frac{ad + bc}{bd}$$

When the fraction on the right is obtained, it should then be simplified as much as possible.

Illustration 1

a $\dfrac{2x - 1}{x + 3} + \dfrac{x^2}{3x - 1} = \dfrac{(2x - 1)(3x - 1) + (x + 3)x^2}{(x + 3)(3x - 1)}$

$= \dfrac{x^3 + 9x^2 - 5x + 1}{3x^2 + 8x - 3}$

b $\dfrac{x}{x + 3} + \dfrac{5x^2}{x^2 - 9} = \dfrac{x(x^2 - 9) + (x + 3)(5x^2)}{(x + 3)(x^2 - 9)}$

$= \dfrac{6x^3 + 15x^2 - 9x}{(x + 3)(x^2 - 9)}$

$= \dfrac{3x(2x - 1)(x + 3)}{(x + 3)(x^2 - 9)}$

$= \dfrac{3x(2x - 1)}{x^2 - 9}$

least common denominator

Although the above process always gives the correct result, it may involve unnecessary complexities. These occur because the process leads to a denominator which is not necessarily the *least common denominator* (L.C.D.). You have doubtless met the notion of a least common denominator in arithmetic. When you added

$$\frac{4}{9} + \frac{7}{12}$$

you learned to write

$$\frac{4}{9} + \frac{7}{12} = \frac{16}{36} + \frac{21}{36} = \frac{37}{36}$$

and you avoided the use of $9 \times 12 = 108$ as a denominator. In arithmetic the least common denominator is the smallest number which contains the given denominators as factors. We found it above by first factoring $9 = 3^2$ and $12 = 2^2 \times 3$ into prime factors. Then we formed a number (the L.C.D.) by multiplying together the several distinct factors we had found (namely, 2 and 3), each raised to the larger of the two powers to which it was raised in the given two numbers. Thus the L.C.D. of $\frac{4}{9}$ and $\frac{7}{12}$ is $3^2 \times 2^2 = 36$.

In algebra we would like to follow the same procedure, but here we run into difficulties of both a theoretical and a practical nature. Our factoring of 9 and 12 above was into *prime* factors, but we can give no definition of what is meant by a prime factor of a general algebraic expression. Even with numbers we get into difficulty in factoring when we leave the domain of integers; for $6 = 2 \times 3$ and also

$$6 = (1 - \sqrt{-5}) \times (1 + \sqrt{-5})$$

and the unique factorization theorem (Sec. 2.6) fails to hold. When, in particular, our denominators are polynomials, it is possible to define prime factors; but then we may well have practical difficulty in finding these factors. Therefore, there may in fact be no L.C.D., or we may be unable to find one even when it exists. The point to remember is that the use of the L.C.D. is a great convenience when it can be found easily but that the method of the L.C.D. cannot be applied universally. The use of the L.C.D. is not, therefore, a fixed requirement of the addition process, but it should be used wherever possible. In most of the problems below its use is recommended.

Illustration 2

a $\dfrac{3x + 4}{x^2 - 16} + \dfrac{x - 3}{x^2 + 8x + 16}$

We write $x^2 - 16 = (x + 4)(x - 4)$ and $x^2 + 8x + 16 = (x + 4) \times (x + 4)$. Hence we form the L.C.D., which is $(x + 4)^2(x - 4)$. Then

$$\frac{3x + 4}{x^2 - 16} + \frac{x - 3}{x^2 + 8x + 16} = \frac{3x + 4}{(x + 4)(x - 4)} + \frac{x - 3}{(x + 4)^2}$$

$$= \frac{(3x + 4)(x + 4)}{(x + 4)^2(x - 4)} + \frac{(x - 3)(x - 4)}{(x + 4)^2(x - 4)}$$

$$= \frac{4x^2 + 9x + 28}{(x + 4)^2(x - 4)}$$

Since this fraction does not simplify, it is the final answer.

b $\dfrac{x}{x + 3} + \dfrac{5x^2}{x^2 - 9}$ (see Illustration 1*b*)

The L.C.D. is $x^2 - 9$. Hence we write

$$\frac{x(x - 3)}{x^2 - 9} + \frac{5x^2}{x^2 - 9} = \frac{6x^2 - 3x}{x^2 - 9} = \frac{3x(2x - 1)}{x^2 - 9}$$

c $\dfrac{x}{x^2 - 2x + 5} + \dfrac{3}{x - 1}$

In this case our procedure breaks down, for we do not know how to factor $x^2 - 2x + 5$. So we forget about the L.C.D. and write

$$\frac{x}{x^2 - 2x + 5} + \frac{3}{x - 1} = \frac{(x^2 - x) + 3(x^2 - 2x + 5)}{(x^2 - 2x + 5)(x - 1)}$$

$$= \frac{4x^2 - 7x + 15}{(x^2 - 2x + 5)(x - 1)}$$

d $\dfrac{x}{x^2 - 2x + 5} + \dfrac{3}{x - 1 + 2i}$

As in Illustration 2c, the only thing you can do is to write

$$\frac{x}{x^2 - 2x + 5} + \frac{3}{x - 1 + 2i} = \frac{(x^2 - x + 2ix) + (3x^2 - 6x + 15)}{(x^2 - 2x + 5)(x - 1 + 2i)}$$

$$= \frac{4x^2 - 7x + 2ix + 15}{(x^2 - 2x + 5)(x - 1 + 2i)}$$

Actually, however, a little more knowledge will give you a better result. If you had been clever, you might have noted that

$$x^2 - 2x + 5 = (x - 1 + 2i)(x - 1 - 2i)$$

So you could have written

$$\frac{x}{x^2 - 2x + 5} + \frac{3}{x - 1 + 2i} = \frac{x + 3(x - 1 - 2i)}{x^2 - 2x + 5}$$

$$= \frac{4x - 6i - 3}{x^2 - 2x + 5}$$

This is a better answer than that given above, but it was obtained by methods which you are not likely to have thought of. We shall not discuss these methods here, but this illustration should be a sufficient hint for a good student.

Problems 4.3

Carry out the indicated operations.

1 $\dfrac{3}{x + 1} + \dfrac{7}{x - 1}$ **2** $\dfrac{4}{3 - x} + \dfrac{5}{6 - x}$

3 $\dfrac{5}{2u - 3v} + \dfrac{4}{3v - 2u}$ **4** $\dfrac{2}{3x + 6y} + \dfrac{8}{x + 2y}$

5 $\dfrac{5}{a} - \dfrac{3}{a - b}$ **6** $\dfrac{6}{x + 4} - \dfrac{9}{x + 3}$

7 $\dfrac{5}{xy + y^2} + \dfrac{3}{x^2 + xy}$ **8** $\dfrac{y}{x - y} + \dfrac{3}{x^2 - y^2}$

9 $\dfrac{3x}{x^2 + 5x + 4} - \dfrac{x}{x^2 - 16}$ **10** $\dfrac{x - 4}{x^2 - 5x + 6} + \dfrac{2x}{x^2 - 2x - 3}$

11 $\dfrac{2x}{x^2 + 2x + 1} - \dfrac{x^2}{x^2 + 4x + 4}$

12 $\dfrac{4x}{(x+y)^2} + \dfrac{5}{x-y}$

13 $\dfrac{x+3}{x^2+3x+3} + \dfrac{2x}{x^2+x+5}$

14 $\dfrac{x^2}{x^2+x-1} - \dfrac{x+3}{x+3}$

15 $\dfrac{2x}{x-1} - \dfrac{3}{x} - \dfrac{2}{x+1}$

16 $2 - \dfrac{3}{a-1} + \dfrac{4}{a^2-1}$

17 $\dfrac{3}{x^2+5x+4} - \dfrac{2}{x^2+4x+3} + \dfrac{7}{x^2+8x+15}$

18 $\dfrac{4}{x^2-3x-4} + \dfrac{3}{x^2-16} - \dfrac{5}{x^2+6x+8}$

19 $\dfrac{2x}{x^2-6x+25} + \dfrac{7}{x-3+4i}$

20 $\dfrac{5x}{x^2+10x+29} - \dfrac{3}{x+5-2i}$

4.4   Multiplication and Division

We saw in Sec. 2.8, Theorem 11, that the multiplication of fractions follows the rule

$$\frac{a}{b} \times \frac{c}{d} = \frac{ac}{bd}$$

Also from Prob. 7, Sec. 2.10, we know that division follows the rule

$$\frac{a}{b} \div \frac{c}{d} = \frac{a}{b} \times \frac{d}{c} = \frac{ad}{bc}$$

These rules apply equally well to algebraic fractions. When a, b, c, and d are polynomials, it is desirable to factor them if possible so that simplifications in the final answer can be made easily. As shown in the illustrations below, it is usually convenient to make these simplifications at an early stage of the work rather than to wait to carry them out after the product has been found.

Illustration 1

a $\dfrac{x^2-y^2}{4y} \times \dfrac{2y}{x+y} = \dfrac{(x-y)(x+y)}{4y} \times \dfrac{2y}{x+y}$

$$= \frac{(x-y)(x+y)(2y)}{4y(x+y)} = \frac{x-y}{2}$$

In the next to the last fraction we have divided numerator and denominator by $(x + y)(2y)$ in order to obtain the final result. You might just as well, however, have carried out this division at the previous stage and thus have shortened the calculation as follows:

$$\frac{x^2 - y^2}{4y} \times \frac{2y}{x + y} = \frac{\overset{\vee}{(x - y)}(x + y)}{\underset{\underset{2}{\vee\vee}}{4y}} \times \frac{\overset{\vee\vee}{2y}}{\underset{\vee}{x + y}} = \frac{x - y}{2}$$

In order to keep track of our divisions, we have placed check marks ($\vee$) above those factors that have been divided out. We have also written a 2 as the quotient of 4 by 2. Instead of using check marks many

cancellation   people cross out these factors and say that they have been "canceled". There is no harm in canceling if it is done with understanding, but too often it is used blindly without an appreciation of the fact that *division* is the true operation involved.

b $\dfrac{x^2 + 3x + 2}{x^2 - x - 2} \times \dfrac{x^2 + x - 2}{x^2 + 4x + 4} \div \dfrac{x^2 - x}{x^2 + x - 6}$

$$= \frac{\overset{\vee}{(x + 2)}\overset{\vee}{(x + 1)}}{\underset{\vee}{(x - 2)}\underset{\vee}{(x + 1)}} \times \frac{\overset{\vee}{(x + 2)}\overset{\vee}{(x - 1)}}{\underset{\vee}{(x + 2)}\underset{\vee}{(x + 2)}} \times \frac{\overset{\vee}{(x - 2)}(x + 3)}{\underset{\vee}{x}(x - 1)} = \frac{x + 3}{x}$$

c $\dfrac{x^2 + 5x + 6}{x^2 - 1} \div \dfrac{x^2 + 1}{x + 4} = \dfrac{(x + 2)(x + 3)}{(x + 1)(x - 1)} \times \dfrac{x + 4}{x^2 + 1}$

$$= \frac{(x + 2)(x + 3)(x + 4)}{(x + 1)(x - 1)(x^2 + 1)}$$

There are no common factors to divide out, and so no simplification is possible.

Do not expect that every problem will simplify. You will make errors if you force yourself to simplify *every* problem just because simplifications do occur in *many* problems proposed in books. Actual problems derived from nature rarely simplify, but you must know the process just in case you are lucky enough to find a problem in your work which does simplify.

Problems 4.4

Carry out the indicated operations, and simplify where possible.

1 $\dfrac{x^2 - 4x + 4}{x^2 - 4} \times \dfrac{x + 2}{3}$ **2** $\dfrac{x^2 - 9}{x^2 + 2x} \times \dfrac{x + 2}{x - 3}$

3 $\dfrac{x^2 - x - 6}{x^2 + 6x + 8} \div \dfrac{x^2 - 4x + 3}{x^2 + 5x + 4}$ **4** $\dfrac{x^2 - 3x - 4}{x^2 - 1} \div \dfrac{x^2 - 16}{x + 3}$

5 $\dfrac{x^2 + 2x + 1}{x^2 + 4x + 4} \times \dfrac{x^2 - 1}{(x + 2)^2}$

6 $\dfrac{x^2 - 25}{x^2 - 36} \div \dfrac{x^2 - 5}{36x^2 - 1}$

7 $\dfrac{x^2 - x - 12}{x^2 + x - 12} \times \dfrac{x^2 + 2x - 15}{x + 1}$

8 $\dfrac{x^2 + 6x}{x^2 - 3x - 10} \times \dfrac{x^2 - 4x - 12}{x^2 + 5x - 6}$

9 $\dfrac{4r + 8s}{3rs} \div \dfrac{3(r + 2s)}{9rs}$

10 $\dfrac{xy + 3x}{7y} \div \dfrac{x^2y + 3x^2}{4y}$

11 $\dfrac{3x^2 + x - 2}{x^2 - 1} \times \dfrac{x^2 - x}{6x - 4}$

12 $\dfrac{8x^2 + 2x - 15}{3x^2 + 13x + 4} \times \dfrac{6x^2 - x - 1}{2x^2 + 5x + 3}$

13 $\dfrac{6x^2 + 7x - 20}{12x^2 + 31x + 7} \div \dfrac{3x^2 + 11x - 20}{3x^2 - 2x - 21}$

14 $\dfrac{6x^2 - x - 35}{4x^2 + 5x - 6} \div \dfrac{3x^2 + 13x + 14}{4x^2 - 3x}$

15 $\dfrac{xy - xz}{xy + xz} \times \dfrac{y}{y - z} \times \dfrac{y + z}{y}$

16 $\dfrac{xy + x}{y^2 - 1} \times \dfrac{y - 1}{x + 3} \times \dfrac{2x + 6}{x^2 + 4x}$

17 $\dfrac{x^3 - y^3}{x^2} \times \dfrac{y^2}{x^2 - y^2} \div \dfrac{x^2 + xy + y^2}{x^2 + 2xy + y^2}$

18 $\dfrac{r^3 + 27}{r^2 + 6r + 9} \times \dfrac{r^2 - 3r}{9 - r^2} \div \dfrac{9r - 3r^2 + r^3}{r^2 - 4r + 3}$

19 $\dfrac{p}{p + q + r} \times \dfrac{(p + q)^2 - r^2}{rp + rq - r^2}$

20 $\dfrac{(a + b)^2 - 36}{a + b - 6} \times \dfrac{a^2 - ab + 6a}{(a + 6)^2 - b^2}$

4.5   Compound Fractions

The operations of adding, subtracting, multiplying, and dividing fractions can be combined in various ways. The only situation which calls for special comment is that of simplifying a fraction whose numerator and denominator are themselves sums of fractions. In this case the numerator and denominator should be simplified separately, and finally the division should be performed.

Illustration 1

$$\frac{\dfrac{1}{x+1} - \dfrac{1}{x-1}}{\dfrac{1}{x+1} + \dfrac{1}{x-1}} = \frac{\dfrac{(x-1)-(x+1)}{(x+1)(x-1)}}{\dfrac{(x-1)+(x+1)}{(x+1)(x-1)}} = \frac{\dfrac{-2}{x^2-1}}{\dfrac{2x}{x^2-1}}$$

$$= \frac{-2}{x^2-1} \times \frac{x^2-1}{2x} = -\frac{1}{x}$$

Problems 4.5

General Review of Fractions Carry out the indicated operations, and simplify where possible.

1 $\left(\dfrac{3}{x-2} - \dfrac{4}{x-1}\right) \times \dfrac{4x}{x-5}$

2 $\left(\dfrac{2}{x+2} + \dfrac{5}{x-3}\right) \times \dfrac{2x-1}{7x+4}$

3 $\left(\dfrac{x}{x-3} + \dfrac{2x}{-2x+4}\right) \div \dfrac{x}{4x+7}$

4 $\left(\dfrac{x}{x+1} - \dfrac{9}{x+9}\right) \div \dfrac{x+3}{2x-1}$

5 $\dfrac{\dfrac{x}{x-y} - \dfrac{y}{x+y}}{\dfrac{x}{x+y} + \dfrac{y}{x-y}}$

6 $\dfrac{3 + \dfrac{4a}{a+5b}}{\dfrac{2a}{a+5b}}$

7 $3x - \dfrac{2}{4x - \dfrac{5}{3x}}$

8 $4x + \dfrac{6}{2x - \dfrac{3}{5x}}$

9 $\left(\dfrac{x^2+x-2}{x^2-4x} \times \dfrac{x^2-3x-4}{x^2+5x+6}\right) + \left(\dfrac{2x+6}{x} \times \dfrac{x+4}{x^2+x-6}\right)$

10 $\left(\dfrac{x^2-9}{9x^2-1} \times \dfrac{1-3x}{x+3}\right) - \left(\dfrac{x-3}{x^2+4x-5} \times \dfrac{x+5}{x^2-2x-3}\right)$

11 $\left(\dfrac{3x-1}{x-1} \times \dfrac{2}{x+2}\right) + \left(\dfrac{4}{x+2} \div \dfrac{x-3}{x-5}\right) - \left(\dfrac{2}{x-1} \times \dfrac{2x+1}{x-3}\right)$

12 $\left(\dfrac{3}{x} \div \dfrac{x+1}{2x+1}\right) - \left(\dfrac{4}{x+1} \times \dfrac{x-1}{x-3}\right) + \left(\dfrac{x-2}{x-3} \div \dfrac{x}{2}\right)$

13 $\dfrac{\dfrac{x+3}{9x^2-1}}{\dfrac{2x-1}{6x^2-5x+1}}$

14 $\dfrac{\dfrac{2x-1}{2x^2+5x+2}}{\dfrac{x+2}{2x^2+11x+5}}$

15 $\left(\dfrac{3x+1}{x} - \dfrac{x}{3x+1}\right) \times \left(\dfrac{5x-2}{x} + \dfrac{x}{5x-2}\right)$

16 $\left(\dfrac{x^2}{x+2} + \dfrac{x+2}{x^2}\right) \times \left(\dfrac{x}{5x+2} - \dfrac{5x+2}{x}\right)$

17 $\dfrac{\dfrac{x^2}{x+2} + \dfrac{x-1}{x+3}}{\dfrac{x^2}{x-2} - \dfrac{2}{x+4}}$

18 $\dfrac{\dfrac{x}{x+2} + \dfrac{3x}{2x-5}}{\dfrac{x^2}{x-3} - \dfrac{x+1}{x+2}}$

19 $\dfrac{x+1}{x^2 - 2x + 5} - \dfrac{2x}{x-1+2i}$

20 $\dfrac{1+2x}{x^2 + 4x + 13} + \dfrac{x+2}{x+2-3i}$

EXPONENTS AND RADICALS

Chapter

5.1 Positive Integral Exponents

By this time you should be well acquainted with the notation a^n, where a is any real or complex number and n is a positive integer.   The simplest approach to its meaning is the statement that a^n stands for the product of n factors, each equal to a.   We call a^n "the nth power of a", and n the exponent.

For our purposes in this chapter we prefer to use instead the following *inductive* definition of a^n.

Definition   The symbol a^n (where a is any number and n is a positive integer) is defined inductively by the formulas $a^1 = a$, $a^{n+1} = a^n \times a$.

inductive definition of a^n

According to this definition

$$a^1 = a$$
$$a^2 = a^1 \times a = a \times a$$
$$a^3 = a^2 \times a = (a \times a) \times a = a \times a \times a$$
$$a^4 = a^3 \times a = (a \times a \times a) \times a = a \times a \times a \times a$$
$$\text{etc.}$$

We must now examine the rules for handling these symbols.   All of them are derived from the definition and from properties R1 to R11 (Sec. 2.5).

Theorem 1   Let m and n be positive integers, and let a be any number.

106

Then

$$a^m \times a^n = a^{m+n}$$

Proof Let us fix m and construct an induction on n. As step I we must verify that $a^m \times a^1 = a^{m+1}$. But this is immediate, for by definition $a^{m+1} = a^m \times a = a^m \times a^1$.

In step II of the induction we must prove that for all $k \geq 1$: If $a^m \times a^k = a^{m+k}$, then $a^m \times a^{k+1} = a^{m+k+1}$. To show this we consider the product $a^m \times a^k \times a$. Now

$$\begin{aligned}
a^m \times a^k \times a &= (a^m \times a^k) \times a \\
&= (a^{m+k}) \times a &&\text{[Hypothesis]} \\
&= a^{m+k+1} &&\text{[Definition]}
\end{aligned}$$

Also $\qquad a^m \times a^k \times a = a^m \times (a^k \times a)$

$$= a^m \times a^{k+1} \qquad\qquad \text{[Definition]}$$

Equating these two expressions for $a^m \times a^k \times a$, we have

$$a^m \times a^{k+1} = a^{m+k+1}$$

which we were to prove.

Exercise

A Where have we used the associative law of multiplication in the above proof?

Corollary I If $m > n$, and $a \neq 0$, then $a^m/a^n = a^{m-n}$

Proof First, $\qquad \dfrac{a^m}{a^n} \times a^n = a^m$

Moreover, $\qquad a^{m-n} \times a^n = a^m \qquad$ [Theorem 1]

Therefore $\qquad \dfrac{a^m}{a^n} \times a^n = a^{m-n} \times a^n$

Dividing both sides by a^n, we complete the proof.

Corollary II If $m < n$ and $a \neq 0$, then $a^m/a^n = 1/a^{n-m}$.

Proof First, $\qquad \dfrac{a^m}{a^n} \times \dfrac{1}{a^m} = \dfrac{1}{a^n}$

Moreover,

$$\frac{1}{a^{n-m}} \times \frac{1}{a^m} = \frac{1}{(a^{n-m}) \times a^m} = \frac{1}{a^n} \qquad \text{[Theorem 1]}$$

The conclusion follows as in Corollary I above.

Illustration 1

a $2^5 \times 2^3 = 2^{5+3} = 2^8 = 256$

b $r^8 \times r^{15} = r^{23}$

c $3^6/3^2 = 3^4$

d $4^3/4^7 = 1/4^4$

e But note that the theorem does not apply to $2^4 \times 3^6$, which cannot be written more simply in terms of exponents.

Theorem 2 Let m and n be positive integers, and let a be any number. Then

$$(a^m)^n = a^{mn}$$

Proof Let us fix m and construct an induction on n. As step I we must verify that $(a^m)^1 = a^{m \times 1}$, a result which is immediately evident.

In step II of the induction we must prove that for all $k \geq 1$: If $(a^m)^k = a^{mk}$, then $(a^m)^{k+1} = a^{m(k+1)}$. To do so we consider the product $(a^m)^k \times a^m$. From the definition

$$(a^m)^k \times a^m = (a^m)^{k+1}$$

On the other hand,

$$
\begin{aligned}
(a^m)^k \times a^m &= a^{mk} \times a^m && \text{[Hypothesis]} \\
&= a^{mk+m} && \text{[Theorem 1]} \\
&= a^{m(k+1)}
\end{aligned}
$$

Equating these two expressions for $(a^m)^k \times a^m$, we have

$$(a^m)^{k+1} = a^{m(k+1)}$$

which we have to prove.

Exercises

B Which of R1 to R11 have we used in the above proof?

C Find a counterexample to the following false relation, which is sometimes confused with Theorem 2:

$$(a^m)^n = a^{m^n}$$

Illustration 2

a $(4^2)^3 = 4^6$

b $(x^4)^2 = x^8$

c $2^3 \times 4^5 = 2^3 \times (2^2)^5 = 2^3 \times 2^{10} = 2^{13}$

Theorem 3 Let n be a positive integer, and let a and b be any numbers. Then

$$(ab)^n = a^n \times b^n$$

Proof   Again we use mathematical induction. In step I, we immediately verify that $(ab)^1 = a^1 \times b^1$. As step II we must prove that for all $k \geq 1$: If $(ab)^k = a^k \times b^k$, then $(ab)^{k+1} = a^{k+1} \times b^{k+1}$. To do so we consider the product $(ab)^k \times (ab)$. On the one hand we have

$$(ab)^k \times (ab) = (ab)^{k+1} \qquad \text{[Definition]}$$

On the other hand

$$\begin{aligned} (ab)^k \times (ab) &= a^k \times b^k \times (ab) \qquad &\text{[Hypothesis]} \\ &= (a^k \times a) \times (b^k \times b) \\ &= a^{k+1} \times b^{k+1} \qquad &\text{[Definition]} \end{aligned}$$

Equating these two expressions for $(ab)^k \times (ab)$, we have

$$(ab)^{k+1} = a^{k+1} \times b^{k+1}$$

which we were to prove.

Exercise

D   Which of R1 to R11 have we used in this proof?

Illustration 3

a $(3 \times 5)^4 = 3^4 \times 5^4$
b $(xy)^7 = x^7 \times y^7$

5.2   Zero and Negative Exponents

So far in our discussion of a^n we have required that n be a positive integer. We now wish to extend the definition of a^n to permit n to be zero or a negative integer. How shall we define a^0 and a^{-p} where p is a positive integer? It would seem reasonable to seek definitions which would permit us to extend Theorem 1 to this more general situation. Thus, for instance, we should choose a definition for a^0 which makes the following equation true:

$a^0 = 1$

$$a^0 \times a^n = a^{0+n} = a^n$$

Thus, provided that $a \neq 0$,

$$a^0 = \frac{a^n}{a^n} = 1$$

We use this argument to motivate the following definition.

Definition   The symbol a^0 (where a is any number $\neq 0$) is equal to 1. We give no meaning to the symbol 0^0.

In a similar fashion we wish to define a^{-p} so that

$$a^{-p} \times a^p = a^{-p+p} = a^0 = 1$$

Provided that $a \neq 0$, this leads to the result

$$a^{-p} = \frac{1}{a^p}$$

a^{-p}
$= 1/a^p$

Definition   The symbol a^{-p} (where p is a positive integer and a is any number $\neq 0$) stands for the quotient $1/a^p$. We give no meaning to the symbol 0^{-p}.

We can now reexamine Theorems 1, 2, and 3 of Sec. 5.1 in order to see how they generalize when the exponents are arbitrary integers (positive, negative, or zero).

Theorem 1'   Let m and n be any integers, and let a be any number $\neq 0$. Then

$$a^m \times a^n = a^{m+n}$$

Proof   To prove this, we must treat various cases separately.
1   $m > 0, n > 0$. This is then Theorem 1 of Sec. 5.1.
2   m arbitrary, $n = 0$. Then

$$a^m \times a^n = a^m \times a^0 = a^m \times 1 = a^m = a^{m+0} = a^{m+n}$$

3   $m > 0, n < 0$. Let $n = -p$, and suppose $m > p$. Then

$$a^m \times a^n = a^m \times a^{-p} = \frac{a^m}{a^p} = a^{m-p} = a^{m+n}$$

Now suppose $m < p$. Then

$$a^m \times a^n = a^m \times a^{-p} = \frac{a^m}{a^p} = \frac{1}{a^{p-m}} = a^{m-p} = a^{m+n}$$

Finally suppose $m = p$. Then

$$a^m \times a^n = a^p \times a^{-p} = \frac{a^p}{a^p} = 1 = a^0 = a^{p-p} = a^{m+n}$$

4   $m < 0, n < 0$. Let $m = -p; n = -q$. Then

$$a^m \times a^n = a^{-p} \times a^{-q} = \frac{1}{a^p} \times \frac{1}{a^q}$$

$$= \frac{1}{a^p \times a^q} = \frac{1}{a^{p+q}} = a^{-(p+q)} = a^{-p-q} = a^{m+n}$$

Theorem 2'   Let m and n be any integers, and let a be any number $\neq 0$. Then

$$(a^m)^n = a^{mn}$$

The proof is by cases as above and is included in the Problems.

Theorem 3' Let n be any integer, and let a and b be any numbers $\neq 0$. Then

$$(ab)^n = a^n \times b^n$$

The proof is again by cases and is included in the Problems.

Exercise

A Why must we exclude $a = 0$ in the statements of Theorems 1', 2', and 3'?

Problems 5.2

In Probs. 1 to 20 perform the indicated operations. Write your answers in a form which uses *positive* exponents only.

1 $4^3 \times 4^7$ **2** $6^2 \times 6^5$

3 $5^{-3} \times 5^4 \times 5^0$ **4** $9^6 \times 9^{-3} \times 9^0$

5 $3^5 \times 6^{-2} \times 6^5 \times 3^{-8}$ **6** $7^4 \times 15^7 \times 7^{-8} \times 15^{-3}$

7 $(-3)^4 \times 9^2 \times (-3)^5 \times 9^4$ **8** $4^5 \times (-2)^3 \times 4^{-2} \times (-2)^4$

9 $\dfrac{(4^{-1})7 + 4 + 4(7^{-1})}{(4^{-1})(7^{-1})}$ **10** $\dfrac{3(5^{-1}) + 4 + (3^{-1})6}{(5^{-1})(3^{-1})}$

11 $\dfrac{xy^{-2} + 2 + x^{-3}y^2}{x^{-3}y^{-4}}$ **12** $\dfrac{p^{-3}q + 4 + p^2q^{-4}}{p^{-5}q^{-7}}$

13 $\dfrac{abc^{-3} - a^{-1}b^{-2}c^2}{a^2b^{-1}c^4}$ **14** $(z + 2z^{-2})(3z^{-1} - z^2)$

15 $(y^2 + 3y^{-1})(2y - y^{-2})$ **16** $(1 - x^{-1} + x^{-2})(1 - x)$

17 $\dfrac{4x^{-1} + 3 - 5x}{3x^{-2} + 4}$ **18** $\dfrac{7x^2 + 2x - 9x^{-4}}{2x^{-2} + 3x}$

19 $\dfrac{3x^{-1} + 5x^{-2} + 4x^{-4}}{7x^2 + x^0}$ **20** $\dfrac{2x^4 - 3x^{-2} + 4x^0 - x^{-2}}{x^3 - x^{-3}}$

21 Prove by induction: For all integers $n \geq 1$, and for any numbers a and b $(\neq 0)$,

$$\left(\frac{a}{b}\right)^n = \frac{a^n}{b^n}$$

22 Prove Theorem 2'. Consider the cases: $(m > 0, n > 0)$; $(m > 0, n < 0)$; $(m < 0, n > 0)$; $(m < 0, n < 0)$; $(m = 0, n$ arbitrary); $(m$ arbitrary, $n = 0)$.

23 Prove Theorem 3'. Consider the cases: $n > 0, n = 0, n < 0$.

5.3 Fractional Exponents

We now come to the matter of taking the square root, the cube root, and fourth root, etc., of a number. In order to avoid some troublesome dif-

a is positive   ficulties, in this section we shall assume that we are dealing with the roots of *positive real* numbers only.

Let *a* be a positive real number, and suppose that $b^n = a$. Then we say that "*b* is *an* nth root of *a*"; we must not say "*the* nth root", for indeed there may be several of these. For instance,

$$2^2 = 4 \quad \text{and} \quad (-2)^2 = 4$$

so that both 2 and (-2) are square roots of 4. As a general principle a mathematical symbol should stand for just one mathematical object rather than for several such objects. For this reason we use $\sqrt[n]{a}$ in the carefully defined sense given below.

Definition   Let *a* be a positive real number and *n* a positive integer. Then the symbols $\sqrt[n]{a}$ and $a^{1/n}$ will be used interchangeably to mean that particular one of the *n*th roots of *a* which is a positive real number.

$\sqrt[n]{a}$ is
positive *n*th
root of *a*

Illustration 1 $\sqrt{4} = 4^{1/2} = +2$; $\sqrt{25} = 25^{1/2} = +5$; $-\sqrt{36} = -36^{1/2} = -6$. We never write $\sqrt{25} = \pm 5$.

Exercise

A   Why is $a^{1/n}$ a desirable notation for $\sqrt[n]{a}$? HINT: Consider a reasonable generalization of Theorem 2.

$a^{p/q}$
$= (a^{1/q})^p$

Now that we know what we mean by $a^{1/n}$, we must extend our definition to symbols of the form $a^{p/q}$.

Definition   Let *a* be a positive real number, and let *p* and *q* be positive integers. Then the symbols $(\sqrt[q]{a})^p$ and $a^{p/q}$ are used interchangeably to mean the *p*th power of $a^{1/q}$. That is, $a^{p/q} = (a^{1/q})^p$.

Another meaning for $a^{p/q}$ is derived from Theorem 4.

Theorem 4 $a^{p/q} = \sqrt[q]{a^p}$, where $\sqrt[q]{a^p}$ denotes the positive *q*th root of a^p. That is, $(a^{p/q})^q = a^p$.

Proof   We must show that $(a^{p/q})^q = a^p$. From the above definition we have

$$
\begin{aligned}
(a^{p/q})^q &= [(a^{1/q})^p]^q \\
&= [a^{1/q}]^{pq} & \text{[Theorem 2]} \\
&= [(a^{1/q})^q]^p & \text{[Theorem 2]} \\
&= a^p & \text{[Definition of } a^{1/q}]
\end{aligned}
$$

Finally we extend our definition of negative exponents to our fractional exponents.

Definition   Let a be a positive real number and p and q be positive integers. Then

$$a^{-p/q} = \frac{1}{a^{p/q}}$$

We have now completely defined the symbol a^r when a is positive and r is any rational number.   Let us see how Theorems 1, 2, and 3 generalize to this situation.

Theorem 1″   Let a be a positive real number and r and s any rational numbers.   Then

$$a^r \times a^s = a^{r+s}$$

Proof   Let $r = p/q$ and $s = u/v$, where p, q, u, and v are integers.   Then we must prove that

$$a^{p/q} \times a^{u/v} = a^{(pv+qu)/qv}$$

Because of Theorem 4 this is equivalent to proving that

$$(a^{p/q} \times a^{u/v})^{qv} = a^{pv+qu}$$

However:

$$
\begin{aligned}
(a^{p/q} \times a^{u/v})^{qv} &= (a^{p/q})^{qv} \times (a^{u/v})^{qv} &&\text{[Theorem 3′]}\\
&= [(a^{p/q})^q]^v \times [(a^{u/v})^v]^q &&\text{[Theorem 2′]}\\
&= (a^p)^v \times (a^u)^q &&\text{[Theorem 4]}\\
&= a^{pv} \times a^{qu} &&\text{[Theorem 2′]}\\
&= a^{pv+qu} &&\text{[Theorem 1′]}
\end{aligned}
$$

Theorem 2″   Let a be a positive real number and r and s any rational numbers.   Then

$$(a^r)^s = a^{rs}$$

Theorem 3″   Let a and b be any positive real numbers and r any rational numbers.   Then

$$(ab)^r = a^r \times b^r$$

The proofs of these theorems are similar to that of Theorem 1″ and are included in the Problems.

Exercise

B   Define $0^r = 0$ where r is a *positive* rational number.   Show that Theorems 1″, 2″, and 3″ can be extended to include the case where $a = 0$ and r and s are *positive* rational numbers.

5.4   Special Problems Concerning Square Roots

In the last section we required that a be a positive real number. Here we relax that restriction and consider two special difficulties that occur when we take square roots.

The Square Root of a^2**, or** $\sqrt{a^2} = (a^2)^{1/2}$ When a is positive, Theorem 2″ tells us that $(a^2)^{1/2} = a$.   Consider, however, $[(-3)^2]^{1/2}$.   We have

$$[(-3)^2]^{1/2} = 9^{1/2} = +3$$

This does not agree with Theorem 2″, and so we see that we cannot extend the validity of this theorem to cover the case where a is negative.   A correct statement which includes all cases is the following theorem.

$\sqrt{a^2}$
$= |a|$

Theorem 5   For any real number a, $(a^2)^{1/2} = |a|$, where $|a|$ denotes the absolute value of a.

Proof   When $a \geq 0$, Theorem 5 is a consequence of Theorem 2″.   When $a < 0$, let $a = -b$, where $b > 0$.   Then

$$(a^2)^{1/2} = [(-b)^2]^{1/2} = (b^2)^{1/2} = b = |a|$$

Illustration 1

a $(-5^2)^{1/2} = \sqrt{(-5)^2} = +5$
b $(x^2)^{1/2} = \sqrt{x^2} = |x|$
c $(x^2 + 2x + 1)^{1/2} = \sqrt{x^2 + 2x + 1} = |x + 1|$

The result of Theorem 5 can be extended at once to any even root.

Theorem 6   For any real number a and any positive integer n, $(a^{2n})^{1/(2n)} = |a|$.

Proof $\qquad (a^{2n})^{1/(2n)} = [(a^2)^n]^{1/(2n)} = (a^2)^{1/2} = |a|$

Exercise

A   State the justification for each step in the proof of Theorem 6.

$\sqrt{-a}$
$= i\sqrt{a}$

The Square Root of a Negative Number   Let a be a positive real number, and consider $(-a)^{1/2}$.   This needs to be defined.

Definition   For any positive real number a,

$$(-a)^{1/2} = i(a^{1/2})$$

Illustration 2 $\qquad \sqrt{-3} = (-3)^{1/2} = i\sqrt{3} = i3^{1/2}$

Expressions of this kind need special care, for Theorem 3 does not hold in this case. Let us consider $(-a)^{1/2} \times (-b)^{1/2}$, where a and b are both positive.

$$(-a)^{1/2} \times (-b)^{1/2} = i(a^{1/2}) \times i(b^{1/2}) = i^2 a^{1/2} b^{1/2} = -(ab)^{1/2}$$

The application of Theorem 3 would have given the incorrect result:

$$(-a)^{1/2} \times (-b)^{1/2} = +(ab)^{1/2}$$

Illustration 3

a $\quad (-3)^{1/2} \times (-5)^{1/2} = i(3)^{1/2} \times i(5)^{1/2} = -15^{1/2}$
b $\quad \sqrt{-10} \times \sqrt{-7} = i\sqrt{10} \times i\sqrt{7} = -\sqrt{70}$
c $\quad \sqrt{13} \times \sqrt{-3} = \sqrt{13} \times i\sqrt{3} = i\sqrt{39}$

5.5 Special Problems Concerning Odd Roots

Again let a be a positive real number, and consider $\sqrt[3]{-a}$. There is always a negative real number $-b$ such that $(-b)^3 = -a$, and we shall write $-b = \sqrt[3]{-a} = (-a)^{1/3}$. In general we proceed as follows.

Definition   Let a be a positive real number and p be any odd positive integer. Then the symbols $\sqrt[p]{-a}$ and $(-a)^{1/p}$ will be used interchangeably to denote the negative real number, $-b$ such that $(-b)^p = -a$.

Illustration 1

a $\quad \sqrt[3]{-27} = (-27)^{1/3} = -3$
b $\quad \sqrt[5]{-32} = (-32)^{1/5} = -2$
c $\quad \sqrt[11]{-x^{11}} = -x^1 = -x$

Remarks

1   The notation $(-a)^{1/p}$ is used by many authors in a fashion different from that defined above. In their usage this symbol denotes a certain complex number called the "principal pth root of $-a$". We shall discuss this notion in Sec. 10.11.

2   The symbol $(-a)^{p/q}$ for a positive and q odd can now be defined in a fashion analogous to the definition of $a^{p/q}$ in Sec. 5.3. These symbols then obey Theorems 1″, 2″, and 3″. The proofs are included in the Problems.

5.6   Unanswered Questions

Although we have discussed the meaning of the symbol a^b, where a and b are certain types of real numbers, there are still several cases which we have omitted. We call your attention to the following situations, which we shall treat later in this book:

1 $(-a)^{p/q}$, where a is positive, p is odd, and q is even. This is included as a special case in Sec. 10.11.

2 a^b, where a is positive and b is irrational. This is discussed in Sec. 11.2.

3 We must omit entirely the complicated, but fascinating story of a^b, where a and b are any complex numbers. For this consult G. H. Hardy, "Pure Mathematics", pages 409 to 410, 457 to 459, Cambridge, New York, 1945.

Problems 5.6

In Probs. 1 to 8 perform the given operations. The universal set for each variable mentioned is the set of positive real numbers.

1 $(4x^{3/2} - 2x^{-1/2})x^{3/2}$

2 $(2x^{1/3} - 3x^{-2/3})x^{-1/3}$

3 $(x^{1/2} + y^{1/2})^2$

4 $(a^{1/3} + b^{1/3})^3$

5 $(p^{1/2} - q^{1/2})(p^{1/2} + q^{1/2})$

6 $(x^{-3/2} + y^{5/2})(x^{1/2} - y^{3/2})$

7 $\dfrac{x^3 + 4x^{3/2}}{2x^{1/2}}$

8 $\dfrac{x^3 + 3x^{-4} - x^{-3/2}}{x^{3/2}}$

In Probs. 9 to 12 change the given expression to an equivalent form in which all exponents are positive and no compound fractions appear. All variables are positive.

9 $\dfrac{x^{2/3} + 2x^{1/3} - 4x^{-4/3}}{3x^{2/3} + x^{-5/3}}$

10 $\dfrac{3y^{1/7} - 4y^{-3/7} + 5y^{2/7}}{2y^2 - y^{-4/7}}$

11 $\dfrac{5a^{1/3}b^{-2/3} + 4a^{-2/3}b^{4/3}}{5a^{-2}b^{1/3} + 3a^{1/3}b}$

12 $\dfrac{4r^{-1/5}s^{3/5} - 3r^{2/5}s^{-1/5}}{5rs + 7r^2s^{-4}t}$

In Probs. 13 to 20 simplify the given expressions. The universal set for each variable mentioned is the set of all real numbers.

13 $(x^2 + 6x + 9)^{1/2}$

14 $(9x^2 + 6x + 1)^{1/2}$

15 $(x^2 + 2x + 1)^{1/2} + (x^2 - 2x + 1)^{1/2}$. Give a counterexample to show that $2x$ is an incorrect answer.

16 $(x^2 + 4x + 4)^{1/2} - (x^2 - 4x + 4)^{1/2}$. Give a counterexample to show that 4 is an incorrect answer.

17 $\dfrac{(4x^2 - 12x + 9)^{1/2}}{2x - 3}$ **18** $\dfrac{(16x^2 + 40x + 25)^{1/2}}{4x + 5}$

19 $x\sqrt{1 + x^{-2}}$ **20** $(x + 2)\sqrt{1 + (x + 2)^{-2}}$

In Probs. 21 to 30 simplify the given expression, assuming that the variables are restricted as indicated.

21 $(x^2 + 10x + 25)^{1/2}$ where $x + 5 \geq 0$
22 $(x^2 - 12x + 36)^{1/2}$ where $x - 6 \geq 0$
23 $(x^2 + 10x + 25)^{1/2}$ where $x + 5 \leq 0$
24 $(x^2 - 12x + 36)^{1/2}$ where $x - 6 \leq 0$
25 $(x^2 + 2x + 1)^{1/2} + (x^2 + 4x + 4)^{1/2}$ where $x + 1 \geq 0$
26 $(x^2 - 8x + 16)^{1/2} - (x^2 - 10x + 25)^{1/2}$ where $x - 5 \geq 0$
27 $(x^2 + 2x + 1)^{1/2} + (x^2 + 4x + 4)^{1/2}$ where $x + 2 \leq 0$
28 $(x^2 - 8x + 16)^{1/2} - (x^2 - 10x + 25)^{1/2}$ where $x - 4 \leq 0$
29 $(x^2 + 2x + 1)^{1/2} + (x^2 + 4x + 4)^{1/2}$ where $-2 \leq x \leq -1$
30 $(x^2 - 8x + 16)^{1/2} - (x^2 - 10x + 25)^{1/2}$ where $4 \leq x \leq 5$

In Probs. 30 to 45 perform the indicated operations:

31 $\sqrt{-5} \times \sqrt{-7}$ **32** $\sqrt{-25} \times \sqrt{-36}$
33 $\sqrt{9} \times \sqrt{-16}$ **34** $\sqrt{4} \times \sqrt{-25}$
35 $\sqrt[3]{-27} \times \sqrt{25}$ **36** $\sqrt{8} \times \sqrt[3]{-64}$
37 $\sqrt[3]{-8} \times \sqrt[4]{16}$ **38** $\sqrt[3]{-125} \times \sqrt[3]{-27}$
39 $\sqrt{-16} \times \sqrt[3]{-216}$ **40** $\sqrt{2} + \sqrt{8}$ HINT: $\sqrt{8} = 2\sqrt{2}$
41 $\sqrt{5} - \sqrt{20}$ **42** $\sqrt{11} - 2\sqrt{17}$
43 $\sqrt{40} - 2\sqrt{10}$ **44** $\sqrt{5} + \sqrt{125} + \sqrt{-5}$
45 $2\sqrt{2} + \sqrt{50} - 3\sqrt{-32}$ **46** Prove Theorem 2″.
47 Prove Theorem 3″.
48 Give a definition for $(-a)^{p/q}$, where a is a positive real number, and p and q are integers with q odd.
49 Review the proofs of Theorems 4, 1″, 2″, and 3″. What changes are needed in these proofs so that these theorems apply to the symbols $(-a)^{p/q}$, where a is positive, p any integer, and q an odd integer?

In Probs. 50 to 54 we consider the symbol a^b where a and b are positive real numbers. We assume that these symbols are defined (Sec. 11.2) and that for them the analogs of Theorems 1, 2, and 3 are true. The symbol a^b can be considered as a binary operation on the two positive real numbers a and b similar to the sum $a + b$ and the product $a \times b$. According to its definition, a^b is itself a positive real number, so that the law of closure holds.

50 Is this operation associative?
51 Is this operation commutative?
52 Is there an identity?
53 Is there an inverse?
54 What distributive law holds between this operation and multiplication?

5.7   Rationalizing Denominators

From time to time you will meet fractions containing square roots in the denominator, such as

$$\frac{1}{\sqrt{2}} \qquad \frac{2}{\sqrt{3} - \sqrt{5}} \qquad \frac{x + 5}{\sqrt{x + 1} + \sqrt{2x - 3}}$$

Let us consider problems which these present. If we wish to express $1/\sqrt{2}$ as a decimal, it is awkward to divide $1/1.414$. A simpler procedure is to write

$$\frac{1}{\sqrt{2}} = \frac{1}{\sqrt{2}} \cdot \frac{\sqrt{2}}{\sqrt{2}} = \frac{\sqrt{2}}{2} \approx \frac{1.414}{2} = 0.707$$

In some textbooks it is required that all answers be written with rational denominators; thus $1/\sqrt{2}$ is incorrect, and $\sqrt{2}/2$ is correct. This is an absurd requirement, and we shall accept either answer as correct. The choice between them depends on how we are to use them. Consider the examples below.

Illustration 1

a Find $1/\sqrt{2} + 1/\sqrt{3}$. Here we find it wise to write

$$\frac{1}{\sqrt{2}} + \frac{1}{\sqrt{3}} = \frac{\sqrt{2}}{2} + \frac{\sqrt{3}}{3} = \frac{3\sqrt{2} + 2\sqrt{3}}{6}$$

b Find $1/\sqrt{2} \times 1/\sqrt{2}$. Here we can write

$$\frac{1}{\sqrt{2}} \times \frac{1}{\sqrt{2}} = \frac{1}{\sqrt{2} \times \sqrt{2}} = \frac{1}{2}$$

It would be silly to write

$$\frac{1}{\sqrt{2}} \times \frac{1}{\sqrt{2}} = \frac{\sqrt{2}}{2} \times \frac{\sqrt{2}}{2} = \frac{2}{4} = \frac{1}{2}$$

Hence, leave your answer in whatever form is most convenient for later use.

When we are faced with $2/(\sqrt{3} - \sqrt{5})$, another technique is needed if we wish a rational denominator. We can rationalize this one as follows:

$$\frac{2}{\sqrt{3} - \sqrt{5}} = \frac{2}{\sqrt{3} - \sqrt{5}} \times \frac{\sqrt{3} + \sqrt{5}}{\sqrt{3} + \sqrt{5}} = \frac{2\sqrt{3} + 2\sqrt{5}}{3 - 5}$$

$$= \frac{2\sqrt{3} + 2\sqrt{5}}{-2} = -\sqrt{3} - \sqrt{5}$$

We can apply this method to various cases as shown in the Illustrations below.

Illustration 2

a $\quad \dfrac{3}{1 + \sqrt{3}} = \dfrac{3}{1 + \sqrt{3}} \times \dfrac{1 - \sqrt{3}}{1 - \sqrt{3}} = \dfrac{3 - 3\sqrt{3}}{1 - 3} = -\dfrac{3 - 3\sqrt{3}}{2}$

b $\quad \dfrac{x + 5}{\sqrt{x + 1} + \sqrt{2x - 3}}$

$$= \dfrac{(x + 5)(\sqrt{x + 1} - \sqrt{2x - 3})}{(\sqrt{x + 1} + \sqrt{2x - 3})(\sqrt{x + 1} - \sqrt{2x - 3})}$$

$$= \dfrac{(x + 5)\sqrt{x + 1} - (x + 5)\sqrt{2x - 3}}{(x + 1) - (2x - 3)}$$

$$= \dfrac{(x + 5)\sqrt{x + 1} - (x + 5)\sqrt{2x - 3}}{-x + 4}$$

c $\quad \dfrac{x}{1 + \sqrt{x}} + \dfrac{2}{\sqrt{x + 1}} = \dfrac{x}{1 + \sqrt{x}} \times \dfrac{1 - \sqrt{x}}{1 - \sqrt{x}} + \dfrac{2\sqrt{x + 1}}{x + 1}$

$$= \dfrac{x - x\sqrt{x}}{1 - x} + \dfrac{2\sqrt{x + 1}}{x + 1}$$

$$= \dfrac{(x - x\sqrt{x})(x + 1) + 2(1 - x)\sqrt{x + 1}}{1 - x^2}$$

Problems 5.7

In Probs. 1 to 10 rationalize the denominator in the given expressions.

1 $\dfrac{3}{\sqrt{7}}$

2 $\dfrac{-2}{\sqrt{11}}$

3 $\dfrac{-5}{\sqrt{2} - \sqrt{3}}$

4 $\dfrac{1}{\sqrt{5} + \sqrt{7}}$

5 $\dfrac{4}{\sqrt{x - 1}}$

6 $\dfrac{x^2}{\sqrt{1 - x^2}}$

7 $\dfrac{3}{\sqrt{x - 2} + \sqrt{x + 1}}$

8 $\dfrac{x}{\sqrt{x^2 - 4} + \sqrt{x - 3}}$

9 $\dfrac{4}{\sqrt{2} - \sqrt{3} + \sqrt{8}}$

10 $\dfrac{5}{\sqrt{5} - \sqrt{20} + \sqrt{3}}$

In Probs. 11 to 18 perform the stated operation. State your answers with rationalized denominators.

11 $\dfrac{1}{\sqrt{3}} \times \dfrac{1}{\sqrt{27}}$

12 $\dfrac{1}{\sqrt{2}} \times \dfrac{1}{\sqrt{8}}$

13 $\dfrac{1}{\sqrt{3}} + \dfrac{1}{\sqrt{5}}$

14 $\dfrac{1}{\sqrt{5}} - \dfrac{1}{\sqrt{7}}$

15 $\dfrac{1}{3 - \sqrt{x}} + \dfrac{1}{x}$

16 $\dfrac{3}{\sqrt{x + 1} - 2} + \dfrac{1}{x^2}$

17 $\dfrac{x^2}{1 - \sqrt{x}} \times \dfrac{4}{1 + \sqrt{x}}$

18 $\dfrac{3}{4 + \sqrt{1 + x}} \times \dfrac{2}{4 - \sqrt{1 + x}}$

In Probs. 19 to 22 use a table of square roots to compute the following to three decimal places.

19 $\dfrac{2}{\sqrt{3}}$

20 $\dfrac{5}{\sqrt{6}}$

21 $\dfrac{1}{\sqrt{5} - \sqrt{3}}$

22 $\dfrac{2}{\sqrt{3} - \sqrt{7}}$

6.1 Solutions of Equations

One of the most common problems in mathematics is *solving an equation*. In this chapter we shall tackle a number of problems of this kind, but first we must know what is meant by the *solution of an equation*. In order to avoid unnecessary complications we shall begin by considering the simple case of polynomial equations in one variable.

Let us recall the definition of a polynomial from Chap. 3.

Definition A *polynomial* $P(x)$ in one variable is an expression of the form

$$P(x) = a_0 x^n + a_1 x^{n-1} + \cdots + a_{n-1} x + a_n$$

where n is an integer which is positive or zero; the coefficients $a_0, a_1, \ldots, a_n$ are elements of a given field F; and x is a variable whose universal set is F. We say that $P(x)$ is a polynomial *over the field F*.

Definition If $a_0 \neq 0$, the polynomial

$$P(x) = a_0 x^n + a_1 x^{n-1} + \cdots + a_{n-1} x + a_n$$

is said to be of *degree n*.

Remarks We shall generally choose the field F to be the field of complex numbers, but on occasion may restrict it to the real numbers. In any case, the field involved must be specified in advance.

121

Illustration 1   The following expressions are polynomials over the field of complex numbers:

   a 2

   b $3x^5 - \pi x^2 - 1$

   c $(2 + i)x^2 - 3x + (-3 + i)$

Illustration 2   The following expressions are not polynomials over the field of complex numbers:

   a $1/x - 3i$

   b 2^x

   c $\sqrt{x}$

   d $|x|$

We can now define a polynomial equation and its solution as follows.

Definition   A polynomial equation is an open sentence of the form $P(x) = Q(x)$, where $P(x)$ and $Q(x)$ are polynomials.

Definition   The *solution set* of a polynomial equation $P(x) = Q(x)$ is the set
solution set $\{x \mid P(x) = Q(x)\}$; i.e., it is the truth set of the given open sentence. Any element of this solution set is called a *solution* or *root* of $P(x) = Q(x)$.

Illustration 3

   a The solution set of $2x = 6$ is $\{3\}$.

   b The solution set of $x^2 - 5x + 6 = 0$ is $\{2, 3\}$. The roots are the integers 2 and 3.

   c The solution set of $(x - 1)^2 = 0$ is $\{1\}$. There is only one root, namely 1.

6.2   Equivalent Polynomial Equations

In Sec. 1.4 we remarked that each step in the process of solving equations amounts to the replacement of an equation by a simpler one which
equivalence   is equivalent to it. In the last step the equation is so simple that we can write down the solution set by inspection. We remind you that two equations are equivalent if and only if they have the same solution set.

Illustration 1   Solve $4x - 5 = 2x + 7$.
   This is equivalent in turn to

$$4x = 2x + 12$$
$$2x = 12$$
$$x = 6$$

So the solution set is $\{6\}$.

So that we can carry out this process efficiently, we must examine various transformations of equations to determine once and for all which of these do replace an equation by one equivalent to it. Such transformations may be used safely at any time; others are dangerous to use and may lead to wrong answers.

One of the most important of these transformations is the addition of the same polynomial to both sides of an equation, as we did in the first two steps of Illustration 1 above. In some textbooks this process is called "transposition".

Theorem 1   Let $P(x) = Q(x)$ be a polynomial equation over a field F, and let $G(x)$ be another polynomial over F. Then the equations

addition to
both sides

$$P(x) = Q(x)$$
$$\text{and} \qquad P(x) + G(x) = Q(x) + G(x)$$

are equivalent; i.e.,

$$\{x \,|\, P(x) = Q(x)\} = \{x \,|\, P(x) + G(x) = Q(x) + G(x)\}$$

Proof   Let a be any element of F for which $P(a) = Q(a)$. Then $P(a) + G(a) = Q(a) + G(a)$, since the two sides are just different names of the same element of F. Therefore

$$\{x \,|\, P(x) = Q(x)\} \subseteq \{x \,|\, P(x) + G(x) = Q(x) + G(x)\}$$

Conversely, if a is such that $P(a) + G(a) = Q(a) + G(a)$, it follows by subtraction of $G(a)$ from both sides that $P(a) = Q(a)$. Therefore

$$\{x \,|\, P(x) + G(x) = Q(x) + G(x)\} \subseteq \{x \,|\, P(x) = Q(x)\}$$

Hence the two sets are identical, and the equations are equivalent.

Remarks

1   This theorem also includes the case of subtraction of $G(x)$ from both sides; for the subtraction of $G(x)$ is the same as the addition of $-G(x)$.
2   By virtue of this theorem we can rewrite any polynomial equation $P(x) = Q(x)$ in the *standard* form $R(x) = 0$, where

$$R(x) = P(x) - Q(x)$$

In later sections of this chapter we shall assume that polynomial equations are written in this form.

Now let us turn to multiplication of both sides by the same expression, as we did in the last step of Illustration 1 above, where we multiplied by ½.

Theorem 2   Let $P(x) = Q(x)$ be a polynomial equation over F, and let a be any nonzero element of F. Then the equations

multiplica-
tion of both
sides

$$P(x) = Q(x)$$
$$\text{and} \qquad aP(x) = aQ(x)$$

are equivalent; i.e.,

$$\{x \,|\, P(x) = Q(x)\} = \{x \,|\, aP(x) = aQ(x)\}$$

The proof is immediate and is included in the problems.

Remarks

1 Since a may be $1/b$, Theorem 2 includes division by a nonzero element of F.
2 If $a = 0$, the two equations in Theorem 2 are not necessarily equivalent, since any x in F is a solution of $0 \cdot P(x) = 0 \cdot Q(x)$. Thus we must exclude $a = 0$ in the statement of the theorem.

When we try to extend Theorem 2 to multiplication by a polynomial of degree ≥ 1, we run into trouble. Indeed, such operations may well lead to wrong answers.

Theorem 3 Let $P(x) = Q(x)$ be a polynomial equation over F, and let $G(x)$ be a polynomial of degree ≥ 1. Then the equations

$$P(x) = Q(x)$$

and
$$G(x) \cdot P(x) = G(x) \cdot Q(x)$$

may not be equivalent. Indeed,

$$\{x \,|\, P(x) = Q(x)\} \subseteq \{x \,|\, G(x) \cdot P(x) = G(x) \cdot Q(x)\}$$

Proof The set on the right contains all the elements of the set on the left, and in addition it contains those elements a for which $G(a) = 0$. If there is an a such that $G(a) = 0$ and $P(a) \neq Q(a)$, the equations are not equivalent. If, however, every a with $G(a) = 0$ also satisfies $P(a) = Q(a)$, or if $G(a) \neq 0$ for all a in F, the equations are equivalent.

Illustration 2

a The equations $2x + 1 = 3$ and $(x - 1)(2x + 1) = 3(x - 1)$ are equivalent.
b The equations $2x + 1 = 3$ and $x(2x + 1) = 3x$ are not equivalent.

The following restatement of Theorem 3 warns us of the dangers of removing a common factor $G(x)$ of degree ≥ 1 from both sides.

Theorem 4 Let $P(x) = A(x) \cdot G(x)$ and $Q(x) = B(x) \cdot G(x)$, where $A(x)$ and $B(x)$ are polynomials, and $G(x)$ is a polynomial of degree ≥ 1. Then the equations

$$P(x) = Q(x) \qquad \text{and} \qquad A(x) = B(x)$$

may not be equivalent. Indeed

$$\{x \,|\, A(x) = B(x)\} \subseteq \{x \,|\, P(x) = Q(x)\}$$

Illustration 3 The solution set of $(x + 4)x = (x + 4)$ is the set $\{1, -4\}$. But the solution set of $x = 1$ is the set $\{1\}$. By dividing by $x + 4$ we have lost the root $x = -4$.

Finally, we shall need the following theorem of a somewhat different type.

Theorem 5 Let $P(x)$ be a polynomial which factors into the product

factoring

$$P(x) = P_1(x) \cdot P_2(x) \cdots P_r(x). \quad \text{Then}$$

$$\{x \,|\, P(x) = 0\} = \{x \,|\, P_1(x) = 0\} \cup \{x \,|\, P_2(x) = 0\} \cup \cdots \cup \{x \,|\, P_r(x) = 0\}$$

Proof If a satisfies any one of the equations $P_1(a) = 0$; $P_2(a) = 0, \ldots$; $P_r(a) = 0$; it is clear that $P(a) = 0$. Hence the right-hand set is a subset of the left-hand side. Conversely, suppose that a satisfies $P(a) = 0$. Then the product

$$P_1(a) \cdot P_2(a) \cdots P_r(a) = 0$$

and at least one of the factors must be zero. Hence the left-hand set is a subset of the right-hand set. Combining these results, we get a proof of the theorem.

As we have noted earlier (Sec. 2.4), this theorem is extremely useful in the solution of polynomial equations. We factor the given polynomial and then set each factor equal to zero. The union of the solution sets of the equations so obtained is the solution set of the given equation.

Illustration 4 Solve $x^2 + 12x + 32 = 0$.

Since $x^2 + 12x + 32 = (x + 8)(x + 4)$, we consider

$$x + 8 = 0 \quad \text{and} \quad x + 4 = 0$$

Since their solution sets are $\{-8\}$ and $\{-4\}$, respectively, the solution set of the given equation is

$$\{-8\} \cup \{-4\} = \{-8, -4\}$$

Problems 6.2

In each of Probs. 1 to 10 one of the following relations is true: $A \subset B$; $A = B$; $A \supset B$. Write the correct relation in each case. All polynomials are defined over the field of real numbers.

	A	B
1	$\{x \,|\, 3x - 2 = x + 4\}$	$\{x \,|\, 4x + 3 = 2x + 9\}$
2	$\{x \,|\, x^2 + 4 = -2x + 3\}$	$\{x \,|\, x^2 + 2x = -1\}$
3	$\{x \,|\, (x + 2)(x - 3) = 0\}$	$\{x \,|\, x + 2 = 0\}$
4	$\{x \,|\, (x + 1)(x - 4) = 0\}$	$\{x \,|\, x + 1 = 0\} \cup \{x \,|\, x - 4 = 0\}$
5	$\{x \,|\, (x + 3)(x + 5) = 0\}$	$\{x \,|\, x + 3 = 0\} \cup \{x \,|\, x + 5 = 0\}$
6	$\{x \,|\, x + 3 = 7\}$	$\{x \,|\, (x + 3)^2 = 49\}$

A B

7 $\{x \mid \sqrt{x+2} = -3\}$ $\qquad$ $\{x \mid x+2 = 9\}$

8 $\{x \mid 2x+5 = 9\}$ $\qquad$ $\{x \mid \sqrt{2x+5} = -3\}$

9 $\{x \mid 3x(x-2) = 4(x-2)\}$ $\qquad$ $\{x \mid x = 2\}$

10 $\{x \mid 6x-3 = 9x+12\}$ $\qquad$ $\{x \mid 2x-1 = 3x+4\}$

11 Show that the equations $P(x) = Q(x)$ and $[P(x)]^2 = [Q(x)]^2$ may not be equivalent.

12 Prove Theorem 2.

6.3   Linear Equations

These have appeared so often in earlier chapters (especially in Chap. 2) that we need not discuss them here. For reference let us state the following theorem, which summarizes the situation.

Theorem 6   The linear equation $ax + b = 0$, where a and b are real or complex numbers and $a \neq 0$, has one and only one solution, namely, $x = -b/a$.

6.4   Quadratic Equations

The next simplest type of equation is the quadratic

(1) $$ax^2 + bx + c = 0$$

where $a \neq 0$. The coefficients a, b, and c are, in general, complex numbers, but in various special circumstances we shall require them to be real. We have seen (Theorem 5, Sec. 6.2) that one way to solve such an equation is to factor it, and in Sec. 3.6 we learned how to factor certain expressions of the form $ax^2 + bx + c$. When we cannot write down the factors at sight in this way, we proceed by a method known as "completing the square". This depends upon the fact that

completing the square

(2) $$x^2 + 2dx + d^2 = (x + d)^2$$

Since $a \neq 0$, let us write Eq. (3), which is equivalent to (1):

(3) $$x^2 + \frac{b}{a}x + \frac{c}{a} = 0 \qquad \text{[Sec. 6.2, Theorem 2]}$$

If we put $d = b/2a$, the first two terms on the left side of (2) are equal to the corresponding terms in (3). In general, however, $d^2 \neq c/a$. Therefore we write (4), which is equivalent to (3):

(4) $$x^2 + \frac{b}{a}x + \left(\frac{b}{2a}\right)^2 = \left(\frac{b}{2a}\right)^2 - \frac{c}{a}$$

Now the left-hand side of (4) is of the same form as the left-hand side of (2).

Thus

(5)
$$\left(x + \frac{b}{2a}\right)^2 = \frac{b^2 - 4ac}{4a^2}$$

or
$$\left(x + \frac{b}{2a}\right)^2 - \left(\frac{\sqrt{b^2 - 4ac}}{2a}\right)^2 = 0$$

This factors easily into

$$\left(x + \frac{b}{2a} - \frac{\sqrt{b^2 - 4ac}}{2a}\right)\left(x + \frac{b}{2a} + \frac{\sqrt{b^2 - 4ac}}{2a}\right) = 0$$

Therefore

$$x = \frac{-b + \sqrt{b^2 - 4ac}}{2a} \quad \text{or} \quad x = \frac{-b - \sqrt{b^2 - 4ac}}{2a}$$

This proves the following theorem.

Theorem 7   The quadratic equation

$$ax^2 + bx + c = 0 \qquad a \neq 0$$

quadratic
formula

where a, b, and c are complex numbers, and where x is an element of the set of all complex numbers, has two solutions, namely,

$$x = \frac{-b \pm \sqrt{b^2 - 4ac}}{2a}$$

In other words, the set $\{x \mid ax^2 + bx + c = 0\}$ is the set

$$\left\{\frac{-b + \sqrt{b^2 - 4ac}}{2a}, \frac{-b - \sqrt{b^2 - 4ac}}{2a}\right\}$$

Remarks

1   Theorem 7 would have been false, even for real a, b, and c, if we had restricted x to belong to the set of real numbers. For $x^2 + 1 = 0$ has no real solutions.

2   If a, b, and c are real, we may easily deduce the following properties of the solutions:
When $b^2 - 4ac$ is positive, the two solutions are real and unequal.
When $b^2 - 4ac$ is zero, the two solutions are real and equal; i.e., there is only one solution.
When $b^2 - 4ac$ is negative, the two solutions are unequal and neither of them is real.

3   Let r_1 and r_2 be the two roots of $ax^2 + bx + c = 0$; that is,

(6) $$r_1 = \frac{-b + \sqrt{b^2 - 4ac}}{2a} \qquad r_2 = \frac{-b - \sqrt{b^2 - 4ac}}{2a}$$

Then

(7) $$r_1 + r_2 = -\frac{b}{a} \qquad r_1 r_2 = \frac{c}{a}$$

Exercise

A By direct calculation verify the above statements.

4 The above formulas permit us to give a general expression for the factors of $ax^2 + bx + c$, namely:

(**8**) $$ax^2 + bx + c = a(x - r_1)(x - r_2)$$

where r_1 and r_2 are defined in (6) above. To prove this correct, we note that

$$a(x - r_1)(x - r_2) = ax^2 - a(r_1 + r_2)x + ar_1r_2$$

$$= ax^2 - a\left(-\frac{b}{a}\right)x + a\left(\frac{c}{a}\right)$$

$$= ax^2 + bx + c$$

The use of formula (8) permits us to factor trinomials which could not be factored by the methods of Chap. 3. It also provides a direct approach to certain trinomials which can be factored by the methods of Chap. 3 but whose factorization by those methods is likely to be long and tedious.

Illustration 1

a Factor $x^2 + 2x + 5$.

$$r_1 = \frac{-2 + \sqrt{-16}}{2} = -1 + 2i$$

$$r_2 = \frac{-2 - \sqrt{-16}}{2} = -1 - 2i$$

So $x^2 + 2x + 5 = (x + 1 - 2i)(x + 1 + 2i)$.

b Factor $35x^2 - 11x - 72$.

$$r_1 = \frac{11 + \sqrt{10,201}}{70} = \frac{11 + 101}{70} = \frac{112}{70} = \frac{8}{5}$$

$$r_2 = \frac{11 - \sqrt{10,201}}{70} = \frac{11 - 101}{70} = -\frac{90}{70} = -\frac{9}{7}$$

So $35x^2 - 11x - 72 = 35(x - \frac{8}{5})(x + \frac{9}{7}) = (5x - 8)(7x + 9)$.

Problems 6.4

In Probs. 1 to 10 find the solutions of the quadratic equations.

1 $x^2 + 4x + 5 = 0$ **2** $x^2 + 2x + 2 = 0$
3 $9x^2 + 6x + 1 = 0$ **4** $x^2 + 10x + 25 = 0$
5 $36x^2 + 17x - 35 = 0$ **6** $18x^2 + 27x - 56 = 0$
7 $6x^2 - 17x - 45 = 0$ **8** $15x^2 + 26x + 8 = 0$
9 $x^2 - (5 + 2i)x + (5 + 5i) = 0$
10 $x^2 + (4 + 3i)x + (7 + i) = 0$ HINT: $(2 + 5i)^2 = -21 + 20i$

In Probs. 11 to 16 add terms to both sides so that the left-hand side becomes a perfect square.

11 $x^2 + 5x + 1 = 0$ **12** $x^2 - 2x - 3 = 0$
13 $x^2 - 8x + 2 = 0$ **14** $x^2 + 7x - 2 = 0$
15 $4x^2 - 10x + 3 = 0$ **16** $3x^2 + 12x + 1 = 0$

In Probs. 17 to 22 find the sum and product of the solutions without solving the equation.

17 $x^2 - 3x + 10 = 0$ **18** $x^2 + 4x - 7 = 0$
19 $4x^2 - 7x + 12 = 0$ **20** $5x^2 + 3x + 6 = 0$
21 $(2 - i)x^2 + (4 + 3i)x + (-1 + i) = 0$
22 $(3 + 2i)x^2 + (1 - 2i)x + (3 + i) = 0$

In Probs. 23 to 28 find the value of k for which the solutions of the given equation are equal.

23 $x^2 + 3x + k = 0$ **24** $x^2 - 5x + k = 0$
25 $2x^2 + 7x + k = 0$ **26** $3x^2 - x + k = 0$
27 $x + 3kx + 4 = 0$ **28** $x^2 - 4kx + 5 = 0$

In Probs. 29 to 34 find a quadratic equation the sum and product of whose solutions have the given values.

29 Sum 3, product 4   **30** Sum 0, product 5
31 Sum ⅓, product ⅔   **32** Sum ⅖, product − ⅘
33 Sum $3 - i$, product $2 + 5i$ **34** Sum $1 + i$, product $2 - i$

In Probs. 35 to 40 factor the given trinomial.

35 $2x^2 + 3x + 5$ **36** $x^2 - x + 1$
37 $x^2 - 2x + 2$ **38** $4x^2 + 2x + 3$
39 $54x^2 - 3x - 35$ **40** $28x^2 + 3x - 40$
41 The sum of two numbers is 14, and their product is 45. Write the quadratic equation of which they are the solutions, and solve.
42 Find the value of k for which the sum of the solutions of the following equation is twice their product:

$$5x^2 + 6x + k = 0$$

43 Find a quadratic equation whose solutions are the reciprocals of those of

$$3x^2 + 5x + 9 = 0$$

HINT: $\dfrac{1}{p} + \dfrac{1}{q} = \dfrac{p + q}{pq}$; $\left(\dfrac{1}{p}\right)\left(\dfrac{1}{q}\right) = \dfrac{1}{pq}$

44 Find a quadratic equation whose solutions are the squares of those of $x^2 + 5x + 3 = 0$.
HINT: $p^2 + q^2 = (p + q)^2 - 2pq$; $(p^2)(q^2) = (pq)^2$

45 Find a quadratic equation whose solutions are respectively the sum and product of the solutions of

$$x^2 + 6x + 11 = 0$$

46 Find a quadratic equation whose solutions are 3 smaller, respectively, than the solutions of

$$x^2 + 7x - 11 = 0$$

6.5 Related Equations in Two Variables

There is a clear relationship between the equations

(1) $\qquad ax + b = 0 \qquad ax^2 + bx + c = 0$

and the polynomial equations in two variables

(2) $\qquad y = ax + b \qquad y = ax^2 + bx + c$

We have seen that the solution sets of equations of the form (1) are sets of numbers:

(3) $\qquad \{x \mid ax + b = 0\} \qquad \{x \mid ax^2 + bx + c = 0\}$

By contrast, however, the solution sets of equations of the form (2) are sets whose elements are ordered pairs of numbers:

(4) $\qquad \{(x,y) \mid y = ax + b\} \qquad \{(x,y) \mid y = ax^2 + bx + c\}$

Illustration 1

a A few of the elements of $\{(x,y) \mid y = 3x + 2\}$ are $(-2,-4), (-1,-1),$ (0,2), (1,5).

b A few of the elements of $\{(x,y) \mid y = x^2 - 6x + 5\}$ are $(-1,12), (0,5),$ (1,0), (2,-3), (3,-4), (4,-3), (5,0), (6,5), (7,12).

The method of finding elements of sets of the form (4) is undoubtedly familiar to you. To do so, choose any value of x, say x_0, and from the given equation calculate the corresponding value of y, say y_0. Then the pair (x_0,y_0) is an element of the corresponding solution set. This method of finding these elements shows us another important difference between the sets (3) and the sets (4), namely: the sets (3) contain a finite number of elements (actually one or two), whereas the sets (4) contain an *infinite number of elements*. For this reason we can never list all the elements of the sets (4), and to describe them we must use another technique, namely graphing. This method requires that we restrict the underlying field F to the set of real numbers; and, hereafter, in this discussion we shall assume that this has been done.

Earlier (Sec. 2.12), we have seen that there is a 1 to 1 correspondence between ordered pairs of real numbers and points in the plane. Using this correspondence, we can define the notion of the graph of an equation of the form (2).

Definition The graph of a polynomial equation $P(x,y) = 0$ (over the reals) is the set of points in the plane whose ordered pairs of coordinates are elements of the solution set of $P(x,y) = 0$.

graphs

The usual procedure for plotting the graph of such an equation is to compute a small number of points on it and then to connect these with a smooth curve.

Illustration 2 Plot the graph of

$$y = 3x + 2$$

We write the following table of pairs (see Illustration 1).

x	-2	-1	0	1
y	-4	-1	2	5

We then draw Fig. 6.1.

Straight line

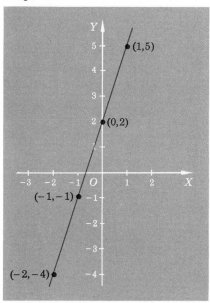

Figure 6.1
The graph of $y = ax + b$ is a straight line.
Here $y = 3x + 2$.

Remarks

1 The graph in Fig. 6.1 is a straight line. It is possible to prove that the graph of any equation of the form $y = ax + b$ is a straight line. Hence you need compute only two points to determine this

graph. To catch errors in computation, however, it is wise to calculate at least three points.

2 The point $(-\frac{2}{3},0)$ where this line cuts the X-axis corresponds to the solution $x = \frac{2}{3}$ of $3x + 2 = 0$. In general, the graph of $y = ax + b$ will cut the X-axis at the point $(-b/a,0)$ which corresponds to the solution $x = -b/a$ of the equation $ax + b = 0$. This suggests a graphical method for solving $ax + b = 0$, namely: Plot the graph of $y = ax + b$, and find the x-coordinate of the point in which this line intersects the X-axis. This coordinate is the desired solution.

Illustration 3 Plot the graph of

$$y = x^2 - 6x + 5$$

We write the following table of pairs.

x	-1	0	1	2	3	4	5	6	7
y	12	5	0	-3	-4	-3	0	5	12

We then draw Fig. 6.2.

Parabola

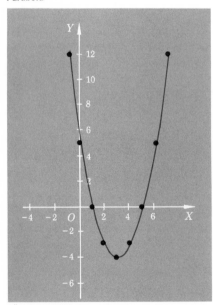

Figure 6.2
The graph of $y = ax^2 + bx + c$ where $a > 0$, is a parabola opening upward. Here $y = x^2 - 6x + 5$.

Illustration 4 Plot the graph of

$$y = -2x^2 + x - 2$$

The table of pairs is as follows.

x	-2	-1	0	1	2	3
y	-12	-5	-2	-3	-8	-17

Then we draw Fig. 6.3.

Parabola

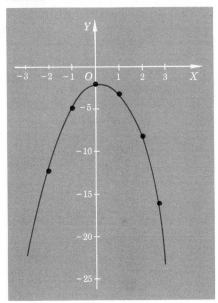

Figure 6.3
When $a < 0$, the graph of $ax^2 + bx + c$ opens downward. Here $y = -2x^2 + x - 2$.

Definition The graph of $y = ax^2 + bx + c$ $(a \neq 0)$ is called a *parabola*.
parabola   Its highest (or lowest) point is called its *vertex*.

Remarks The graphs of these equations have the following important properties:

1 If $a > 0$, the graph opens upward.
2 If $a < 0$, the graph opens downward.
3 If $ax^2 + bx + c = 0$ has the real solutions r_1 and r_2, where $r_1 \neq r_2$, the graph crosses the X-axis at $x = r_1$ and $x = r_2$.
4 If $ax^2 + bx + c = 0$ has two equal real solutions r, the graph is tangent to the X-axis at $x = r$.
5 If $ax^2 + bx + c = 0$ has nonreal solutions, the graph does not cross the X-axis.
6 The graph is symmetric about the line

$$x = -\frac{b}{2a}$$

7 The vertex of the parabola is the point

$$\left(-\frac{b}{2a}, \frac{-b^2 + 4ac}{4a} \right)$$

Exercise

A Verify as many as possible of these properties by examining the graphs in Illustrations 3 and 4 above.

Problems 6.5

In Probs. 1 to 16 plot the graphs of the given equations, and from these find the solutions of the corresponding linear or quadratic equations.

1	$y = 2x + 8$	**2**	$y = 3x + 6$
3	$y = -4x + 12$	**4**	$y = -2x + 4$
5	$y = -x - 3$	**6**	$y = -5x - 10$
7	$y = x^2 + 9x + 8$	**8**	$y = x^2 - 8x + 12$
9	$y = -x^2 + 9x - 18$	**10**	$y = -x^2 + 10x - 24$
11	$y = 4x^2 + 4x + 1$	**12**	$y = 9x^2 + 6x + 1$
13	$y = 2x^2 - x + 8$	**14**	$y = -3x^2 + 2x - 2$
15	$y = -x^2 - 2x - 1$	**16**	$y = -16x^2 - 8x - 1$

Problems 17 to 23 refer to properties 1 to 7 of parabolas, which are stated on page 133.

17 Why is property 1 true?   **18** Why is property 2 true?
19 Why is property 3 true?   **20** Why is property 4 true?
21 Why is property 5 true?
22 Why is property 6 true?   HINT: Write $y = ax^2 + bx + c$ in the form

$$y + \frac{b^2 - 4ac}{4a} = a\left(x + \frac{b}{2a}\right)^2$$

23 Why is property 7 true?   HINT: Write $y = ax^2 + bx + c$ in the form

$$y = \frac{-b^2 + 4ac}{4a} + a\left(x + \frac{b}{2a}\right)^2$$

6.6   Equations Containing Fractions

It is not uncommon for you to meet equations like

$$\frac{1}{x} + \frac{2}{x + 1} = \frac{3}{x - 1} \qquad \text{or} \qquad \frac{x - 1}{x + 4} - \frac{x + 2}{x - 3} = \frac{5}{x + 2}$$

in which algebraic fractions appear. In order to solve these, we first express all the given fractions in terms of a common denominator. Then by addition and/or subtraction we convert the given equation into one of the form

$$\frac{P(x)}{Q(x)} = 0$$

where $P(x)$ and $Q(x)$ are polynomials. By the methods of Sec. 6.2 we can show that this equation is equivalent to the given equation, but we omit the details here.

From Sec. 2.3 it follows that a solution of $P(x)/Q(x) = 0$ is a number a such that $P(a) = 0$ and $Q(a) \neq 0$. To find such solutions we, therefore, solve the equation $P(x) = 0$, and from the set of its solutions we eliminate those for which $Q(x) = 0$. The remaining set is the solution set of the given equation.

Illustration 1 Solve $\dfrac{1}{x} + \dfrac{6}{x + 4} = 1$.

In terms of a common denominator this becomes

$$\frac{(x + 4) + 6x}{x(x + 4)} = \frac{x(x + 4)}{x(x + 4)}$$

or

$$\frac{-x^2 + 3x + 4}{x(x + 4)} = 0$$

Solving $-x^2 + 3x + 4 = 0$ or $(-x + 4)(x + 1) = 0$

we obtain $x = 4, -1$

Since neither of these satisfies $x(x + 4) = 0$, the solution set of the given equation is $\{4, -1\}$.

Illustration 2 Solve $\dfrac{7}{x - 1} - \dfrac{6}{x^2 - 1} = 5$.

In terms of the common denominator $(x - 1)(x^2 - 1)$, we obtain after addition and rearrangement:

$$\frac{(x - 1)(x - 2)(5x + 3)}{(x - 1)(x^2 - 1)} = 0$$

Hence, the tentative solution set is $\{1, 2, -\%\}$. We see, however, that the denominator is zero for $x = 1$. Hence we must discard $x = 1$; and the final solution set is $\{2, -\%\}$.

The above method is not as elegant as it might have been, for we did not use the L.C.D. If we had observed that $x^2 - 1$ serves as the L.C.D., we would have arrived at the equation

$$\frac{(x - 2)(5x + 3)}{x^2 - 1} = 0$$

Thus our tentative and final solution sets are both $\{2, -\%\}$. But even the consistent use of the L.C.D. will not excuse you from testing every tentative solution. See the next illustration.

Illustration 3 Solve $\dfrac{x}{x + 2} - \dfrac{4}{x + 1} = \dfrac{-2}{x + 2}$.

Using the L.C.D. $(x + 2)(x + 1)$, we finally transform this into

$$\frac{(x - 3)(x + 2)}{(x + 1)(x + 2)} = 0$$

The solution set of the numerator is $\{3, -2\}$, but $x = -2$ makes the denominator zero. Hence the correct solution set is $\{3\}$.

This illustration suggests removing common factors from the numerator and denominator before proceeding with the solution. This is usually a wise move, but it has its pitfalls. See the next illustration.

Illustration 4   Solve $\dfrac{1}{x-1} + x = 1 + \dfrac{1}{x-1}$.

By the usual process we convert this into

$$\frac{(x-1)^2}{x-1} = 0$$

Removing the common factor in numerator and denominator, we obtain

$$x - 1 = 0 \qquad \text{or} \qquad x = 1$$

Unfortunately this is not a solution, for $x = 1$ makes the denominator of $(x-1)^2/x - 1$ equal to zero. This equation, therefore, has no solutions.

6.7   Equations Containing Radicals

In this section we are interested in equations like

$$\sqrt{x+13} - \sqrt{7-x} = 2 \qquad \text{or} \qquad 2\sqrt{x+4} - x = 1$$

in which x appears under a radical. For simplicity we shall consider square roots only. The only possible method of procedure involves squaring both sides, but this does not necessarily lead to an equivalent equation. The difficulty is that although from $F(x) = G(x)$ we can derive $[F(x)]^2 = [G(x)]^2$, we cannot go backward. For $[F(x)^2] = [G(x)]^2$ implies that $F(x) = G(x)$ or that $F(x) = -G(x)$. Therefore, all solutions of $[F(x)]^2 = [G(x)]^2$ must be tested in the given equation to make certain that they actually do solve the problem which is posed.

When there is only one radical in the given equation, write the equivalent equation in which the radical is on one side and all the other terms on the other side. Then squaring both sides removes the radical and leaves an equation without radicals to be solved. Since this equation may not be equivalent to the given equation, all solutions must be checked in the given equation.

solutions must be tested

Illustration 1   Solve $2\sqrt{x+4} - x = 1$.

$$\begin{aligned}
2\sqrt{x+4} &= x + 1 \\
4(x+4) &= x^2 + 2x + 1 \\
x^2 - 2x - 15 &= 0 \\
(x-5)(x+3) &= 0 \\
x &= 5, -3
\end{aligned}$$

Checking $x = 5$, we have $2\sqrt{9} - 5 = 1$, or $6 - 5 = 1$, which is true. Checking $x = -3$, we have $2\sqrt{1} + 3 = 1$, or $2 + 3 = 1$, which is false. The solution set of the given equation is, therefore, $\{5\}$.

When there are two radicals, the method is similar; but two squarings are required. Proceed as in the illustration below.

Illustration 2   Solve $\sqrt{x + 13} - \sqrt{7 - x} = 2$.

$$\sqrt{x + 13} = 2 + \sqrt{7 - x}$$
$$x + 13 = 4 + 4\sqrt{7 - x} + 7 - x$$
$$2x + 2 = 4\sqrt{7 - x}$$
$$x + 1 = 2\sqrt{7 - x}$$
$$x^2 + 2x + 1 = 28 - 4x$$
$$x^2 + 6x - 27 = 0$$
$$(x - 3)(x + 9) = 0$$
$$x = 3, -9$$

Checking $x = 3$, we have $\sqrt{16} - \sqrt{4} = 2$, or $4 - 2 = 2$, which is true.

Checking $x = -9$, we have $\sqrt{4} - \sqrt{16} = 2$, or $2 - 4 = 2$, which is false.

Hence the correct solution set is $\{3\}$.

Illustration 3   Solve $\sqrt{x + 1} - \sqrt{x + 6} = 1$.

$$\sqrt{x + 1} = 1 + \sqrt{x + 6}$$
$$x + 1 = 1 + 2\sqrt{x + 6} + x + 6$$
$$-6 = 2\sqrt{x + 6}$$
$$36 = 4(x + 6)$$
$$4x = 12$$
$$x = 3$$

Testing, we find

$$\sqrt{3 + 1} - \sqrt{3 + 6} \neq 1$$

Therefore the equation has no solution.

Problems 6.7

Solve:

1 $\dfrac{6}{x + 2} + \dfrac{7}{x} = 9$

2 $\dfrac{8}{x - 2} - \dfrac{4}{x} = 3$

3 $\dfrac{6}{x - 1} + \dfrac{16}{x^2 - 1} = 5$

4 $\dfrac{4}{x - 3} - \dfrac{16}{x^2 - 9} = 1$

5 $\dfrac{6}{x} - \dfrac{1}{x - 1} - \dfrac{10}{x + 3} = 0$

6 $-\dfrac{6}{x} - \dfrac{1}{x + 1} + \dfrac{20}{x - 3} = 0$

7 $\dfrac{-4}{x + 1} + \dfrac{5x - 5}{x^2 + 1} = 0$

8 $\dfrac{-2}{x + 2} + \dfrac{3x - 6}{x^2 + 2} = 0$

9 $\dfrac{2x}{x + 2} - \dfrac{2}{x - 2} = \dfrac{-4}{x + 2}$

10 $\dfrac{-3x}{x + 1} + \dfrac{6}{x} = \dfrac{3}{x + 1}$

11 $\sqrt{x + 7} + 5x - 13 = 0$

12 $\sqrt{x - 2} - 3x + 16 = 0$

13 $\sqrt{x + 3} + \sqrt{x + 8} = 0$

14 $\sqrt{x - 11} + \sqrt{x + 4} = 0$

15 $\sqrt{x + 21} - \sqrt{x + 14} = 1$

16 $\sqrt{x + 3} + \sqrt{x + 15} = 6$

17 $\sqrt{2x + 9} + \sqrt{3x + 16} = 7$

18 $\sqrt{5x - 1} - \sqrt{2x - 3} = 2$

19 $\dfrac{1}{\sqrt{x}} - \dfrac{2}{\sqrt{x + 27}} = 0$

20 $\dfrac{x}{\sqrt{x + 1}} + \dfrac{2x}{\sqrt{x + 3}} = 0$

6.8   Simultaneous Linear Equations

Here we complicate the situation a little by considering a pair of simultaneous linear equations in two variables. The general expression for such a system is

(1) $\begin{cases} a_1x + b_1y + c_1 = 0 & a_1 \neq 0 \text{ or } b_1 \neq 0 \\ a_2x + b_2y + c_2 = 0 & a_2 \neq 0 \text{ or } b_2 \neq 0 \end{cases}$

where the coefficients are real numbers, and the universal set of the variables is also the set of real numbers. By a solution of (1) we mean an ordered pair (x,y) which satisfies both equations.

In set language, the solution set of (1) is the set

$$\{(x,y) \mid a_1x + b_1y + c_1 = 0\} \cap \{(x,y) \mid a_2x + b_2y + c_2 = 0\}$$

Since the graph of each equation in (1) is a straight line, we are looking for the points which lie on both these lines.

method of addition and subtraction

The method of solution is to transform (1) into an equivalent system whose solution is obvious. You are undoubtedly familiar with this procedure as the method of elimination by *addition and subtraction*. Let us state it formally.

Theorem 8   The linear systems

(1) $\begin{cases} a_1x + b_1y + c_1 = 0 & a_1 \neq 0 \text{ or } b_1 \neq 0 \\ a_2x + b_2y + c_2 = 0 & a_2 \neq 0 \text{ or } b_2 \neq 0 \end{cases}$

and

(2) $\begin{cases} a_1x + b_1y + c_1 = 0 \\ k_1(a_1x + b_1y + c_1) + k_2(a_2x + b_2y + c_2) = 0 \end{cases}$

with $k_2 \neq 0$ are equivalent.

Proof   It is clear that every solution of (1) satisfies (2). Conversely, if $(\bar{x},\bar{y})$ satisfies (2), it follows that $a_1\bar{x} + b_1\bar{y} + c_1 = 0$, and

$$k_2(a_2\bar{x} + b_2\bar{y} + c_2) = 0$$

Since $k_2 \neq 0$, $(\bar{x},\bar{y})$ satisfies (1).

Using this theorem, we can eliminate x and y in turn by choosing suitable values of k_1 and k_2 and thus solve the system. Let us do this systematically. By hypothesis, either $a_1 \neq 0$ or $b_1 \neq 0$. Let us suppose that

$a_1 \neq 0$. Then, choosing $k_1 = -a_2$ and $k_2 = a_1$ in the system (2), we replace (1) by

(3)
$$\begin{cases} ax_1 \qquad\qquad +b_1 y \qquad\qquad +c_1 = 0 \\ \qquad (a_1 b_2 - a_2 b_1)y + (a_1 c_2 - a_2 c_1) = 0 \end{cases}$$

If $a_1 b_2 - a_2 b_1 \neq 0$, we solve the second equation of (3) for y and obtain

$$y = -\frac{a_1 c_2 - a_2 c_1}{a_1 b_2 - a_2 b_1}$$

We may substitute this value of y in the first equation of (3) and solve for x:

$$x = -\frac{b_2 c_1 - b_1 c_2}{a_1 b_2 - a_2 b_1}$$

If we suppose that $b_1 \neq 0$, we choose $k_1 = b_2$ and $k_2 = -b_1$ and replace (1) by

(4)
$$\begin{cases} a_1 x \qquad\qquad +b_1 y \qquad\qquad +c_1 = 0 \\ (a_1 b_2 - a_2 b_1)x \qquad\qquad +(b_2 c_1 - b_1 c_2) = 0 \end{cases}$$

Solving as before, we obtain the values for x and y derived above.

If, on the other hand, $a_1 b_2 - a_2 b_1 = 0$, there are two possibilities:

1 $a_1 c_2 - a_2 c_1 = 0$, and $b_2 c_1 - b_1 c_2 = 0$. Then the system is equivalent to the single equation $a_1 x + b_1 y + c_1 = 0$, which has infinitely many solutions.

2 At least one of $(a_1 c_2 - a_2 c_1)$ and $(b_2 c_1 - b_1 c_2)$ is not zero. Then (3) or (4) or both contains a contradiction and there is no solution.

Exercise

A Show that if $a_1 b_2 - a_2 b_1 = 0$ and $a_1 c_2 - a_2 c_1 = 0$, then

$$b_2 c_1 - b_1 c_2 = 0$$

We summarize these results in the following theorem.

Theorem 9 The simultaneous equations

$$\begin{aligned} a_1 x + b_1 y + c_1 = 0 \qquad & a_1 \neq 0 \text{ or } b_1 \neq 0 \\ a_2 x + b_2 y + c_2 = 0 \qquad & a_2 \neq 0 \text{ or } b_2 \neq 0 \end{aligned}$$

a Have a unique solution if $a_1 b_2 - a_2 b_1 \neq 0$.
b Have no solution if $a_1 b_2 - a_2 b_1 = 0$ and at least one of $(a_1 c_2 - a_2 c_1)$ and $(b_2 c_1 - b_1 c_2)$ is not zero.
c Have infinitely many solutions if $a_1 b_2 - a_2 b_1 = 0$, $a_1 c_2 - a_2 c_1 = 0$, and $b_2 c_1 - b_1 c_2 = 0$.

Remark Since the coefficients are real, case a of Theorem 9 corresponds to two intersecting lines, case b to two parallel lines, and case c to a single line.

Illustration 1   Solve

$$2x - 5y - 19 = 0$$
$$3x + 4y + 6 = 0 \qquad \text{(Fig. 6.4)}$$

Intersecting lines; Theorem 9, case *a*

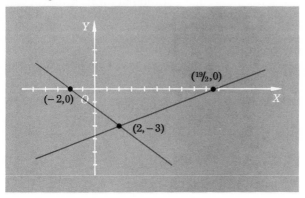

Figure 6.4 $2x - 5y - 19 = 0$ $3x + 4y + 6 = 0$

To eliminate x, we multiply the first equation by 3 and the second by 2. This gives us

$$6x - 15y - 57 = 0$$
$$6x + 8y + 12 = 0$$

Subtracting, we have

$$-23y - 69 = 0 \qquad \text{or} \qquad y + 3 = 0$$

This equation, combined with the first equation of the stated system, gives us the equivalent system

$$2x - 5y - 19 = 0$$
$$y + 3 = 0$$

We solve the second equation for y and get $y = -3$.   Putting $y = -3$ in the first equation and solving for x, we have $x = 2$.   Hence the solution is the pair $(2, -3)$.

Illustration 2   Solve

$$3x + 2y + 5 = 0$$
$$6x + 4y - 4 = 0 \qquad \text{(Fig. 6.5)}$$

Elimination of x as in Illustration 1 gives us the equivalent system

$$3x + 2y + 5 = 0$$
$$14 = 0$$

Since the last equation is, in fact, not an equality, there can be no solution. The lines are parallel.

Parallel lines; Theorem 9, case *b*

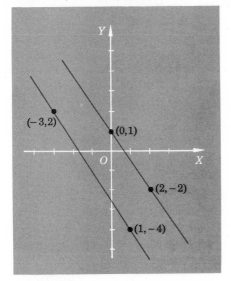

Figure 6.5
$3x + 2y + 5 = 0$
$6x + 4y - 4 = 0$

Illustration 3   Solve

$$4x - y + 3 = 0$$
$$8x - 2y + 6 = 0 \qquad \text{(Fig. 6.6)}$$

Single line; Theorem 9, case *c*

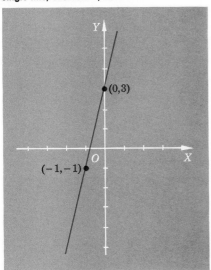

Figure 6.6
$4x - y + 3 = 0$
$8x - 2y + 6 = 0$

Elimination of x as above gives us the equivalent system

$$4x - y + 3 = 0$$
$$0 = 0$$

Hence the system reduces to a single equation: i.e., the lines are coincident, and there are infinitely many solutions.

6.9   Simultaneous Linear Equations (Continued)

In this section we consider a further idea which will help you to understand what we have just been discussing. Let us multiply the left members of the system (1) of Sec. 6.8 by k_1 and k_2, respectively, where k_1 and k_2 are not both zero, and then add. The result is

$$(5) \qquad k_1(a_1x + b_1y + c_1) + k_2(a_2x + b_2y + c_2) = 0$$

For all values of k_1 and k_2 (not both zero) this is the equation of some line. What line is it?

To simplify matters let us assume that the two lines intersect (case a of Sec. 6.8) and call the point of intersection $P(x_0,y_0)$. Then it follows that line (5) passes through P. To see this substitute (x_0,y_0) in (5); the result is zero since each parenthesis is zero by hypothesis. As k_1 and k_2 take different values, we get a family of lines all passing through P.

Theorem 10   If the lines $a_1x + b_1y + c_1 = 0$ and $a_2x + b_2y + c_2 = 0$ intersect at a point P, Eq. (5) represents a family of lines, each of which passes through P.

family of
lines
through P

Let us not forget our original problem—to solve the simultaneous system (1) of Sec 6.8. The point P of intersection of the two given lines can be found equally well by solving the equations of any two other lines through P. In other words, we will get the same point P if we solve any pair of equations chosen from the family (5). In the terminology of

Equivalent linear systems

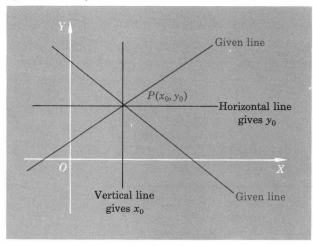

Figure 6.7   If the lines corresponding to a system of linear equations intersect at P, then the lines corresponding to any equivalent system also intersect at P.

Sec. 6.2, the given system of two equations is equivalent to any system of two distinct equations obtained from (5). So let us pick the simplest possible pair of equations from (5). These will correspond to the horizontal and vertical lines through P. To find the horizontal line, choose k_1 and k_2 so that the coefficient of x in (5) is zero; i.e., eliminate x. This gives the solution for y. Similarly, to find the vertical line through P, choose k_1 and k_2 so that y disappears, and solve for x. This solves the problem.

Incidentally, Eq. (5) permits us to obtain easy solutions to a number of other problems. The illustration below will give you the idea.

Illustration 1   Find the equation of the line passing through the point $(2, -1)$ and through the point of intersection of the lines $2x + y - 5 = 0$ and $x + 2y - 7 = 0$.

Using (5), we obtain the equation of the family of lines through this point of intersection:

$$k_1(2x + y - 5) + k_2(x + 2y - 7) = 0$$

We want to pick out the one passing through $(2, -1)$. So substitute $(2, -1)$ for (x,y) in the above equation. This gives

$$k_1(-2) + k_2(-7) = 0$$

Choose any k_1 and k_2 for which this is true, say $k_1 = 7$, $k_2 = -2$. This gives the required equation, namely:

$$7(2x + y - 5) + (-2)(x + 2y - 7) = 0$$
or
$$12x + 3y - 21 = 0$$

which is the answer. Observe that we never found the point $(1,3)$, which is the point of intersection of the two given lines.

6.10 Simultaneous Linear Equations in Three Unknowns

The method of Sec. 6.8 can be applied without substantial change to simultaneous systems of three equations. The general expression for such a system is

(1)
$$\begin{aligned} a_1x + b_1y + c_1z + d_1 = 0 \qquad &\text{Not all } a_1, b_1, c_1 = 0 \\ a_2x + b_2y + c_2z + d_2 = 0 \qquad &\text{Not all } a_2, b_2, c_2 = 0 \\ a_3x + b_3y + c_3z + d_3 = 0 \qquad &\text{Not all } a_3, b_3, c_3 = 0 \end{aligned}$$

A solution is an ordered triple (x,y,z) which satisfies all three equations. If we wish to plot ordered triples, we need three dimensions and so we shall not draw the graphs of these equations. Their geometric interpretation, however, is helpful. It can be proved that the equation $ax + by + cz + d = 0$ corresponds to a plane in 3-space; so the system (1) represents three

the graph of each equation is a plane

planes. The number of possible configurations for three planes is a little large, but here they are:

1 The three planes intersect in a point; hence (1) has a unique solution.
2 The three planes are mutually parallel; hence (1) has no solutions.
3 Two planes coincide, and the third plane is parallel to this common plane; hence (1) has no solutions.
4 All three planes coincide; hence (1) has a plane of solutions.
5 The three planes intersect in three parallel lines; hence (1) has no solutions.
6 Two planes are parallel, and the third intersects them in two parallel lines; hence (1) has no solutions.
7 The three planes intersect in a common line; hence (1) has a line of solutions.

In summary, (1) may have a unique solution, a line of solutions, a plane of solutions, or no solution.

The method of solution, which handles all these cases, is best explained by the illustrations below.

Illustration 1 Solve

$$2x - y + 3z + 9 = 0$$
$$x + 3y - z - 10 = 0$$
$$3x + y - z - 8 = 0$$

unique
solution

First we eliminate x between the first and second and between the first and third equations. The result, together with the unchanged first equation, is the equivalent system

$$2x - y + 3z + 9 = 0$$
$$7y - 5z - 29 = 0$$
$$5y - 11z - 43 = 0$$

Next eliminate y between the second and third equations. Leaving the first two equations unchanged, we have the equivalent system

$$2x - y + 3z + 9 = 0$$
$$7y - 5z - 29 = 0$$
$$z + 3 = 0$$

From the last equation, $z = -3$. Putting $z = -3$ in the second equation enables us to find $y = 2$; and putting $y = 2$ and $z = -3$ in the first equation gives us $x = 1$.

Hence the solution is $(1, 2, -3)$.

Illustration 2 Solve

$$x + 2y - z + 3 = 0$$
$$2x + y + z - 1 = 0$$
$$3x + 3y + 2 = 0$$

The first elimination (of x) gives us the equivalent system

$$x + 2y - z + 3 = 0$$
$$3y - 3z + 7 = 0$$
$$3y - 3z + 7 = 0$$

The final elimination (of y) gives the equivalent system

$$x + 2y - z + 3 = 0$$
$$3y - 3z + 7 = 0$$
$$0 = 0$$

Thus the system really reduces to two equations, i.e., to two planes which meet in a line. Hence there is a line of solutions.

Illustration 3 Solve

$$x - y + 2z + 1 = 0$$
$$2x - 2y + 4z + 2 = 0$$
$$3x - 3y + 6z + 3 = 0$$

The first elimination (of x) gives

$$x - y + 2z + 1 = 0$$
$$0 = 0$$
$$0 = 0$$

Thus the system reduces to a single equation, i.e., to a single plane all of whose points are solutions. The three planes are coincident, and there is a plane of solutions.

Illustration 4 Solve

$$x + y + z - 1 = 0$$
$$2x - 3y - 2z + 4 = 0$$
$$3x - 2y - z + 2 = 0$$

The first elimination (of x) gives the system

$$x + y + z - 1 = 0$$
$$5y + 4z - 6 = 0$$
$$5y + 4z - 5 = 0$$

The final elimination (of y) gives the system

$$x + y + z - 1 = 0$$
$$5y + 4z - 6 = 0$$
$$-1 = 0$$

Since the last equation is not, in fact, an equality, the system has no solution. There is no need to look further into the geometry of the case.

Problems 6.10

In Probs. 1 to 10 solve the given pair of equations algebraically. Then plot the graph of the two lines, and check your solution graphically.

1   $3x - 4y + 2 = 0$
    $x + 2y - 6 = 0$

2   $3x - 2y - 11 = 0$
    $5x + 2y + 3 = 0$

3   $4x + y - 3 = 0$
    $8x - y + 3 = 0$

4   $6x + y + 10 = 0$
    $3x - 2y + 10 = 0$

5    $x + 3y - 6 = 0$
    $-2x + 4y + 2 = 0$

6   $5x + 2y + 2 = 0$
    $x - y - 8 = 0$

7    $x + y - 2 = 0$
    $2x + 2y + 1 = 0$

8   $3x + 5y + 15 = 0$
    $6x + 10y - 2 = 0$

9   $-2x + 7y + 19 = 0$
    $4x - 14y - 38 = 0$

10 $3x - 2y + 2 = 0$
    $9x - 6y + 6 = 0$

In Probs. 11 to 20 solve for x, y, and z.

11   $x + 2y - z + 3 = 0$
    $2x - y + z - 5 = 0$
    $3x + 2y - 2z + 3 = 0$

12   $5x - y + 2z - 5 = 0$
    $-3x + 2y - z = 0$
    $4x + y + z - 7 = 0$

13 $2x + z + 4 = 0$
    $3y + 4z - 11 = 0$
    $x + y + 2 = 0$

14 $3x + 4y - 5z + 2 = 0$
    $-2x + y + z - 9 = 0$
    $4x - 3z + 13 = 0$

15 $3x + y - 4z + 6 = 0$
    $2x + 3y + z - 5 = 0$
    $5x + 4y - 3z + 2 = 0$

16 $2x - 3y + 5z - 5 = 0$
    $-x + y + 3z + 6 = 0$
    $x - 2y + 8z - 1 = 0$

17 $3x + 2y + 7z - 5 = 0$
    $4x - y + 5z + 4 = 0$
    $7x + y + 12z - 1 = 0$

18   $2x + 5y - 3z + 1 = 0$
    $-4x - 10y + 6z - 2 = 0$
    $6x + 15y - 9z + 3 = 0$

19   $x + 2y - z + 3 = 0$
    $-2x - 4y + 2z - 6 = 0$
    $3x + 6y - 3z + 9 = 0$

20 $5x - y - z + 1 = 0$
    $2x + 3y - z + 4 = 0$
    $x - 7y + z - 7 = 0$

In Probs. 21 to 25 find an equation of the line passing through the intersection of the two given lines and the given point.

21   $2x + y - 3 = 0$; $x - 3y + 2 = 0$; $(2,4)$
22   $2x + y = 0$; $3x - y - 5 = 0$; $(-1,3)$
23   $4x - y - 6 = 0$; $x - 3y + 4 = 0$; $(5,5)$
24   $x = 3$; $y = 1$; $(-2,-4)$
25   $x = -2$; $y = 4$; $(1,5)$
26   Find an equation of the line through $(1,3)$ and $(2,7)$. HINT: $x - 1 = 0$, $y - 3 = 0$ pass through $(1,3)$.
27   Find an equation of the line through $(-1,4)$ and $(2,6)$. See hint for Prob. 26.
28   Find an equation of the line through (x_1,y_1) and (x_2,y_2).

29   What does the equation

$$k_1(a_1x + b_1y + c_1) + k_2(a_2x + b_2y + c_2) = 0$$

represent when the two given lines are parallel?

30   What does the equation

$$k_1(a_1x + b_1y + c_1) + k_2(a_2x + b_2y + c_2) = 0$$

represent when the two given lines are coincident?

6.11   Simultaneous Linear and Quadratic Equations

Suppose that we are given a mixed system consisting of one linear equation and one quadratic equation, each involving two variables, such as

(1)
$$\begin{cases} y = ax + b \\ y = cx^2 + dx + e \end{cases}$$

We are looking for those ordered pairs (x,y) which satisfy both equations. When x and y are real, this amounts to finding the points of intersection (if any) of the line and the parabola, which are the graphs of the given equations.

The method of solution is quite straightforward: Set the two right-hand sides equal and solve the resulting quadratic equation for x.   Then substitute the solution (or solutions) into the linear equation and find the cor-

Straight line and parabola

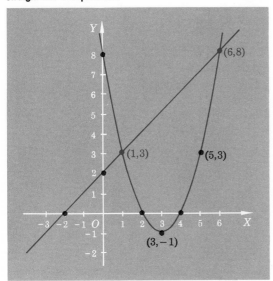

Figure 6.8
There are two points of intersection.
$y = x + 2$
$y = x^2 - 6x + 8$

responding values for y. The various possibilities are suggested by the following illustrations.

Illustration 1   Solve

$$\begin{cases} y = x + 2 \\ y = x^2 - 6x + 8 \end{cases}$$

Putting the right-hand sides equal, we obtain

$$x + 2 = x^2 - 6x + 8 \qquad \text{or} \qquad x^2 - 7x + 6 = 0$$
$$(x - 6)(x - 1) = 0 \qquad \text{or} \qquad x = 1, 6$$

When $x = 1$, $y = 3$ and when $x = 6$, $y = 8$. So the solutions are $(1,3)$ and $(6,8)$. The line intersects the parabola in two points (Fig. 6.8).

Illustration 2   Solve

$$\begin{cases} y = 2x - 4 \\ y = x^2 - 4x + 5 \end{cases}$$

We obtain

$$2x - 4 = x^2 - 4x + 5$$

or $\qquad\qquad x^2 - 6x + 9 = 0 \qquad \text{or} \qquad x = 3$

Thus the only solution is $(3,2)$. The line is tangent to the parabola (Fig. 6.9).

Straight line and parabola

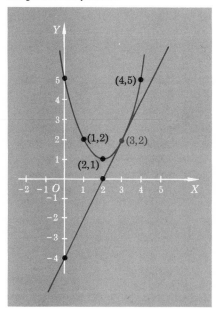

Figure 6.9
The line is tangent to the parabola.
$y = 2x - 4$
$y = x^2 - 4x + 5$

Illustration 3 Solve

$$\begin{cases} y = -4x + 2 \\ y = x^2 - 4x + 3 \end{cases}$$

We obtain

$$-4x + 2 = x^2 - 4x + 3 \qquad \text{or} \qquad x^2 + 1 = 0$$

Hence $x = +i, -i$, and $y = 2 - 4i, 2 + 4i$. The solutions are $(i, 2 - 4i)$, $(-i, 2 + 4i)$. The line and the parabola do not intersect (Fig. 6.10).

Straight line and parabola

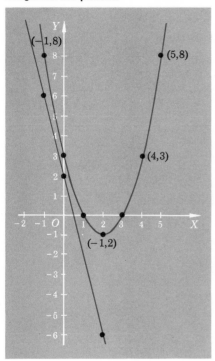

Figure 6.10
The line and the parabola do not intersect.
$y = -4x + 2$
$y = x^2 - 4x + 3$

The same methods apply to simultaneous systems such as

(2)
$$\begin{cases} y = ax^2 + bx + c \\ y = px^2 + qx + r \end{cases}$$

and

(3)
$$\begin{cases} y = ax + b \\ y = \dfrac{k}{cx + d} \end{cases}$$

Simple transformations will often convert other simultaneous systems to one of these types.

Illustration 4 Solve

$$\begin{cases} 2x - y - 2 = 0 \\ \qquad\quad xy = 4 \end{cases}$$

This is equivalent to

$$\begin{cases} y = 2x - 2 \\ y = \dfrac{4}{x} \end{cases}$$

So we solve

$$2x - 2 = \frac{4}{x}$$

or $$2x^2 - 2x - 4 = 0$$

or $$x^2 - x - 2 = 0$$

or $$(x - 2)(x + 1) = 0$$

Thus $x = 2$, or -1. The corresponding values of y are 2 and -4. Hence the solutions are $(2, 2)$ and $(-1, -4)$ (Fig. 6.11).

Straight line and hyperbola

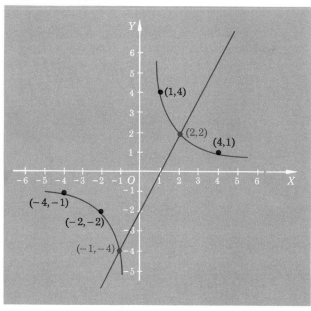

Figure 6.11 There are two points of intersection.
$$y = 2x - 2 \qquad y = 4/x$$

Problems 6.11

In Probs. 1 to 12 solve algebraically and graphically.

1 $y = 2x - 1$
$y = x^2 + 4x - 4$

2 $y = 3x - 11$
$y = x^2 - 5x + 4$

3 $y = 3x + 3$
$y = -x^2 + 2x + 15$

4 $y = -2x + 2$
$y = -x^2 + x + 6$

5 $y = 2x - 14$
$y = x^2 - 4x - 5$

6 $y = -2x + 5$
$y = -x^2 + 4$

7 $y = x - 10$
$y = x^2 + 2x - 8$

8 $y = -x + 7$
$y = -x^2 - x + 6$

9 $y = x^2 - 3x - 18$
$y = -x^2 + 4x - 3$

10 $y = x^2 + 4x + 4$
$y = x^2 + 8x + 16$

11 $y = x + 1$

$y = \dfrac{6}{x + 2}$

12 $y = x - 2$

$y = \dfrac{6}{x + 3}$

In Probs. 13 to 20 solve algebraically. To solve these you will need to invent your own methods.

13 $x^2 + y^2 = 4$
$x^2 + y^2 - 4y = 0$

14 $x + y = 5$
$xy + 36 = 0$

15 $x^2 + y^2 = 200$
$x + y = 12$

16 $x^2 + 3y^2 = 13$
$6x^2 - y^2 = 2$

17 $x^2 + 3xy + y^2 - 5 = 0$
$x = y$

18 $x^2 + xy + y^2 = 7$
$x^2 - xy + y^2 = 3$

19 $x + y = 3$
$y + z = 1$
$x^2 + yz = -1$

20 $xy = 6$
$yz = -3$
$xz = -2$

6.12 Word Problems

As we pointed out in Chap. 1, the ability to solve a given set of equations is not the only mathematical skill which a person requires. He must also be able to translate his practical problems into mathematical terms. In order to develop this ability of translation from nature to mathematics, we include a set of "word problems" here. Since the real problems which you are likely to meet in practice are too complicated for you to handle at this stage, these word problems represent situations which have been greatly simplified. They are worth your attention, however, for by solving them you will be preparing yourself to handle less artificial problems.

The method of solution involves the following steps:

1 Let variables x, y, z, etc., represent the unknown quantities. If the problem permits, draw a sketch labeling the known and unknown parts.

2 Translate the statement of the problem into a set of equations involving the chosen variables.

3 Solve these equations for the required quantities.

4 Check the solution to make sure that it fits the conditions as stated in the original problem. In particular be sure that the solution makes sense practically; i.e., distances cannot be negative, numbers of people must be positive integers, etc.

Illustration 1   Find the dimensions of a rectangle whose perimeter is 24 ft and whose area is 35 ft².

1   Let x be the length and y the width in feet.   Draw Fig. 6.12.

Figure 6.12

2   Then
$$2x + 2y = 24$$
$$xy = 35$$

3   Solving, we obtain $y = 12 - x$ and $y = 35/x$.   Hence
$$(12 - x)x = 35$$
or
$$x^2 - 12x + 35 = 0$$
$$(x - 5)(x - 7) = 0$$
$$x = 5 \text{ or } 7$$

Thus the solutions are $x = 5$, $y = 7$ and $x = 7$, $y = 5$.   *Ans.:* 5 by 7 ft.

4   These check in the original problem.

Problems 6.12

1   The sum of the digits of a certain two-digit number is 12.   If the order of the digits is reversed, the number is increased by 36.   What is the number?

2   The sum of two numbers is 24, and their product is 135.   Find the numbers.

3   A father is now four times as old as his son.   If both he and his son live 20 years longer, he will be twice as old as his son.   What are their present ages?

4   Two persons, A and B, are now respectively 30 years old and 10 years old.   In how many years will A be twice as old as B?

5   If 2 ft are added to each side of a certain square, its area is increased by 100 ft².   What is the original area of the square?

6   A farmer wishes to fence a rectangular field with 1,200 yd of wire fence.   If a river runs along one side and no fence is required there, and if the area of the field is to be 160,000 yd², what are its dimensions?

7   In an election for the mayor of Oxbridge the Labour candidate received 6,412 more votes than the Conservative candidate.   A total of 22,838 votes was cast.   How many people voted for the winner?

8 An eastbound, nonstop flight of 3,500 mi requires 5 hr. A similar westbound flight requires 7 hr. Assuming a constant westerly wind throughout, find the speed of the wind and the airspeed of the plane.

9 When two bricklayers, A and B, are working separately, A lays 2 more bricks/min than B. When they work together, each of their rates of laying drops to three-fourths of what it was when they worked alone, and together they lay 12 bricks/min. What are their rates of laying when they work separately?

10 A jet plane flies twice as fast as a propeller plane. Over a route of 3,000 mi the propeller plane takes 5 hr longer than the jet plane. Find the speeds of the two planes.

11 The total cost of 4 milk shakes and 2 coffees is $1.40. The cost of 1 milk shake and 3 coffees is $0.60. Find the price of a milk shake and of a coffee.

12 The annual cost C of operating a new car is $C = f + cm$, where f is the fixed cost (depreciation, insurance, license, etc.), c is the operating cost per mile, and m is the number of miles driven. The total cost for 10,000 mi is $2,100, and the cost for 15,000 mi is $2,700. Find the fixed cost and the cost per mile.

13 At supermarkets in Suburbia, the price of a pack of cigarettes includes a tax of 20 cents, which is the same for all brands. In Suburbia, 4 packs of Notar cigarettes cost the same as 3 packs of Green Grass cigarettes. In the free port of Utopia, there are no taxes, and hence the price of a pack of cigarettes is 20 cents lower than that in Suburbia. In Utopia, 2 packs of Notar cigarettes cost the same as 1 pack of Green Grass cigarettes. Find the prices of the cigarettes in Suburbia.

14 A citizen of the nation of Nancago has an annual taxable income of $9,900. The income tax rate in Nancago is 20 per cent. Moreover, the province of Camford also imposes an income tax of 5 per cent. The arrangement is that the national tax is based upon the annual taxable income less the provincial tax paid, and the provincial tax is based upon the annual taxable income less the national tax paid. Find the tax payable to Nancago and also that payable to its province, Camford.

15 On certain days of the week a family of father, mother, and teen-aged children traveling by first-class rail can take advantage of "family-plan" rates. Under one version of this scheme the father pays full fare and his wife and teen-aged children pay half fare. On the other hand the family could travel by coach, in which case each member would pay the full coach fare, which is two-thirds of the first-class fare. For what number of children is the total cost of first-class family plan equal to the total cost of coach?

16 This problem is the same as Prob. 15, except that the family now consists of father, mother, teen-aged children, and one eight-year-old child. The eight-year-old pays half the full first-class fare under family plan or half the full coach fare if they travel by coach. For what number of teen-aged children are the two costs equal?

17 A tourist has a collection of 21 coins consisting of Belgian francs (worth 2 cents each), British shillings (14 cents each), and German marks (25 cents each). The total value of the collection is $3.32. He has twice as many marks as shillings. How many coins of each kind has he?

18 A man has two *major medical* insurance policies. Policy A will pay him 80 per cent of the difference between the cost of an illness and the portion of this cost paid by policy B. Policy B will pay him 90 per cent of the difference between the cost of an illness and the portion of the cost paid by policy A. If the total cost of an illness is $2,800, how much of this cost is paid by each policy? How much must the man pay himself?

19 Neverstart automobiles use ordinary gasoline at 30 cents/gal and get 20 mi/gal. Everknock automobiles use premium gasoline at 34 cents/gal and get 17 mi/gal. In one day the sum of the distances traveled by a Neverstart and an Everknock was 200 mi, and the total cost for gasoline was $3.40. How far did each go?

20 In Redtape College all courses carry one *unit* credit, and *grade points* for each course are assigned as follows: A, 4 grade points; B, 3 points; C, 2 points; D, 1 point; E, 0 points. A student's *grade point average* is the quotient of the sum of his grade points divided by the number of units completed. After spending some time at Redtape, a student is enrolled for the current term in algebra and German (only). If he receives a C in algebra and an A in German, his grade point average at the end of this term will be 2.25. If he receives a D in algebra and a B in German, it will be 2.125. Find his grade point average at the beginning of the current term.

In Probs. 21 to 26 the following formulas from physics will be helpful.

$s = vt$	$s =$ distance, $v =$ velocity (constant), $t =$ time
$s = s_0 + \frac{1}{2} at^2$	$s =$ distance for general t, $s_0 =$ distance at $t = 0$, $a =$ acceleration (constant), $t =$ time
$v = v_0 + at$	$v =$ velocity for general t, $v_0 =$ velocity at $t = 0$, $a =$ acceleration (constant), $t =$ time
$f = ma$	$f =$ force, $m =$ mass, $a =$ acceleration
$E = IR$	$E =$ voltage, $I =$ current, $R =$ resistance
$PV = KT$	$P =$ pressure of a gas, $V =$ volume, $K =$ universal gas constant, $T =$ absolute temperature

21 A force of 5 dynes acts on a body A whose mass is 10 g. A force of 2 dynes acts on a body B whose mass is 15 g. Bodies C and D, the sum of whose masses is 11 g, are now fastened to bodies A and B, respectively, but the forces acting remain unchanged. Find the masses of C and D so that the acceleration produced on $A + C$ is twice that produced on $B + D$.

22 In an electric circuit A, the impressed voltage is 12 volts, and the resistance is 3 ohms. In circuit B, the voltage is 20 volts, and the resistance is 7 ohms. Additional batteries with a total voltage of 38 volts are to be added to these two circuits so that after the addition the currents in the two circuits will be equal. How much voltage should be added to each circuit?

23 In an electric circuit, the voltage is 15 volts. If the current is increased by 2 amp and the resistance decreased by 1 ohm, the voltage is increased by 5 volts. Find the original current and resistance.

24 In a certain gas, the product of the pressure (pounds per square inch) and the volume (cubic inches) is 24 in.-lb. If the pressure is increased by 5 lb/in.^{2} and the volume is decreased by 5 in.^{3}, the temperature is unchanged. Find the original pressure and volume.

25 Two different rockets are fired vertically at the same instant. The acceleration (constant) of one rocket is twice that of the other. After 3 sec one rocket is 81 ft higher than the other. Find their accelerations.

26 The second stage of a rocket is fired vertically with a constant acceleration a at a time $(t = 0)$ when the first stage has a vertical velocity v_0. One second after $t = 0$ the velocity of the second stage is 1,100 ft/sec, and after 3 sec it is 1,900 ft/sec. Find v_0 and a.

6.13 Transformation of Coordinates

If we have a coordinate system on a line with coordinates x, we can obtain a new coordinate system by means of the linear transformation

(1) $$x' = ax + b \qquad \text{where} \qquad a \neq 0$$

The effect of this is to give new names, x', to the points in place of their old names, x. In Probs. 2.13 we developed certain properties of such linear transformations and discussed the following two special cases:

$$x' = x + a \qquad\qquad \text{[Translation]}$$
$$x' = ax \qquad a \neq 0 \qquad \text{[Dilatation]}$$

We observed that a translation merely shifts the origin, and that a dilatation multiplies lengths of segments by $|a|$. A general linear transformation combines these two operations into one.

linear transformation

Now we turn to the similar problem in the plane. If we have a coordinate system (x,y), we can define a new coordinate system (x',y') by means of the *linear transformation*.

$$x' = a_1x + b_1y + c_1$$
$$y' = a_2x + b_2y + c_2$$

where we assume $a_1b_2 - a_2b_1 \neq 0$. The new pairs (x',y') serve as new labels for the points in the plane. The point O', where $x' = 0$, $y' = 0$, is the new origin, the X'-axis is the line $y' = 0$, and the Y'-axis is the line $x' = 0$. These axes, however, need not be at right angles (Fig. 6.13).

A point P is called a *fixed point* if its coordinates in both systems are equal, i.e., if $x' = x$, $y' = y$ at P. If every point is a fixed point, the trans-

Oblique axes

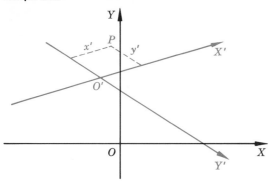

Figure 6.13
A linear transformation
results in a new set of axes.

formation has the equations $x' = x$, $y' = y$ and is called the *identity transformation*.

We shall develop the properties of these linear transformations in Probs. 6.13 which is the section below. In working some of these, you will need the following result. Let d be the distance between $P_1(x_1,y_1)$ and $P_2(x_2,y_2)$. Then

(2) $$d^2 = (x_2 - x_1)^2 + (y_2 - y_1)^2$$

This is really nothing but the Pythagorean Theorem.

Problems 6.13

Probs. 1 to 4 refer to linear transformations on a line.

1 Find the linear transformation which relabels $x = 1$ with $x' = 5$, and $x = 2$ with $x' = 7$. HINT: Solve for a and b in Eq. (1).

2 Given that $0°C$ corresponds to $32°F$ and $100°C$ corresponds to $212°F$, find the linear transformation which expresses F in terms of C.

3 In one grading system 70 is passing and 90 perfect. In a second grading system 60 is passing and 100 is perfect. Find a linear transformation which takes the first grading system into the second which takes passing into passing and perfect into perfect. What grade remains unchanged?

4 A faulty speedometer reads 15 mi/hr for an actual speed of 10 mi/hr, and 60 mi/hr for a speed of 70 mi/hr. Assuming a linear relationship, find a formula by means of which the correct speed can be calculated from the speedometer reading. For what speed is the speedometer correct?

Problems 5 to 26 refer to linear transformations in the plane. In Probs. 5 to 8 we take $x' = x + a$, $y' = y + b$. This transformation is called a *translation*.

translation

5 Prove that a translation leaves the lengths of segments unchanged.
6 Prove: if a translation has a fixed point, then it is the identity transformation.

7 Show that the correspondence $(x,y) \leftrightarrow (x',y')$ defined by a translation is 1 to 1.

8 For the translation $x' = x - 2$, $y' = y - 5$, find the new origin and sketch the new axes.

In Probs. 9 to 12 we take $x' = ax$, $y' = ay$, where $a \neq 0$.   This transformation is called a *dilatation*.

dilatation

9 Prove that a dilatation multiplies lengths of segments by $|a|$ and areas of rectangles by a^2.   Hence show that a triangle is transformed into a similar triangle.

10 Prove that the origin is a fixed point under a dilatation.

11 Prove that, if a dilatation has a fixed point in addition to the origin, then it is the identity transformation.

12 Show that the correspondence $(x,y) \leftrightarrow (x',y')$ defined by a dilatation is 1 to 1.

In Probs. 13 to 15 we take $x' = -x$, $y' = y$.   This transformation is called a *reflection* in the Y-axis.

reflection

13 Prove that a reflection leaves the lengths of segments unchanged.

14 Find the fixed points for the above reflection.

15 Prove that the correspondence $(x,y) \leftrightarrow (x',y')$ defined by a reflection is 1 to 1.

In Probs. 16 to 20 we take

rotation

$$\left. \begin{array}{l} x' = \quad ax + by \\ y' = -bx + ay \end{array} \right\} \quad \text{where } a^2 + b^2 = 1$$

This transformation is called a *rotation*.

16 Sketch the new axes when

$$x' = \quad \frac{1}{\sqrt{2}} x + \frac{1}{\sqrt{2}} y$$

$$y' = - \frac{1}{\sqrt{2}} x + \frac{1}{\sqrt{2}} y$$

17 Prove that a rotation leaves the lengths of segments unchanged.

18 Prove that the origin is a fixed point under a rotation.

19 Prove that, if a rotation has a fixed point other than the origin, then it is the identity transformation.   HINT: Solve

$$\left. \begin{array}{l} x = \quad ax + by \\ y = -bx + ay \end{array} \right\} \quad \text{for } a \text{ and } b$$

assuming $(x,y) \neq (0,0)$.

20 Prove that the correspondence $(x,y) \leftrightarrow (x',y')$ defined by a rotation is 1 to 1.

In Probs. 21 to 26 we consider the *centered* linear transformation

$$\left.\begin{aligned} x' &= a_1x + b_1y \\ y' &= a_2x + b_2y \end{aligned}\right\} \quad \text{where } a_1b_2 - a_2b_1 \neq 0$$

21 Prove that every point on the line $x + y = 0$ is a fixed point for the transformation

$$\begin{aligned} x' &= 2x + \ y \\ y' &= \ x + 2y \end{aligned}$$

HINT: Solve

$$\left.\begin{aligned} x &= 2x + \ y \\ y &= \ x + 2y \end{aligned}\right\} \quad \text{for } x \text{ and } y$$

22 Find the fixed points of the transformation

$$\begin{aligned} x' &= 3x - y \\ y' &= 2x \end{aligned}$$

23 Prove that the origin is the only fixed point of the general centered linear transformation unless $a_1b_2 - a_2b_1 - b_2 - a_1 + 1 = 0$.

24 Solve the equations of the general centered linear transformation for (x,y) in terms of (x',y'). This is the *inverse* transformation.

25 Prove that the correspondence $(x,y) \leftrightarrow (x',y')$ defined by a centered linear transformation is 1 to 1.

26 Consider the pair of transformations

$$\begin{aligned} x' &= ax + by & x'' &= px' + qy' \\ y' &= cx + dy & y'' &= rx' + sy' \end{aligned}$$

Find (x'',y'') in terms of (x,y). This formula will be used in Sec. 7.5 to motivate our definition for the product of two matrices.

VECTORS AND MATRICES

Chapter **7**

7.1 Introduction

In this chapter we shall study a new kind of algebra in which the elements are as follows:

vector

1 Ordered pairs like (x,y), ordered triples like (x,y,z). We call these vectors.

2 Rectangular arrays like those in the coefficients of our simultaneous equations, such as

$$\begin{pmatrix} a_1 & b_1 \\ a_2 & b_2 \end{pmatrix} \begin{pmatrix} a_1 & b_1 & c_1 \\ a_2 & b_2 & c_2 \end{pmatrix}$$

$$\begin{pmatrix} a_1 & b_1 & c_1 \\ a_2 & b_2 & c_2 \\ a_3 & b_3 & c_3 \end{pmatrix} \begin{pmatrix} a_1 & b_1 & c_1 & d_1 \\ a_2 & b_2 & c_2 & d_2 \\ a_3 & b_3 & c_3 & d_3 \end{pmatrix}$$

We call these matrices.

matrix

Definition A *matrix* is any rectangular (or square) array of numbers, and a *vector* is a special case of a matrix which has only one row or one column.

7.2 Vectors

Let us begin by restricting ourselves to vectors. You have probably met vectors before in your study of physics, and may wonder about the connection between the vectors of physics and those defined above. In physics a vector is represented in the plane as a directed distance $\overrightarrow{PQ}$ and is said to have magnitude and direction. The magnitude is represented by the length of the line segment in the plane, and the direction is given by the angle which this line makes with the horizontal and by the sense in which the arrow points. Common examples of vectors in physics are velocity, acceleration, and force.

Corresponding to the vector $\overrightarrow{PQ}$, we may draw a right triangle PQR (Fig. 7.1) with PR horizontal and QR vertical. The length of PR is the

Components of a vector

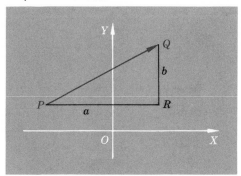

Figure 7.1
A vector can be resolved into horizontal and vertical components.

"x-component a" of $\overrightarrow{PQ}$; a is positive if $\overrightarrow{PQ}$ points to the right and is negative if $\overrightarrow{PQ}$ points to the left. Similarly RQ is the "y-component b" of $\overrightarrow{PQ}$; b is positive if $\overrightarrow{PQ}$ points up and negative if $\overrightarrow{PQ}$ points down. Clearly these components are known if the vector is known, and, conversely, a pair of components determines a vector. To simplify the discussion, we shall suppose that all of our vectors have the origin O as their initial point, so that the coordinates of their end points are equal to the components of the vectors. Then any vector is determined by the ordered pair of numbers (a,b). In the same way vectors in space have three components and are determined by a triple (a,b,c).

This gives us the connection between our vector and the vectors of physics. You should note, however, that every physical vector can be represented by a pair or a triple, but that vectors as we have defined them do not necessarily have physical interpretations. This is a good example of a mathematical concept which has arisen as a generalization of a concrete physical object.

Notation for Vectors   We shall write our vectors as "row-vectors": (a,b) or (a,b,c) or as "column-vectors": $\begin{pmatrix} a \\ b \end{pmatrix}$ or $\begin{pmatrix} a \\ b \\ c \end{pmatrix}$. There is no real distinction between row-vectors and column-vectors, but it will be convenient to use both notations in the applications which follow. Sometimes we use a single boldface letter such as **a** or **i** to stand for a vector. An equation of the form **a** $= (a,b,c)$ means that **a** and (a,b,c) are two different names for the same vector.

addition **Addition of Vectors** Since vectors are not numbers, the sum of two vectors is a new idea and must be defined.

Definition The *sum of two vectors* is defined by the formulas

$$(a,b) + (c,d) = (a + c, b + d)$$
$$(a,b,c) + (d,e,f) = (a + d, b + e, c + f)$$

Thus, to add two vectors of the same dimension we add their corresponding components. This has an important geometric interpretation, which we illustrate in the plane (Fig. 7.2). In order to add $\overrightarrow{OP}$ to $\overrightarrow{OQ}$, we

Addition of vectors

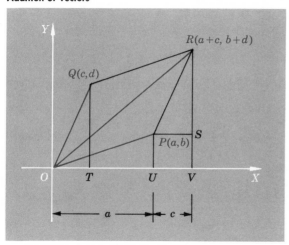

Figure 7.2
The sum of two vectors is the vector obtained by completing the parallelogram.

find point R, which is the fourth vertex of the parallelogram having O, P, and Q as its other vertices. Then triangle PRS is congruent to triangle OQT so that $PS = c$ and $RS = d$. Now $OU = a$, and OV is the x-component of $\overrightarrow{OR}$. From the figure

$$
\begin{aligned}
OV &= OU + UV \\
&= OU + PS \\
&= a + c
\end{aligned}
$$

Similarly $RV = b + d$. Thus $\overrightarrow{OR} = \overrightarrow{OP} + \overrightarrow{OQ}$. This interpretation is the source of the graphical method for adding vectors which is used widely in physics and navigation.

You will observe that this process is identical with that which we described in Sec. 2.15 for the graphical addition of two complex numbers. Although this shows that complex numbers and two-dimensional vectors have the same rules for addition, we must warn you that their rules for multiplication are completely different. Do not be misled into the common error of confusing vectors with complex numbers.

Illustration 1

a $(1,-3,2) + (3,4,-1) = (4,1,1)$
b $(5,2,-1) + (-5,-2,1) = (0,0,0)$, the "zero vector"
c $(-4,7,3) - (2,-1,4) = (-6,8,-1)$
d $(6,4,3) - (6,4,3) = (0,0,0)$

7.3   Products of Vectors

When we are speaking of vectors, we shall refer to an ordinary real number as a *scalar*. We now define the product of a scalar times a vector.

Definition   If (a,b,c) is a vector and k is a scalar, the *product* $k(a,b,c)$ is defined to be the vector (ka,kb,kc).

Illustration 1

a $2(1,3,-4) = (2,6,-8)$
b $-1(2,1,3) = (-2,-1,-3)$
c $0(a,b,c) = (0,0,0)$

It is often useful to define three *base vectors* **i**, **j**, and **k** as follows. These are vectors of length 1 drawn along the positive directions of the three coordinate axes.

Definition $\mathbf{i} = (1,0,0); \mathbf{j} = (0,1,0); \mathbf{k} = (0,0,1)$

In terms of these any vector (a,b,c) can be written

$$(a,b,c) = a\mathbf{i} + b\mathbf{j} + c\mathbf{k}$$

Although this notation is quite common in physics and engineering, we shall not use it regularly in this book.

Exercise

A   Using the definitions of **i**, **j**, and **k**, show that (a,b,c) is correctly expressed as $a\mathbf{i} + b\mathbf{j} + c\mathbf{k}$.

inner
product

The product of a vector by a vector is another concept which needs definition. There are, in fact, three kinds of products in common use; but here we shall discuss only the inner (or "scalar", or "dot") product.

Definition The *inner product* of two vectors (a_1,b_1,c_1) and (a_2,b_2,c_2) is defined to be the scalar $a_1a_2 + b_1b_2 + c_1c_2$. This product is denoted by a dot, so that

$$(a_1,b_1,c_1) \cdot (a_2,b_2,c_2) = a_1a_2 + b_1b_2 + c_1c_2$$

Illustration 2

a $(3,1,-2) \cdot (1,3,4) = (3)(1) + (1)(3) + (-2)(4) = -2$
b $(5,2,6) \cdot (1,1,1) = 5 + 2 + 6 = 13$
c $(-4,1,7) \cdot (0,0,0) = 0 + 0 + 0 = 0$
d $\mathbf{i} \cdot \mathbf{j} = (1,0,0) \cdot (0,1,0) = 0 + 0 + 0 = 0$

In terms of inner products we can define the *length* of a vector.

Definition The *length* of a vector (a,b,c) is the positive square root of the inner product $(a,b,c) \cdot (a,b,c)$. That is:

length $\text{Length of } (a,b,c) = \sqrt{(a,b,c) \cdot (a,b,c)} = \sqrt{a^2 + b^2 + c^2}$

Illustration 3

a The length of $(3,2,4) = \sqrt{9 + 4 + 16} = \sqrt{29}$
b The length of $\mathbf{i} = \sqrt{1 + 0 + 0} = 1$
c The length of $(0,0,0) = \sqrt{0 + 0 + 0} = 0$

The importance of the inner product in physics lies in the following geometrical interpretation, which you can understand if you have an elementary knowledge of trigonometry:

The inner product $(a_1,b_1,c_1) \cdot (a_2,b_2,c_2)$ is equal to the length of (a_1,b_1,c_1) times the length of (a_2,b_2,c_2) times the cosine of the angle between these two vectors. This statement is equivalent to the Law of Cosines for a triangle.

Physical concepts are frequently defined in terms of the inner product. For example, if a force $\mathbf{F} = (f_1,f_2,f_3)$ in pounds acts during a displacement $\mathbf{s} = (s_1,s_2,s_3)$ in feet, the work W which is done is defined to be

$$W = \mathbf{F} \cdot \mathbf{s} = (f_1,f_2,f_3) \cdot (s_1,s_2,s_3) \text{ ft-lb}$$

Problems 7.3

In Probs. 1 to 6 add the given vectors algebraically, and check your result graphically.

1 $(2,1) + (-3,4)$ 2 $(-3,7) + (2,2)$
3 $(-4,5) + (2,-3)$ 4 $(3,5) + (-4,1)$
5 $(2,6) - (1,3)$ 6 $(7,-3) - (-4,-2)$

In Probs. 7 to 12 write a vector equal to the given expression.

7 $(1,2,2) + (-4,5,3) - (1,8,2)$
8 $(0,3,-1) - (8,-2,4) + (1,3,-4)$
9 $3(1,4,2) + 2(0,3,5) + 4(1,-3,2)$
10 $-4(3,2,-1) + 3(1,0,2) + 2(5,0,-1)$
11 $4(3i - 2j + 4k) + 5(-2i + 7j - 4k)$
12 $-3(2i + j - k) + 6(i - j + 2k)$

In Probs. 13 to 16 prove the given statement.

13 Addition of vectors is commutative.
14 Addition of vectors is associative.
15 The zero vector $(0,0,0)$ is the additive identity for vectors.
16 The vector $(-a,-b,-c)$ is the additive inverse of (a,b,c).

In Probs. 17 to 22 compute the indicated inner products.

17 $(2,1,3) \cdot (1,-2,4)$ **18** $(6,1,-2) \cdot (-2,1,0)$
19 $(1,2,3) \cdot (2,1,-4)$ **20** $(4,2,-1) \cdot (1,-2,0)$
21 $i \cdot k$ **22** $j \cdot k$

In Probs. 23 to 30 find the length of the given vector.

23 $(1,2,2)$ **24** $(1,3,-4)$
25 $(2,-1,0)$ **26** $(-3,2,1)$
27 j **28** k
29 $3i - 2j + 4k$ **30** $4i + 3j - k$

In Probs. 31 to 33 prove the given statement.

31 In the multiplication of a scalar times a vector the following distributive laws hold:

$$k[(a_1,b_1,c_1) + (a_2,b_2,c_2)] = k(a_1,b_1,c_1) + k(a_2,b_2,c_2)$$
and $$(k_1 + k_2)(a,b,c) = k_1(a,b,c) + k_2(a,b,c)$$

32 The inner product is commutative.
33 For the inner product the following distributive law holds:

$$[(a_1,b_1,c_1) + (a_2,b_2,c_2)] \cdot (a_3,b_3,c_3)$$
$$= (a_1,b_1,c_1) \cdot (a_3,b_3,c_3) + (a_2,b_2,c_2) \cdot (a_3,b_3,c_3)$$

7.4 Matrices

As we have said above, a matrix is a square or rectangular array of numbers. The numbers of which a matrix is composed are called its elements. You are already familiar with many examples of them, such as the statistical tables which compose the bulk of the "World Almanac". In mathematics, matrices first appeared as the arrays of coefficients in simultaneous linear equations. In physics they are widely used in quantum theory and appear in elementary physics as (1) the set of moments and products of inertia of a rigid body or (2) the set of pressures at a point in a viscous fluid.

Although matrices may be of any dimensions, in this book we shall deal only with those of dimensions 2×2 (that is, two rows and two columns), 2×3, 3×2, and 3×3. As special cases we have already discussed vectors, which are matrices of dimensions 2×1, 3×1, 1×2, and 1×3. We shall now develop the elementary algebra of matrices.

Definition Two matrices are *equal* if and only if they have the same dimensions and are *identical*.

That is, for example,

$$\begin{pmatrix} a & b \\ c & d \end{pmatrix} = \begin{pmatrix} x & y \\ z & w \end{pmatrix}$$

if and only if

$$a = x \qquad b = y \qquad c = z \qquad d = w$$

addition The sum of two matrices is analogous to the sum of two vectors.

Definition The *sum* of two matrices of the same dimensions is a matrix whose elements are the sums of the corresponding elements of the given matrices.

For example,

$$\begin{pmatrix} a & b \\ c & d \end{pmatrix} + \begin{pmatrix} x & y \\ z & w \end{pmatrix} = \begin{pmatrix} a + x & b + y \\ c + z & d + w \end{pmatrix}$$

Definition A matrix (of any pair of dimensions) each of whose elements is zero is called the *zero matrix* (for that pair of dimensions).

zero
matrix

Examples of zero matrices are

$$(0) \qquad \begin{pmatrix} 0 \\ 0 \\ 0 \end{pmatrix} \qquad \begin{pmatrix} 0 & 0 \\ 0 & 0 \\ 0 & 0 \end{pmatrix} \qquad \begin{pmatrix} 0 & 0 \\ 0 & 0 \end{pmatrix}$$

Again, as for vectors, we can define the product of a scalar times a matrix.

Definition The *product* of a scalar k times a matrix is a matrix whose elements are k times the corresponding elements of the given matrix.

For example,

$$k \begin{pmatrix} a & b \\ c & d \end{pmatrix} = \begin{pmatrix} ka & kb \\ kc & kd \end{pmatrix}$$

7.5 Products of Matrices

We now turn our attention to the product of two matrices, when this can be defined. This concept is a generalization of the inner product of two vectors.

In order to motivate this definition, let us consider the pair of 2×2 matrices which are formed from the coefficients of the linear transformations (Sec. 6.13):

$$\begin{aligned} x' &= ax + by & x'' &= px' + qy' \\ y' &= cx + dy & y'' &= rx' + sy' \end{aligned}$$

namely:
$$\begin{pmatrix} a & b \\ c & d \end{pmatrix} \quad \text{and} \quad \begin{pmatrix} p & q \\ r & s \end{pmatrix}$$

In Sec. 6.13, Prob. 26, we saw that the combination of these two linear transformations is

$$\begin{aligned} x'' &= (ap + cq)x + (bp + dq)y \\ y'' &= (ar + cs)x + (br + ds)y \end{aligned}$$

If you did not work this problem previously, do so now, to verify this computation. The matrix of the coefficients of the combined linear transformation is then

$$\begin{pmatrix} ap + cq & bp + dq \\ ar + cs & br + ds \end{pmatrix}$$

The question, now, is: "How are these three matrices related?" We shall say that the third matrix is the *product* of the first two and shall write

$$\begin{pmatrix} p & q \\ r & s \end{pmatrix}\begin{pmatrix} a & b \\ c & d \end{pmatrix} = \begin{pmatrix} ap + cq & bp + dq \\ ar + cs & br + ds \end{pmatrix}$$

From this equation we observe that:

1 The element $ap + cq$ in the first row and first column of the product is equal to the inner product of the first row-vector (p,q) of the first factor with the first column-vector $\begin{pmatrix} a \\ c \end{pmatrix}$ of the second factor.

2 The element $bp + dq$ in the first row and second column of the product is equal to the inner product of the first row-vector (p,q) of the first factor with the second column-vector $\begin{pmatrix} b \\ d \end{pmatrix}$ of the second factor.

3 Similarly, $\qquad ar + cs = (r,s) \cdot \begin{pmatrix} a \\ c \end{pmatrix}$

and $\qquad\qquad br + ds = (r,s) \cdot \begin{pmatrix} b \\ d \end{pmatrix}$

As another piece of motivation, let us see how we can combine (by multiplication) the matrices

$$\begin{pmatrix} a & b \\ c & d \end{pmatrix} \quad \text{and} \quad \begin{pmatrix} x \\ y \end{pmatrix}$$

to obtain the column-vector $\begin{pmatrix} ax + by \\ cx + dy \end{pmatrix}$.

We shall write

$$\begin{pmatrix} a & b \\ c & d \end{pmatrix}\begin{pmatrix} x \\ y \end{pmatrix} = \begin{pmatrix} ax + by \\ cx + dy \end{pmatrix}$$

Notice that

$$ax + by = (a,b) \cdot \begin{pmatrix} x \\ y \end{pmatrix}$$

$$cx + dy = (c,d) \cdot \begin{pmatrix} x \\ y \end{pmatrix}$$

Perhaps the pattern is becoming clear, but before giving a precise definition of this product, let us consider some more examples.

Illustration 1

a
$$\begin{pmatrix} a_1 & b_1 & c_1 \\ a_2 & b_2 & c_2 \\ a_3 & b_3 & c_3 \end{pmatrix}\begin{pmatrix} x \\ y \\ z \end{pmatrix} = \begin{pmatrix} a_1 x + b_1 y + c_1 z \\ a_2 x + b_2 y + c_2 z \\ a_3 x + b_3 y + c_3 z \end{pmatrix}$$

b
$$\begin{pmatrix} 1 & -2 & 3 \\ 2 & 1 & 4 \end{pmatrix}\begin{pmatrix} 2 \\ 1 \\ 5 \end{pmatrix} = \begin{pmatrix} 15 \\ 25 \end{pmatrix}$$

c
$$\begin{pmatrix} 2 & -3 \\ 1 & 4 \end{pmatrix}\begin{pmatrix} 3 \\ -2 \end{pmatrix} = \begin{pmatrix} 12 \\ -5 \end{pmatrix}$$

d
$$\begin{pmatrix} 4 & 1 & -2 \\ 3 & 2 & 5 \\ -1 & 2 & 1 \end{pmatrix}\begin{pmatrix} 1 \\ 3 \\ -1 \end{pmatrix} = \begin{pmatrix} 9 \\ 4 \\ 4 \end{pmatrix}$$

Illustration 2   In view of the above,

$$\begin{pmatrix} a_1 & b_1 \\ a_2 & b_2 \end{pmatrix}\begin{pmatrix} x \\ y \end{pmatrix} = \begin{pmatrix} a_1 x + b_1 y \\ a_2 x + b_2 y \end{pmatrix}$$

and
$$\begin{pmatrix} a_1 & b_1 \\ a_2 & b_2 \end{pmatrix}\begin{pmatrix} x \\ y \end{pmatrix} + \begin{pmatrix} c_1 \\ c_2 \end{pmatrix} = \begin{pmatrix} a_1 x + b_1 y + c_1 \\ a_2 x + b_2 y + c_2 \end{pmatrix}$$

We can, therefore, write the system of simultaneous equations

$$a_1 x + b_1 y + c_1 = 0$$
$$a_2 x + b_2 y + c_2 = 0$$

in the compact form

$$AX + C = 0$$

where

$$A = \begin{pmatrix} a_1 & b_1 \\ a_2 & b_2 \end{pmatrix} \quad X = \begin{pmatrix} x \\ y \end{pmatrix} \quad C = \begin{pmatrix} c_1 \\ c_2 \end{pmatrix} \quad 0 = \begin{pmatrix} 0 \\ 0 \end{pmatrix}$$

A Show that the simultaneous system

$$a_1x + b_1y + c_1z + d_1 = 0$$
$$a_2x + b_2y + c_2z + d_2 = 0$$
$$a_3x + b_3y + c_3z + d_3 = 0$$

can be written in the form

$$AX + D = 0$$

where
$$A = \begin{pmatrix} a_1 & b_1 & c_1 \\ a_2 & b_2 & c_2 \\ a_3 & b_3 & c_3 \end{pmatrix} \quad X = \begin{pmatrix} x \\ y \\ z \end{pmatrix}$$

$$D = \begin{pmatrix} d_1 \\ d_2 \\ d_3 \end{pmatrix} \quad 0 = \begin{pmatrix} 0 \\ 0 \\ 0 \end{pmatrix}$$

We are now ready to define the product $AB = C$ of a $p \times q$-dimensional matrix A and a $q \times r$-dimensional matrix B. A consists of p q-dimensional row-vectors, and B consists of r q-dimensional column-vectors. The elements of C are the inner products of the row-vectors of A times the column-vectors of B.

Definition Let A be a $p \times q$-dimensional matrix and B a $q \times r$-dimensional matrix. Their *product* $AB = C$ is a $p \times r$-dimensional matrix whose elements are as follows: The element in the ith row and the jth column of C is the inner product of the ith row-vector of A with the jth column-vector of B.

product

Illustration 3

a $\begin{pmatrix} 2 & -1 & 4 \\ 1 & 3 & 5 \end{pmatrix}\begin{pmatrix} 4 & 2 \\ -1 & 3 \\ 2 & 1 \end{pmatrix} = \begin{pmatrix} 17 & 5 \\ 11 & 16 \end{pmatrix}$

b $(3 \quad -2)\begin{pmatrix} 2 & -1 & -2 \\ 6 & 3 & 4 \end{pmatrix} = (-6 \quad -9 \quad -14)$

c $\begin{pmatrix} 5 & 1 \\ -2 & 3 \end{pmatrix}\begin{pmatrix} 2 & 4 \\ -1 & 1 \end{pmatrix} = \begin{pmatrix} 9 & 21 \\ -7 & -5 \end{pmatrix}$

Remarks

1 The product AB is in this order: A first, B second. The necessity for this follows from the definition in which A and B are treated differently; we multiply the *rows* of A by the *columns* of B.

2 The product AB is defined only when the dimension of the row-vectors of A equals that of the column-vectors of B. That is, the number of columns in A must equal the number of rows in B.

3 When A and B are square and of the same dimension, both AB and BA are defined. However, in general, AB does not equal BA; that is, multiplication of square matrices is *not* commutative.

Illustration 4

$$\begin{pmatrix} 2 & 1 \\ -3 & 4 \end{pmatrix}\begin{pmatrix} 1 & 3 \\ 5 & -1 \end{pmatrix} = \begin{pmatrix} 7 & 5 \\ 17 & -13 \end{pmatrix}$$

but

$$\begin{pmatrix} 1 & 3 \\ 5 & -1 \end{pmatrix}\begin{pmatrix} 2 & 1 \\ -3 & 4 \end{pmatrix} = \begin{pmatrix} -7 & 13 \\ 13 & 1 \end{pmatrix}$$

On the other hand, it can be shown that matrix multiplication *is* associative; that is, $A(BC) = (AB)C$.

Problems 7.5

In Probs. 1 to 6 carry out the indicated operations.

1 $\begin{pmatrix} 2 & -4 \\ 1 & 3 \end{pmatrix} + \begin{pmatrix} 1 & 5 \\ 3 & 6 \end{pmatrix}$
    2 $\begin{pmatrix} 4 & 5 \\ 3 & -1 \end{pmatrix} + \begin{pmatrix} 2 & 2 \\ 1 & 4 \end{pmatrix}$

3 $\begin{pmatrix} 3 & -1 \\ 2 & -3 \end{pmatrix} - \begin{pmatrix} 2 & 4 \\ 1 & 6 \end{pmatrix}$
    4 $\begin{pmatrix} 2 & 1 \\ 0 & -3 \end{pmatrix} - \begin{pmatrix} 2 & 5 \\ 0 & 2 \end{pmatrix}$

5 $3\begin{pmatrix} 2 & 4 \\ -1 & 2 \end{pmatrix} + 4\begin{pmatrix} 1 & 0 \\ 2 & 1 \end{pmatrix}$
    6 $2\begin{pmatrix} 1 & 5 \\ -1 & 1 \end{pmatrix} - 3\begin{pmatrix} 2 & 2 \\ 4 & 7 \end{pmatrix}$

In Probs. 7 to 10 prove the given statement.

7 Addition of matrices is commutative.
8 Addition of matrices is associative.

9 The zero matrix $\begin{pmatrix} 0 & 0 \\ 0 & 0 \end{pmatrix}$ is the additive identity for 2×2 matrices.

10 The matrix $\begin{pmatrix} -a & -b \\ -c & -d \end{pmatrix}$ is the additive inverse of the matrix $\begin{pmatrix} a & b \\ c & d \end{pmatrix}$.

In Probs. 11 to 32 find the products of the given matrices.

11 $(2 \quad -1 \quad 4)\begin{pmatrix} 1 \\ 3 \\ 5 \end{pmatrix}$
    12 $(4 \quad 1 \quad -2)\begin{pmatrix} 3 \\ 6 \\ -2 \end{pmatrix}$

13 $\begin{pmatrix} 2 & 1 & 5 \\ 4 & -1 & 3 \end{pmatrix}\begin{pmatrix} 3 \\ -1 \\ 2 \end{pmatrix}$
    14 $\begin{pmatrix} -3 & 0 & 2 \\ 1 & 5 & 3 \end{pmatrix}\begin{pmatrix} 4 \\ 2 \\ -3 \end{pmatrix}$

15 $\begin{pmatrix} 4 & -1 & 1 \\ 2 & 3 & 0 \end{pmatrix}\begin{pmatrix} -3 & 2 \\ 5 & 2 \\ 1 & 1 \end{pmatrix}$
    16 $\begin{pmatrix} 2 & 8 & -3 \\ 6 & 0 & 2 \end{pmatrix}\begin{pmatrix} 7 & -3 \\ 5 & 4 \\ 1 & 2 \end{pmatrix}$

17 $\begin{pmatrix} 4 & -1 \\ 2 & 5 \end{pmatrix}\begin{pmatrix} 4 & 5 \\ -6 & 2 \end{pmatrix}$
    18 $\begin{pmatrix} 3 & -2 \\ 1 & 4 \end{pmatrix}\begin{pmatrix} 2 & 4 \\ 1 & 6 \end{pmatrix}$

19 $\begin{pmatrix} 4 & 1 \\ -2 & 3 \end{pmatrix}\begin{pmatrix} 3 & 6 \\ 1 & -2 \end{pmatrix}$ **20** $\begin{pmatrix} -5 & 2 \\ 1 & 3 \end{pmatrix}\begin{pmatrix} 2 & 4 \\ 1 & -3 \end{pmatrix}$

21 $\begin{pmatrix} 5 & 3 \\ 2 & 4 \\ 1 & 7 \end{pmatrix}\begin{pmatrix} 2 & -3 \\ 1 & 4 \end{pmatrix}$ **22** $\begin{pmatrix} 4 & 1 \\ 0 & -2 \\ -7 & 3 \end{pmatrix}\begin{pmatrix} 2 & -1 \\ 4 & 0 \end{pmatrix}$

23 $\begin{pmatrix} 2 \\ 1 \\ 4 \end{pmatrix}(1 \quad -5 \quad 2)$ (The resulting matrix is the *tensor product* of the two vectors.)

24 $\begin{pmatrix} -2 \\ 0 \\ 4 \end{pmatrix}(3 \quad 1 \quad -2)$ **25** $\begin{pmatrix} 5 & 2 & 1 \\ -3 & 1 & 7 \\ 0 & 1 & 2 \end{pmatrix}\begin{pmatrix} 1 & 2 & -3 \\ 0 & -2 & 4 \\ 1 & -3 & 1 \end{pmatrix}$

26 $\begin{pmatrix} 3 & -1 & 5 \\ 2 & 0 & 2 \\ 1 & 1 & 1 \end{pmatrix}\begin{pmatrix} 2 & -1 & 4 \\ 4 & -3 & 1 \\ 1 & 2 & 1 \end{pmatrix}$ **27** $\begin{pmatrix} -2 & -3 & 4 \\ 2 & 2 & -3 \\ 1 & 2 & -2 \end{pmatrix}\begin{pmatrix} 2 & 2 & 1 \\ 1 & 0 & 2 \\ 2 & 1 & 2 \end{pmatrix}$

28 $\begin{pmatrix} -9 & 29 & 10 \\ -14 & 45 & 16 \\ 1 & -3 & -1 \end{pmatrix}\begin{pmatrix} 3 & -1 & 14 \\ 2 & -1 & 4 \\ -3 & 2 & 1 \end{pmatrix}$

29 $\begin{pmatrix} 2 & -2 \\ -2 & 2 \end{pmatrix}\begin{pmatrix} 1 & 1 \\ 1 & 1 \end{pmatrix}$ Could this happen for the product of real numbers?

30 $\begin{pmatrix} 1 & 5 \\ 2 & 10 \end{pmatrix}\begin{pmatrix} 5 & 5 \\ -1 & -1 \end{pmatrix}$ Why is this result surprising?

31 $\begin{pmatrix} 2 & -1 & 1 \\ 3 & 0 & 2 \end{pmatrix}\begin{pmatrix} 3 & 2 \\ 1 & 0 \\ 2 & 1 \end{pmatrix}$ and $\begin{pmatrix} 3 & 2 \\ 1 & 0 \\ 2 & 1 \end{pmatrix}\begin{pmatrix} 2 & -1 & 1 \\ 3 & 0 & 2 \end{pmatrix}$

32 $\begin{pmatrix} -1 & 2 & 3 \\ 0 & 4 & 1 \end{pmatrix}\begin{pmatrix} 2 & 3 \\ 2 & -1 \\ 1 & 4 \end{pmatrix}$ and $\begin{pmatrix} 2 & 3 \\ 2 & -1 \\ 1 & 4 \end{pmatrix}\begin{pmatrix} -1 & 2 & 3 \\ 0 & 4 & 1 \end{pmatrix}$

*33 Why do $A = \begin{pmatrix} a & b & c \\ d & e & f \end{pmatrix}$ and $B = \begin{pmatrix} r & u \\ s & v \\ t & w \end{pmatrix}$ not commute under multiplication? Do not perform the multiplication.

*34 Why do $(a \quad b \quad c)$ and $\begin{pmatrix} d \\ e \\ f \end{pmatrix}$ not commute under multiplication?

In Probs. 35 to 37 carry out the stated operations where $I = \begin{pmatrix} 1 & 0 \\ 0 & 1 \end{pmatrix}$.

35 Let $A = \begin{pmatrix} 3 & 1 \\ 2 & 5 \end{pmatrix}$. Compute $A^2 - 8A + 13I$.

36 Let $A = \begin{pmatrix} 2 & -1 \\ 1 & 3 \end{pmatrix}$. Compute $A^2 - 5A + 7I$.

37 Let $A = \begin{pmatrix} 1 & 0 \\ 0 & -1 \end{pmatrix}$. Compute $A^4 - I$.

*38 Let $A = \begin{pmatrix} a & -b \\ b & a \end{pmatrix}$ and $B = \begin{pmatrix} c & -d \\ d & c \end{pmatrix}$.

Compute $A + B$ and AB. Compare your results with the sum and product of the complex numbers $a + bi$ and $c + di$.

In Probs. 39 to 42 write the system of linear equations in matrix form.

39 $3x + 4y - 5 = 0$
$x - 2y + 6 = 0$

40 $x + 5y + 7 = 0$
$2x - 3y + 1 = 0$

41 $2x - y + 3z - 4 = 0$
$x + 2y + z + 6 = 0$
$3x + y - z - 1 = 0$

42 $x + y + z = 0$
$y + 2z - 1 = 0$
$2x + y + 5 = 0$

43 Write the systems

$$\begin{array}{ll} x' = ax + by & \quad x'' = px' + qy' \\ y' = cx + dy & \text{and} \quad y'' = rx' + sy' \end{array}$$

in the matrix forms $X' = AX$; $X'' = BX'$. Hence find X'' in terms of X.

44 Compute the product

$$(x \quad y)\begin{pmatrix} 3 & 1 \\ 1 & 4 \end{pmatrix}\begin{pmatrix} x \\ y \end{pmatrix}$$

45 Compute the product

$$(x \quad y)\begin{pmatrix} 1 & 1 \\ 1 & 1 \end{pmatrix}\begin{pmatrix} x \\ y \end{pmatrix}$$

46 Compute the product

$$(x \quad y \quad z)\begin{pmatrix} 2 & 0 & 0 \\ 0 & 4 & 0 \\ 0 & 0 & -3 \end{pmatrix}\begin{pmatrix} x \\ y \\ z \end{pmatrix}$$

47 Consider the matrix $\begin{pmatrix} a & b \\ -b & a \end{pmatrix}$ where $a^2 + b^2 = 1$. This is the matrix associated with a rotation (see Probs. 16 to 20 of Sec. 6.13). Show that

$$\begin{pmatrix} a & -b \\ b & a \end{pmatrix}\begin{pmatrix} a & b \\ -b & a \end{pmatrix} = \begin{pmatrix} 1 & 0 \\ 0 & 1 \end{pmatrix}$$

48 For the rotation (see Prob. 47):

$$\begin{array}{l} x' = ax + by \\ y' = -bx + ay \end{array} \quad \text{where } a^2 + b^2 = 1$$

show that
$$\begin{pmatrix} x' \\ y' \end{pmatrix} = \begin{pmatrix} a & b \\ -b & a \end{pmatrix}\begin{pmatrix} x \\ y \end{pmatrix}$$

and
$$(x',y') = (x,y)\begin{pmatrix} a & -b \\ b & a \end{pmatrix}$$

49   Show that

$$d^2 = (x_1 - x_2)^2 + (y_1 - y_2)^2$$

$$= (x_1 - x_2, y_1 - y_2)\begin{pmatrix} x_1 - x_2 \\ y_1 - y_2 \end{pmatrix}$$

50   Using the results of Probs. 47, 48, and 49, show that

$$d'^2 = (x_1' - x_2')^2 + (y_1' - y_2')^2$$

$$= (x_1' - x_2', y_1' - y_2')\begin{pmatrix} x_1' - x_2' \\ y_1' - y_2' \end{pmatrix}$$

$$= (x_1 - x_2, y_1 - y_2)\begin{pmatrix} a & -b \\ b & a \end{pmatrix}\begin{pmatrix} a & b \\ -b & a \end{pmatrix}\begin{pmatrix} x_1 - x_2 \\ y_1 - y_2 \end{pmatrix}$$

$$= (x_1 - x_2)^2 + (y_1 - y_2)^2$$

$$= d^2$$

7.6   Inverse of a Square Matrix

The multiplication of square matrices has many, but not all, of the properties R6 to R10 of ordinary multiplication.   For instance, we have seen that it is associative, but not commutative.   We now show that it has an identity element.

Theorem 1   There exists an identity for the multiplication of 3×3 square matrices, namely, the matrix

identity
matrix

$$\begin{pmatrix} 1 & 0 & 0 \\ 0 & 1 & 0 \\ 0 & 0 & 1 \end{pmatrix}$$

Proof

$$\begin{pmatrix} 1 & 0 & 0 \\ 0 & 1 & 0 \\ 0 & 0 & 1 \end{pmatrix}\begin{pmatrix} a_1 & b_1 & c_1 \\ a_2 & b_2 & c_2 \\ a_3 & b_3 & c_3 \end{pmatrix} = \begin{pmatrix} a_1 & b_1 & c_1 \\ a_2 & b_2 & c_2 \\ a_3 & b_3 & c_3 \end{pmatrix}$$

and moreover

$$\begin{pmatrix} a_1 & b_1 & c_1 \\ a_2 & b_2 & c_2 \\ a_3 & b_3 & c_3 \end{pmatrix}\begin{pmatrix} 1 & 0 & 0 \\ 0 & 1 & 0 \\ 0 & 0 & 1 \end{pmatrix} = \begin{pmatrix} a_1 & b_1 & c_1 \\ a_2 & b_2 & c_2 \\ a_3 & b_3 & c_3 \end{pmatrix}$$

Remarks

1   For 2×2 matrices, the identity matrix is similarly $\begin{pmatrix} 1 & 0 \\ 0 & 1 \end{pmatrix}$.

2   We denote these identity matrices by the common symbol I.

Finally, we ask whether there is a multiplicative inverse for square matrices.

inverse

Definition If A is a square matrix, its *inverse* is a square matrix A^{-1} (read "A inverse") which satisfies the equations

$$AA^{-1} = I \quad \text{and} \quad A^{-1}A = I$$

Remarks

1 In the notation A^{-1} the -1 is not an exponent; it is merely a symbol indicating the inverse. Do *not* write $A^{-1} = 1/A$. We shall not define the quotient of two matrices.

2 Some square matrices do not have inverses. See Theorems 2 and 3 below.

In the 2×2 case we are given a matrix A and are looking for an A^{-1} which satisfies the definition. First let us require that

$$AA^{-1} = I$$

If A is $\begin{pmatrix} a & b \\ c & d \end{pmatrix}$ and $A^{-1} = \begin{pmatrix} w & x \\ y & z \end{pmatrix}$, we are asked to solve the matrix equation

$$\begin{pmatrix} a & b \\ c & d \end{pmatrix}\begin{pmatrix} w & x \\ y & z \end{pmatrix} = \begin{pmatrix} 1 & 0 \\ 0 & 1 \end{pmatrix}$$

Taking the product on the left, we have

$$\begin{pmatrix} aw + by & ax + bz \\ cw + dy & cx + dz \end{pmatrix} = \begin{pmatrix} 1 & 0 \\ 0 & 1 \end{pmatrix}$$

From the definition of the equality of two matrices, this gives us the two simultaneous systems:

(1)
$$\begin{cases} aw + by = 1 \\ cw + dy = 0 \end{cases}$$

(2)
$$\begin{cases} ax + bz = 0 \\ cx + dz = 1 \end{cases}$$

Writing $\Delta = ad - bc$, and supposing this not to be zero, we find that the solution is

$$w = \frac{d}{\Delta} \qquad y = -\frac{c}{\Delta} \qquad x = -\frac{b}{\Delta} \qquad z = \frac{a}{\Delta}$$

Therefore
$$A^{-1} = \frac{1}{\Delta}\begin{pmatrix} d & -b \\ -c & a \end{pmatrix}$$

As a bonus we find that

$$A^{-1}A = \frac{1}{\Delta}\begin{pmatrix} d & -b \\ -c & a \end{pmatrix}\begin{pmatrix} a & b \\ c & d \end{pmatrix} = \begin{pmatrix} 1 & 0 \\ 0 & 1 \end{pmatrix} = I$$

This gives us Theorem 2, in which $A = \begin{pmatrix} a & b \\ c & d \end{pmatrix}$.

Theorem 2 If $ad - bc \neq 0$, the matrix A has an inverse A^{-1} such that

$$AA^{-1} = A^{-1}A = I$$

Theorem 3 If $ad - bc = 0$, the inverse of A does not exist.

Proof If an inverse exists, the simultaneous systems (1) and (2) must have solutions. But when $ad - bc = 0$, Theorem 9 of Chap. 6 tells us that no such solutions exist.

To find the inverse of a particular matrix, you may either use the formula just given or you may solve Eqs. (1) and (2).

Illustration 1

a If $A = \begin{pmatrix} 2 & 5 \\ -1 & 4 \end{pmatrix}$, then by the formula

$$A^{-1} = \frac{1}{13}\begin{pmatrix} 4 & -5 \\ 1 & 2 \end{pmatrix} = \begin{pmatrix} {}^{4}\!/_{13} & -{}^{5}\!/_{13} \\ {}^{1}\!/_{13} & {}^{2}\!/_{13} \end{pmatrix}$$

b To find the inverse of $\begin{pmatrix} 2 & 5 \\ -1 & 4 \end{pmatrix}$ from first principles we write

$$\begin{pmatrix} 2 & 5 \\ -1 & 4 \end{pmatrix}\begin{pmatrix} w & x \\ y & z \end{pmatrix} = \begin{pmatrix} 1 & 0 \\ 0 & 1 \end{pmatrix}$$

Thus we derive

$$\begin{cases} 2w + 5y = 1 \\ -w + 4y = 0 \end{cases} \qquad \begin{cases} 2x + 5z = 0 \\ -x + 4z = 1 \end{cases}$$

These are equivalent to

$$\begin{cases} 2w + 5y = 1 \\ -2w + 8y = 0 \end{cases} \qquad \begin{cases} 2x + 5z = 0 \\ -2x + 8z = 2 \end{cases}$$

Hence $13y = 1$, $y = {}^{1}\!/_{13}$, $w = {}^{4}\!/_{13}$ and $13z = 2$, $z = {}^{2}\!/_{13}$, $x = -{}^{5}\!/_{13}$. Therefore

$$A^{-1} = \begin{pmatrix} {}^{4}\!/_{13} & -{}^{5}\!/_{13} \\ {}^{1}\!/_{13} & {}^{2}\!/_{13} \end{pmatrix}$$

c If $A = \begin{pmatrix} 2 & -3 \\ 4 & -6 \end{pmatrix}$, A^{-1} does not exist.

nonsingular matrix

A 2×2 matrix for which $ad - bc \neq 0$ is called *nonsingular;* if $ad - bc = 0$, it is *singular.* Hence a *matrix has an inverse if and only if it is nonsingular.*

In the case of 3×3 matrices, inverses can be computed by the method just described, but the computations are very tedious. We shall approach this problem from a simpler point of view in the next section.

7.7 Determinants

In finding the inverse of a 2×2 matrix and in solving a system of two simultaneous equations, we have run across expressions like $ad - bc$ and $a_1b_2 - a_2b_1$ in critical places. It is time we gave these a formal discussion. This brings us to determinants, which is the name given to expressions of this kind.

Definition Let A be the 2×2 matrix $\begin{pmatrix} a_1 & b_1 \\ a_2 & b_2 \end{pmatrix}$. Then we define the expression $a_1b_2 - a_2b_1$ to be the *determinant* of A and write

$$\det A = \begin{vmatrix} a_1 & b_1 \\ a_2 & b_2 \end{vmatrix} = a_1b_2 - a_2b_1$$

Exercise

A Prove that the determinant of a 2×2 matrix is zero if the two columns (rows) are proportional or equal.

Remarks

1 We use parentheses for matrices and parallel lines for the corresponding determinants.

2 A determinant is a single number associated with a square matrix. The determinant is *not* the array; the array is the matrix.

For 3×3 matrices we define the determinant in the following fashion. Let $A = \begin{pmatrix} a_1 & b_1 & c_1 \\ a_2 & b_2 & c_2 \\ a_3 & b_3 & c_3 \end{pmatrix}$ be a given 3×3 matrix. If we strike out the row and column containing any element, we are left with a 2×2 matrix whose determinant has already been defined. This determinant is called the *minor* of the corresponding element. We list a few examples of these:

minor

Element	Minor
a_1	$b_2c_3 - b_3c_2$
b_1	$a_2c_3 - a_3c_2$
c_2	$a_1b_3 - a_3b_1$
c_3	$a_1b_2 - a_2b_1$

We now attach an algebraic sign to each minor in the following way: Consider the corresponding element, and move it by a series of horizontal and/or vertical steps to the upper left-hand corner. The sign is $+$ if the number of steps required is even, $-$ if this number is odd. The product of the minor times this sign is called the *cofactor* of the corresponding

cofactor

element. The cofactor of any element will be denoted by the corresponding capital letter; for instance, the cofactor of a_1 is A_1. We list a few examples:

Element	Cofactor
a_1	$A_1 = \quad b_2c_3 - b_3c_2$
b_1	$B_1 = -(a_2c_3 - a_3c_2)$
c_2	$C_2 = -(a_1b_3 - a_3b_1)$
c_3	$C_3 = \quad a_1b_2 - a_2b_1$

To define the determinant, we now consider the first row and define

$$\begin{aligned} \det A &= a_1A_1 + b_1B_1 + c_1C_1 \\ &= a_1(b_2c_3 - b_3c_2) - b_1(a_2c_3 - a_3c_2) + c_1(a_2b_3 - a_3b_2) \\ &= a_1b_2c_3 + a_2b_3c_1 + a_3b_1c_2 - a_1b_3c_2 - a_2b_1c_3 - a_3b_2c_1 \end{aligned}$$

We might equally well have done this for any row or column, and at first sight you would expect the results to be six different numbers. They are, in fact, all equal.

Exercise

B By direct computation show that

$$\det A = b_1B_1 + b_2B_2 + b_3B_3$$

Definition The *determinant* of a 3×3 matrix is equal to the inner product of any row-vector (or column-vector) with the vector of its corresponding cofactors.

Illustration 1 Find

$$\det \begin{pmatrix} 1 & -2 & 3 \\ 4 & 1 & -1 \\ 1 & 2 & 1 \end{pmatrix} = \begin{vmatrix} 1 & -2 & 3 \\ 4 & 1 & -1 \\ 1 & 2 & 1 \end{vmatrix}$$

Choosing the first row, we find that the cofactors are

$$\begin{vmatrix} 1 & -1 \\ 2 & 1 \end{vmatrix} = 3 \qquad -\begin{vmatrix} 4 & -1 \\ 1 & 1 \end{vmatrix} = -5 \qquad \begin{vmatrix} 4 & 1 \\ 1 & 2 \end{vmatrix} = 7$$

So the answer is $(1)(3) + (-2)(-5) + (3)(7) = 34$.

As an alternative solution, choose the first column. The cofactors are respectively

$$\begin{vmatrix} 1 & -1 \\ 2 & 1 \end{vmatrix} = 3 \qquad -\begin{vmatrix} -2 & 3 \\ 2 & 1 \end{vmatrix} = 8 \qquad \begin{vmatrix} -2 & 3 \\ 1 & -1 \end{vmatrix} = -1$$

So the answer is $(1)(3) + 4(8) + (1)(-1) = 34$.

The following two theorems about determinants are of great utility.

Theorem 4 If two rows (columns) of a matrix are proportional (or equal), its determinant is zero.

The proof is immediate. First choose the third row (column) not involved in the proportionality. Then all the corresponding cofactors are zero (see Exercise A).

Theorem 5 The inner product of any row- (column-) vector and the vector of cofactors of a different row (column) is zero.

Proof Consider for example $a_1B_1 + a_2B_2 + a_3B_3$. This, however, is the determinant of the matrix

$$\begin{pmatrix} a_1 & c_1 & a_1 \\ a_2 & c_2 & a_2 \\ a_3 & c_3 & a_3 \end{pmatrix}$$

which is zero since two columns are equal.

Finally we can calculate the inverse of a 3×3 matrix.

Theorem 6 Let $A = \begin{pmatrix} a_1 & b_1 & c_1 \\ a_2 & b_2 & c_2 \\ a_3 & b_3 & c_3 \end{pmatrix}$, and let $\det A = \Delta$. Then, if $\Delta \neq 0$,

formula for
A^{-1}

$$A^{-1} = \frac{1}{\Delta}\begin{pmatrix} A_1 & A_2 & A_3 \\ B_1 & B_2 & B_3 \\ C_1 & C_2 & C_3 \end{pmatrix}$$

Proof We must show that $AA^{-1} = A^{-1}A = I$.

$$AA^{-1} = \frac{1}{\Delta}\begin{pmatrix} a_1 & b_1 & c_1 \\ a_2 & b_2 & c_2 \\ a_3 & b_3 & c_3 \end{pmatrix}\begin{pmatrix} A_1 & A_2 & A_3 \\ B_1 & B_2 & B_3 \\ C_1 & C_2 & C_3 \end{pmatrix} = \frac{1}{\Delta}\begin{pmatrix} \Delta & 0 & 0 \\ 0 & \Delta & 0 \\ 0 & 0 & \Delta \end{pmatrix}$$

$$= \begin{pmatrix} 1 & 0 & 0 \\ 0 & 1 & 0 \\ 0 & 0 & 1 \end{pmatrix} = I$$

because of the definition of the determinant and Theorem 5. A similar proof gives $A^{-1}A = I$.

Determinants of square matrices of higher dimension are defined inductively in a similar fashion. For a 4×4 matrix, for example, the cofactors are $\pm$ determinants of 3×3 matrices. The determinant, in an obvious notation, is defined to be $a_1A_1 + a_2A_2 + a_3A_3 + a_4A_4$. We can continue step by step to define determinants of square matrices of any size.

We conclude by giving without proof the following theorem for any square matrices.

Theorem 7 If $AB = C$, then $(\det A) \times (\det B) = \det C$.

Problems 7.7

In Probs. 1 to 10 find the determinants and the inverses of the given matrices when they exist. Check your inverses in the formulas $AA^{-1} = I$, $A^{-1}A = I$.

1 $\begin{pmatrix} 3 & 4 \\ -1 & 2 \end{pmatrix}$ **2** $\begin{pmatrix} 2 & 4 \\ 1 & 3 \end{pmatrix}$

3 $\begin{pmatrix} -3 & 2 \\ 5 & 1 \end{pmatrix}$ **4** $\begin{pmatrix} 1 & 0 \\ -2 & 3 \end{pmatrix}$

5 $\begin{pmatrix} 3 & 6 \\ -1 & -2 \end{pmatrix}$ **6** $\begin{pmatrix} 2 & -4 \\ -4 & 8 \end{pmatrix}$

7 $\begin{pmatrix} 1 & 0 \\ 0 & 1 \end{pmatrix}$ **8** $\begin{pmatrix} 0 & 0 \\ 0 & 0 \end{pmatrix}$

9 $\begin{pmatrix} \dfrac{1}{\sqrt{2}} & -\dfrac{1}{\sqrt{2}} \\ \dfrac{1}{\sqrt{2}} & \dfrac{1}{\sqrt{2}} \end{pmatrix}$ **10** $\begin{pmatrix} 0 & 1 \\ 1 & 0 \end{pmatrix}$

In Probs. 11 to 20 find the determinants of the given matrices.

11 $\begin{pmatrix} 3 & -1 & 4 \\ 1 & 0 & 2 \\ 5 & 3 & -2 \end{pmatrix}$ **12** $\begin{pmatrix} 0 & 1 & 5 \\ -2 & 8 & 3 \\ 4 & -1 & 2 \end{pmatrix}$

13 $\begin{pmatrix} 2 & 2 & 1 \\ -3 & 1 & -4 \\ 1 & 2 & 6 \end{pmatrix}$ **14** $\begin{pmatrix} 3 & -4 & 0 \\ 1 & 5 & 2 \\ -3 & 1 & 3 \end{pmatrix}$

15 $\begin{pmatrix} 2 & 4 & -1 \\ 1 & 6 & 2 \\ 5 & 14 & 0 \end{pmatrix}$ **16** $\begin{pmatrix} -1 & 5 & 3 \\ 4 & 0 & 8 \\ 2 & 1 & 5 \end{pmatrix}$

17 $\begin{pmatrix} 2 & 1 & -3 \\ 2 & 4 & -1 \\ 1 & 5 & 0 \end{pmatrix}$ **18** $\begin{pmatrix} 4 & 2 & 3 \\ -1 & 4 & 2 \\ 3 & 0 & 5 \end{pmatrix}$

19 $\begin{pmatrix} x & y & 1 \\ 3 & 1 & 1 \\ -2 & 4 & 1 \end{pmatrix}$ **20** $\begin{pmatrix} x & y & 1 \\ -2 & 4 & 1 \\ 3 & -2 & 1 \end{pmatrix}$

*21 Prove that $\det \begin{pmatrix} a_1 & b_1 & c_1 \\ a_2 & b_2 & c_2 \\ a_3 & b_3 & c_3 \end{pmatrix} = \det \begin{pmatrix} a_1 + kb_1 & b_1 & c_1 \\ a_2 + kb_2 & b_2 & c_2 \\ a_3 + kb_3 & b_3 & c_3 \end{pmatrix}$

HINT: Expand the second determinant in terms of its first column.

*22 Prove that $\det \begin{pmatrix} a_1 & b_1 & c_1 \\ a_2 & b_2 & c_2 \\ a_3 & b_3 & c_3 \end{pmatrix} = -\det \begin{pmatrix} b_1 & a_1 & c_1 \\ b_2 & a_2 & c_2 \\ b_3 & a_3 & c_3 \end{pmatrix}$

HINT: Expand both determinants in terms of their third columns.

In Probs. 23 to 30 find the inverse of the given matrix, if it exists, and check your inverses in the formula $AA^{-1} = I$.

23 $\begin{pmatrix} 2 & 1 & -2 \\ 0 & 1 & 1 \\ 3 & 0 & -2 \end{pmatrix}$

24 $\begin{pmatrix} 2 & -3 & 11 \\ -2 & 4 & -13 \\ 1 & -2 & 7 \end{pmatrix}$

25 $\begin{pmatrix} 2 & -1 & -5 \\ 1 & 3 & 4 \\ 0 & 1 & 2 \end{pmatrix}$

26 $\begin{pmatrix} -2 & 2 & 3 \\ 3 & 2 & -2 \\ -3 & 3 & 2 \end{pmatrix}$

27 $\begin{pmatrix} 1 & 2 & -3 \\ 3 & -1 & 0 \\ 5 & 3 & -6 \end{pmatrix}$

28 $\begin{pmatrix} 3 & 2 & 7 \\ -1 & 0 & -1 \\ 5 & 3 & 11 \end{pmatrix}$

29 $\begin{pmatrix} 2 & 1 & -8 \\ 1 & 1 & -2 \\ 1 & 2 & 3 \end{pmatrix}$

30 $\begin{pmatrix} 7 & -19 & 6 \\ -5 & 14 & -4 \\ 1 & -3 & 1 \end{pmatrix}$

31 Prove that $\det A = 1/(\det A^{-1})$. HINT: Use Theorem 7, and the fact that $AA^{-1} = I$.

32 Verify the steps in the following derivation: Let the vertices of a triangle be labeled as in Fig. 7.3. The subscripts are numbered in the counterclockwise direction around the triangle. Then the areas of the triangle and the three trapezoids satisfy

$$P_1P_2P_3 = P_1P_3Q_1Q_3 - P_1P_2Q_1Q_2 - P_2P_3Q_2Q_3$$
$$= \tfrac{1}{2}[(x_3 - x_1)(y_1 + y_3) - (x_2 - x_1)(y_1 + y_2)$$
$$- (x_3 - x_2)(y_2 + y_3)]$$
$$= \tfrac{1}{2}[(x_2y_3 - x_3y_2) - (x_1y_3 - x_3y_1) + (x_1y_2 - x_2y_1)]$$

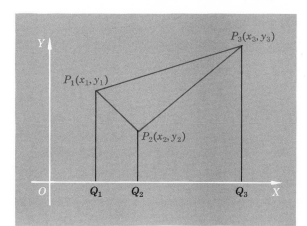

Figure 7.3

Therefore,   Area of $P_1P_2P_3 = \dfrac{1}{2}\begin{vmatrix} x_1 & y_1 & 1 \\ x_2 & y_2 & 1 \\ x_3 & y_3 & 1 \end{vmatrix}$

How is this result altered if we number the vertices in the clockwise direction?

33 Use the method of this section to find the inverse of $\begin{pmatrix} a & b \\ c & d \end{pmatrix}$. Show that the result agrees with the formula of Sec. 7.6.

34 Prove that $\det \begin{pmatrix} x & y & 1 \\ x_1 & y_1 & 1 \\ x_2 & y_2 & 1 \end{pmatrix} = 0$ is an equation of the line passing through (x_1, y_1) and (x_2, y_2). HINT: (a) Expand by the first row to show that the equation is linear. (b) Use Theorem 4 to show that (x_1, y_1) and (x_2, y_2) satisfy this equation.

35 Verify Theorem 7 for the product

$$\begin{pmatrix} 2 & 5 \\ 1 & 4 \end{pmatrix}\begin{pmatrix} 1 & -3 \\ 2 & 5 \end{pmatrix}$$

36 Verify Theorem 7 for the product

$$\begin{pmatrix} 1 & 5 \\ -3 & 2 \end{pmatrix}\begin{pmatrix} 2 & -4 \\ 3 & 7 \end{pmatrix}$$

7.8 Applications of Matrices to Simultaneous Equations

We have seen above that the simultaneous system

$$a_1 x + b_1 y + d_1 = 0$$
$$a_2 x + b_2 y + d_2 = 0$$

can be written in the compact form

$$AX + D = 0$$

where

$$A = \begin{pmatrix} a_1 & b_1 \\ a_2 & b_2 \end{pmatrix} \quad X = \begin{pmatrix} x \\ y \end{pmatrix} \quad D = \begin{pmatrix} d_1 \\ d_2 \end{pmatrix} \quad 0 = \begin{pmatrix} 0 \\ 0 \end{pmatrix}$$

Similarly

$$a_1 x + b_1 y + c_1 z + d_1 = 0$$
$$a_2 x + b_2 y + c_2 z + d_2 = 0$$
$$a_3 x + b_3 y + c_3 z + d_3 = 0$$

can be written in the form $AX + D = 0$, where

$$A = \begin{pmatrix} a_1 & b_1 & c_1 \\ a_2 & b_2 & c_2 \\ a_3 & b_3 & c_3 \end{pmatrix} \quad X = \begin{pmatrix} x \\ y \\ z \end{pmatrix} \quad D = \begin{pmatrix} d_1 \\ d_2 \\ d_3 \end{pmatrix} \quad 0 = \begin{pmatrix} 0 \\ 0 \\ 0 \end{pmatrix}$$

This suggests that the problem of solving simultaneous equations is really that of solving the matrix equation

$$AX + D = 0$$

But this is now an easy problem for us. For $AX + D = 0$ is equivalent to

$$AX = -D$$

Multiplying both sides, on the left, by A^{-1}, we have

$$A^{-1}AX = -A^{-1}D$$
or $$IX = -A^{-1}D$$

since $A^{-1}A = I$, or

$$X = -A^{-1}D \qquad \text{the solution}$$

A possible method of solution, therefore, is to compute A^{-1} by the methods of Secs. 7.6 and 7.7 and then to find $-A^{-1}D$. Although this is not the best method in practice, the above formula is of considerable theoretical value. It is known as "Cramer's Rule".

Cramer's Rule

Illustration 1 Solve

$$2x + 5y - 6 = 0$$
$$x - 2y + 5 = 0$$

$$A = \begin{pmatrix} 2 & 5 \\ 1 & -2 \end{pmatrix} \qquad \det A = -9$$

$$A^{-1} = -\frac{1}{9}\begin{pmatrix} -2 & -5 \\ -1 & 2 \end{pmatrix} \qquad D = \begin{pmatrix} -6 \\ 5 \end{pmatrix}$$

$$X = -A^{-1}D = \frac{1}{9}\begin{pmatrix} -2 & -5 \\ -1 & 2 \end{pmatrix}\begin{pmatrix} -6 \\ 5 \end{pmatrix} = \frac{1}{9}\begin{pmatrix} -13 \\ 16 \end{pmatrix} = \begin{pmatrix} -13/9 \\ 16/9 \end{pmatrix}$$

So $x = -13/9;\ y = 16/9$.

Cramer's Rule is sometimes written in a different form. To illustrate this, let us consider the system

$$a_1x + b_1y + d_1 = 0$$
$$a_2x + b_2y + d_2 = 0$$

Then $A^{-1} = \dfrac{1}{\Delta}\begin{pmatrix} b_2 & -b_1 \\ -a_2 & a_1 \end{pmatrix}$, where $\Delta = a_1b_2 - a_2b_1$. Hence

$$X = -A^{-1}D = -\frac{1}{\Delta}\begin{pmatrix} b_2 & -b_1 \\ -a_2 & a_1 \end{pmatrix}\begin{pmatrix} d_1 \\ d_2 \end{pmatrix} = -\frac{1}{\Delta}\begin{pmatrix} b_2d_1 - b_1d_2 \\ -a_2d_1 + a_1d_2 \end{pmatrix}$$

Therefore

$$x = -\frac{b_2d_1 - b_1d_2}{\Delta} = -\frac{\begin{vmatrix} d_1 & b_1 \\ d_2 & b_2 \end{vmatrix}}{\begin{vmatrix} a_1 & b_1 \\ a_2 & b_2 \end{vmatrix}}$$

and

$$y = -\frac{a_1d_2 - a_2d_1}{\Delta} = -\frac{\begin{vmatrix} a_1 & d_1 \\ a_2 & d_2 \end{vmatrix}}{\begin{vmatrix} a_1 & b_1 \\ a_2 & b_2 \end{vmatrix}}$$

You will observe that the numerator for x is obtained from the denominator by replacing the column of a's (the coefficients of x) by the column of d's. Similarly to obtain the numerator of y we replace the column of b's in the denominator by the column of d's. For the corresponding formula for three equations in three variables see Sec. 7.8, Prob. 15.

Illustration 2 Applying this method to the system in Illustration 1, we obtain

$$x = -\frac{\begin{vmatrix} -6 & 5 \\ 5 & -2 \end{vmatrix}}{\begin{vmatrix} 2 & 5 \\ 1 & -2 \end{vmatrix}} = -\frac{-13}{-9} = -\frac{13}{9}$$

$$y = -\frac{\begin{vmatrix} 2 & -6 \\ 1 & 5 \end{vmatrix}}{\begin{vmatrix} 2 & 5 \\ 1 & -2 \end{vmatrix}} = -\frac{16}{-9} = \frac{16}{9}$$

The problem of solving systems of linear equations is theoretically handled by Cramer's Rule, and we have a practical means for their solution in simple cases. In applied mathematics, however, one meets systems of linear equations containing 100 or more unknowns with coefficients which are 5- to 10-place decimals. The practical problem of solution is quite formidable, even on a high-speed machine. Modern research has developed elaborate techniques for tackling this problem, but improvements in these are currently under study.

two
homogeneous
equations

Finally let us consider the following "homogeneous" system of two equations in three unknowns:

$$a_1 x + b_1 y + c_1 z = 0$$
$$a_2 x + b_2 y + c_2 z = 0$$

Geometrically these equations represent two planes through the origin, and so we expect to find a *line of solutions*. By the use of determinants we can express this solution in a very elegant form.

Theorem 8 The solutions of

$$a_1 x + b_1 y + c_1 z = 0$$
$$a_2 x + b_2 y + c_2 z = 0$$

are

$$x = k\begin{vmatrix} b_1 & c_1 \\ b_2 & c_2 \end{vmatrix} \qquad y = -k\begin{vmatrix} a_1 & c_1 \\ a_2 & c_2 \end{vmatrix} \qquad z = k\begin{vmatrix} a_1 & b_1 \\ a_2 & b_2 \end{vmatrix}$$

where k is an arbitrary scalar (provided that at least one of these is different from zero).

Proof If we substitute these values of x, y, and z into the left-hand side of the first equation, we get

$$k \left\{ a_1 \begin{vmatrix} b_1 & c_1 \\ b_2 & c_2 \end{vmatrix} - b_1 \begin{vmatrix} a_1 & c_1 \\ a_2 & c_2 \end{vmatrix} + c_1 \begin{vmatrix} a_1 & b_1 \\ a_2 & b_2 \end{vmatrix} \right\}$$

The expression in braces, however, is precisely the determinant of the matrix

$$\begin{pmatrix} a_1 & b_1 & c_1 \\ a_1 & b_1 & c_1 \\ a_2 & b_2 & c_2 \end{pmatrix}$$

which is zero by Theorem 4. Therefore the first equation is satisfied. A similar argument shows that the second equation is satisfied.

Illustration 3   Solve

$$3x - 2y + z = 0$$
$$x + 4y + 2z = 0$$

By Theorem 8:

$$x = k \begin{vmatrix} -2 & 1 \\ 4 & 2 \end{vmatrix} \qquad y = -k \begin{vmatrix} 3 & 1 \\ 1 & 2 \end{vmatrix} \qquad z = k \begin{vmatrix} 3 & -2 \\ 1 & 4 \end{vmatrix}$$

or $\qquad\qquad x = -8k \qquad y = -5k \qquad z = 14k$

Illustration 4   Find the vector (x,y,z) whose inner products with each of the vectors $(4,1,-2)$ and $(2,1,3)$ are zero.
The required conditions are

$$4x + y - 2z = 0$$
$$2x + y + 3z = 0$$

Hence $x = 5k$, $y = -16k$, $z = 2k$. The required vector is $k(5,-16,2)$.

Illustration 4 motivates the following definition of the outer (or *vector* or *cross*) product of two vectors.

Definition   The *outer product* of the two vectors (a_1,b_1,c_1) and (a_2,b_2,c_2) is the vector

outer
product

$$\left(\begin{vmatrix} b_1 & c_1 \\ b_2 & c_2 \end{vmatrix}, \ -\begin{vmatrix} a_1 & c_1 \\ a_2 & c_2 \end{vmatrix}, \ \begin{vmatrix} a_1 & b_1 \\ a_2 & b_2 \end{vmatrix} \right)$$

The notation for this product is $(a_1,b_1,c_1) \wedge (a_2,b_2,c_2)$.†

Illustration 5 $(2,4,-3) \wedge (1,-2,6) = (18,-15,-8)$

† The $\wedge$ symbol used here has no connection with the similar symbol which sometimes denotes *conjunction* in symbolic logic. Physicists and engineers commonly use a cross, ✗, to denote this product and write $(a_1,b_1,c_1) \times (a_2,b_2,c_2)$.

Remarks

1   The inner product of $(a_1,b_1,c_1) \wedge (a_2,b_2,c_2)$ with each of its factors is zero.

2 $(a_1,b_1,c_1) \wedge (a_2,b_2,c_2) = -(a_2,b_2,c_2) \wedge (a_1,b_1,c_1)$. The "wedge" symbol $\wedge$ is commonly used in higher mathematics to denote "skew-commutative" multiplication, that is to say, multiplication for which

$$a \wedge b = -b \wedge a$$

3   A convenient way of expressing this product is the expansion of the following symbolic determinant by means of its first row:

$$\begin{vmatrix} \mathbf{i} & \mathbf{j} & \mathbf{k} \\ a_1 & b_1 & c_1 \\ a_2 & b_2 & c_2 \end{vmatrix}$$

Illustration 6

$$(1,-3,2) \wedge (4,1,3) = \begin{vmatrix} \mathbf{i} & \mathbf{j} & \mathbf{k} \\ 1 & -3 & 2 \\ 4 & 1 & 3 \end{vmatrix}$$
$$= -11\mathbf{i} + 5\mathbf{j} + 13\mathbf{k}$$
$$= (-11,5,13)$$

4   Let $\mathbf{A} = (a_1,b_1,c_1)$ and $\mathbf{B} = (a_2,b_2,c_2)$. Then the length of $\mathbf{A} \wedge \mathbf{B}$ is equal to the length of $\mathbf{A}$ times the length of $\mathbf{B}$ times the absolute value of the sine of the angle between $\mathbf{A}$ and $\mathbf{B}$.

Vector moment

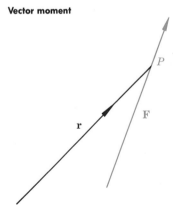

Figure 7.4
To find the vector moment of $\mathbf{F}$ about O, choose P on $\mathbf{F}$, draw $\mathbf{r} = \overrightarrow{OP}$, and compute $\mathbf{M} = \mathbf{r} \wedge \mathbf{F}$. The result does not depend upon the choice of P.

5   Let $\mathbf{F}$ be a force acting on a body at point P; let O be a reference point and $OP = \mathbf{r}$. Then the vector moment $\mathbf{M}$ is defined to be

$$\mathbf{M} = \mathbf{r} \wedge \mathbf{F}$$

Problems 7.8

In Probs. 1 to 14 use Cramer's Rule to solve the given system of equations (if possible).

1 $3x + 4y - 10 = 0$
$-x + 2y \qquad = 0$

[See Prob. 1, Sec. 7.7]

2 $2x + 4y - 14 = 0$
$x + 3y - 10 = 0$

[See Prob. 2, Sec. 7.7]

3 $-3x + 2y - 14 = 0$
$5x + y + 6 = 0$

[See Prob. 3, Sec. 7.7]

4 $x \qquad - 5 = 0$
$-2x + 3y - 11 = 0$

[See Prob. 4, Sec. 7.7]

5 $3x + 6y - 5 = 0$
$-x - 2y + 4 = 0$

[See Prob. 5, Sec. 7.7]

6 $2x - 4y + 3 = 0$
$-4x + 8y - 6 = 0$

[See Prob. 6, Sec. 7.7]

7 $2x + y - 2z + 2 = 0$
$y + z - 2 = 0$
$3x \qquad - 2z + 1 = 0$

[See Prob. 23, Sec. 7.7]

8 $2x - 3y + 11z + 3 = 0$
$-2x + 4y - 13z - 4 = 0$
$x - 2y + 7z + 2 = 0$

[See Prob. 24, Sec. 7.7]

9 $2x - y - 5z - 1 = 0$
$x + 3y + 4z - 4 = 0$
$y + 2z - 1 = 0$

[See Prob. 25, Sec. 7.7]

10 $-2x + 2y + 3z + 8 = 0$
$3x + 2y - 2z - 12 = 0$
$-3x + 3y + 2z + 7 = 0$

[See Prob. 26, Sec. 7.7]

11 $x + 2y - 3z + 1 = 0$
$3x - y \qquad - 2 = 0$
$5x + 3y - 6z \qquad = 0$

[See Prob. 27, Sec. 7.7]

12 $3x + 2y + 7z + 1 = 0$
$-x \qquad - z + 2 = 0$
$5x + 3y + 11z - 1 = 0$

[See Prob. 28, Sec. 7.7]

13 $2x + y - 8z + 5 = 0$
$x + y - 2z \qquad = 0$
$x + 2y + 3z - 6 = 0$

[See Prob. 29, Sec. 7.7]

14 $7x - 19y + 6z - 13 = 0$
$-5x + 14y - 4z + 9 = 0$
$x - 3y + z - 2 = 0$

[See Prob. 30, Sec. 7.7]

*15 Given the system

$$a_1x + b_1y + c_1z + d_1 = 0$$
$$a_2x + b_2y + c_2z + d_2 = 0$$
$$a_3x + b_3y + c_3z + d_3 = 0$$

Use Cramer's Rule to show that

$$x = -\frac{d_1A_1 + d_2A_2 + d_3A_3}{\Delta}$$

$$y = -\frac{d_1B_1 + d_2B_2 + d_3B_3}{\Delta}$$

$$z = -\frac{d_1C_1 + d_2C_2 + d_3C_3}{\Delta}$$

Hence write formulas for x, y, and z as quotients of determinants.

16 Solve Prob. 14 above by using the formulas derived in Prob. 15.

In Probs. 17 to 20 solve the given system of equations.

17 $3x - 5y + 6z = 0$
$x + 4y - z = 0$

18 $2x + 5y - 4z = 0$
$x + 6z = 0$

19 $x + 3y + z = 0$
$2x - y + 5z = 0$

20 $3x - y + z = 0$
$2x + y = 0$

In Probs. 21 to 24 find a vector whose inner products with each of the two given vectors are zero.

21 $(1,2,1); (-3,2,4)$
22 $(3,1,-5); (1,3,0)$
23 $(-2,1,1); (2,5,7)$
24 $(4,0,2); (3,-2,4)$
25 Why does the method of Theorem 8 fail for the system

$$3x - y + 2z = 0$$
$$-6x + 2y - 4z = 0$$

26 Why does the method of Theorem 8 fail for the system

$$x + 3y - z - 6 = 0$$
$$2x + y + z - 5 = 0$$

In Probs. 27 to 36 find the given outer products.

27 $(2,1,-3) \wedge (1,3,4)$
28 $(1,-2,1) \wedge (2,2,2)$
29 $(3,-1,6) \wedge (-2,5,-2)$
30 $(5,0,1) \wedge (1,0,5)$
31 $\mathbf{i} \wedge \mathbf{j}$
32 $\mathbf{j} \wedge \mathbf{k}$
33 $\mathbf{k} \wedge \mathbf{i}$
34 $(-1,2,4) \wedge (-2,4,8)$
35 $(2,-1,5) \wedge (4,-2,10)$
36 $\mathbf{i} \wedge \mathbf{i}$

7.9 Simple Applications to Business

It is often helpful to organize the facts of a given business situation as entries in a vector or a matrix. Then by suitable vector or matrix multiplication, it is possible to combine these data in various desirable ways. This procedure does nothing that could not be accomplished by standard methods of bookkeeping, but it does enable the businessman to organize his information in a very compact and convenient fashion. The process is best explained by examples.

Illustration 1 John Doe goes shopping for new clothes and decides that he needs two suits, six shirts, and one pair of shoes. In the Men's Shop he finds the following prices: suit, $80; shirt, $3; shoes, $15. How much must he pay for a complete outfit?

Solution Let his outfit be represented by the row vector

$$
\begin{array}{ccc}
\text{Suits} & \text{Shirts} & \text{Shoes} \\
(\quad 2 & 6 & 1 \quad)
\end{array}
$$

Let the costs be represented by the column vector

$$
\begin{array}{cc}
\text{Suit} & \\
\text{Shirt} & \begin{pmatrix} 80 \\ 3 \\ 15 \end{pmatrix} \\
\text{Shoes (pair)} &
\end{array}
$$

Then the total cost is the inner product

$$
(2 \quad 6 \quad 1)\begin{pmatrix} 80 \\ 3 \\ 15 \end{pmatrix} = \$193
$$

If Richard Roe decides to purchase one suit, four shirts, and two pairs of shoes at the same store, his total cost is

$$
(1 \quad 4 \quad 2)\begin{pmatrix} 80 \\ 3 \\ 15 \end{pmatrix} = \$122
$$

Illustration 2 John Doe and Richard Roe have the same needs as in Illustration 1, but they wish to shop around. They find the following costs at three stores:

	Men's Shop	*Esquire*	*Smith's Basement*
Suit	$80	$100	$50
Shirt	3	5	2
Shoes	15	20	10

What is the total cost to each man at each store?

Solution The needs of the men are represented by the matrix

$$
\begin{array}{c}
\text{Doe} \\
\text{Roe}
\end{array}
\begin{pmatrix} 2 & 6 & 1 \\ 1 & 4 & 2 \end{pmatrix}
$$

The costs at the several stores are represented by the matrix

$$
\begin{array}{ccc}
\text{Men's} & & \text{Smith's} \\
\text{Shop} & \text{Esquire} & \text{Basement} \\
\begin{pmatrix} 80 & 100 & 50 \\ 3 & 5 & 2 \\ 15 & 20 & 10 \end{pmatrix}
\end{array}
$$

The problem is solved by computing the matrix product

$$
\begin{pmatrix} 2 & 6 & 1 \\ 1 & 4 & 2 \end{pmatrix}
\begin{pmatrix} 80 & 100 & 50 \\ 3 & 5 & 2 \\ 15 & 20 & 10 \end{pmatrix} =
$$

$$
\begin{array}{cccc}
 & \text{Men's} & & \text{Smith's} \\
 & \text{Shop} & \text{Esquire} & \text{Basement} \\
\text{Doe} & \begin{pmatrix} 193 & 250 & 122 \\ 122 & 160 & 78 \end{pmatrix} \\
\text{Roe}
\end{array}
$$

Problems 7.9

1 A small corporation is entering into a merger. For administrative purposes, it retains the services of a law firm and an accounting firm and calls a meeting of its stockholders. The law firm charges $25 an hour for its services; the accounting firm charges $20 an hour; and the stockholders present at the meeting are each paid a fee of $30. The complete merger requires the following:
25 hr by the law firm
56 hr by the accounting firm
1 meeting with 200 stockholders present
Write a three-component row vector giving the requirements for the merger. Write a three-component column vector giving the prices for the various requirements. Using vector multiplication, find the administrative cost of the merger.

2 Using the data from Prob. 1, find the administrative cost of the merger if two stockholders' meetings must be called, each attended by 200 stockholders.

3 The unit costs of producing a newspaper vary with the size of circulation. In Roseville there are three competing papers. *The Herald* has a circulation of 5,000 per day; *The Tribune*, 20,000 per day; and *The Times*, 85,000 per day. *The Herald* spends $14.12 per 1,000 copies on editorial work, $19.70 per 1,000 copies on printing, and $5.28 per 1,000 copies on circulation. *The Tribune* spends $6.64 per 1,000 copies on editorial work, $7.50 per 1,000 copies on printing, and $3.60 per 1,000 copies on circulation. *The Times* spends $4.26 per 1,000 copies on editorial work, $3.40 per 1,000 copies on printing, and $3.48 per 1,000 copies on circulation. Find the total amount per day spent by the three papers combined on each of editorial work, printing and circulation.

4 Each of the papers in Prob. 3 plans to issue a special edition which

will require three times the normal cost of editorial work, two times the normal cost of printing, and no change in the circulation cost.

a What will be the cost per 1,000 copies of this special edition for each paper?

b Assuming the circulation figures of Prob. 3, what will be the total cost of all three special editions?

5 An oil corporation operates three oil refineries; each is located in a different area of the country and thus processes a different quality of crude oil. From the same quantity input of crude oil,

Refinery A produces: 40,000 barrels of kerosene/day
20,000 barrels of gasoline/day
7,000 barrels of low-grade motor oil/day

Refinery B produces: 30,000 barrels of kerosene/day
35,000 barrels of gasoline/day
2,000 barrels of low-grade motor oil/day

Refinery C produces: 27,000 barrels of kerosene/day
25,000 barrels of gasoline/day
15,000 barrels of low-grade motor oil/day

The prevailing wholesale prices of these oil products are:

Kerosene $11 per barrel
Gasoline   19 per barrel
Low-grade motor oil   4 per barrel

Find the total revenue per day of each of the three refineries.

6 Because of labor difficulties the production of the refineries in Prob. 5 is reduced as follows: Refinery A to ½ normal, B to ⅔ normal, and C to ¼ normal.

a What will be the total production of kerosene, gasoline, and oil in this case?

b What will be the revenue of the three refineries combined in this case?

7 A West Coast firm sends representatives to South Bend for yearly meetings with other members of the industry. The expense allowances for the representatives while at the meeting vary with the rank of the individual within the firm.

The president is allowed:

$25 per day hotel expense
20 per day entertainment expense
0.10 per mile traveling expense

All other executives are allowed:

$15 per day hotel expense
20 per day entertainment expense
0.08 per mile traveling expense

The salesmen are allowed:

$10 per day hotel expense
15 per day entertainment expense
0.06 per mile traveling expense

a Find the total allowance for each rank for a week-long meeting in South Bend, Indiana (2,000 miles from the home city).

b What would be the firm's total expense allowance obligation if it sent the president, two junior executives, and four salesmen to South Bend?

8 The expense allowances for employees of a West Coast firm are given in Prob. 7.

a Find the total allowance for each rank for a 5-day meeting in Providence, Rhode Island (3,000 miles from the home city).

b What would be the firm's total expense allowance obligation if it sent the president, five junior executives, and no salesmen to Providence?

9 For the oil refineries of Prob. 5 the following additional information is now available. For all three refineries:

Production costs of kerosene are $10.90 per barrel.

Production costs of gasoline are $18.80 per barrel.

Production costs of low-grade motor oil are $4 per barrel.

a Calculate a matrix with three rows and two columns such that the first column gives the total revenue for each refinery (as in Prob. 5) and the second column gives the total cost for each refinery.

b Multiply the matrix in **a** by the vector $\begin{pmatrix} 1 \\ -1 \end{pmatrix}$. The product is a column vector giving the total profit of each refinery.

10 The expense allowances for employees of a West Coast firm are given in Prob. 7. During the year, representatives of the firm are sent to three national meetings of the industry. The first is for 3 days, 800 mi away in San Francisco. The second is for 8 days, 1,300 mi away in Denver. The third is for 6 days, 3,400 mi away in Miami. What is the firm's total expense allowance obligation for the year if it sends the president, three junior executives, and two salesmen to each meeting?

7.10 Consumption Matrices

In the economic theory of production, a key role is played by certain matrices called *consumption* or *input-output* matrices. This theory was invented by Wassily W. Leontief in 1951 and is one of the most interesting developments in the field of economics in recent years. In this section we present a relatively simple portion of this theory as an illustration of how matrices are applied in modern economic theory.

Let us begin by considering a portion of the economy of the United States which involves the three interrelated industries: coal, steel, and railroad transportation. In order to describe how these are interrelated we write the following *consumption matrix*:

Consumption matrix

	Steel	Coal	R.R.
Steel	0.20	0.30	0.10
Coal	0.30	0.10	0.30
R.R.	0.20	0.40	0.10

The first column says that in the production of $1 worth of steel there is consumed 20 cents worth of steel, 30 cents worth of coal, and 20 cents worth of railroad transport. The second and third columns give, respectively, the values of the three commodities consumed in producing $1 worth of coal and $1 worth of railroad transport. Notice that the entries in this matrix are positive and less than 1. In other examples the entries on the main diagonal (upper left to lower right) may be zero. So our general assumptions are that the entries are all nonnegative and less than 1 and that the off-diagonal entries are not zero.

The sum of the elements in the first row, $0.20 + 0.30 + 0.10 = 0.60$, is the value of the steel consumed in the production of $1 worth of each of the three commodities. Unless steel is to be provided from some other source, such as accumulated inventory, it is reasonable to assume that the value of the steel consumed is not greater than the value of that produced, i.e., that the sum of the entries in the first row is less than or equal to 1. Similarly, we have chosen the entries in the second and third rows in such a way that the sum of each row does not exceed 1.

If the sum of the elements in each row is equal to 1 (rather than being less than 1), the values of the three commodities produced are exactly equal to the amounts consumed, and there is nothing left over for the use of other consumers such as the producers of automobiles. Such a situation is rather pointless, and so we further assume that, after this round-robin of production and consumption, there is a surplus of production over consumption for at least one commodity. That is, we assume that the sum of the entries in at least one row is strictly less than 1; i.e., the consumption of it is less than its production.

Finally, we assume that this model of our economy is valid at any level of production. This means, for example, that the production of $1,000 worth of steel requires the consumption of $200 worth of steel, $300 worth of coal, and $200 worth of railroad transport. Now we shall see how this enables us to solve certain economic problems.

First let us summarize by defining a consumption matrix.

Definition A *consumption matrix* is a square matrix such that (1) its elements are nonnegative and less than 1, (2) the off-diagonal elements are not zero, (3) the sum of the elements in each row is ≤ 1, and (4) there is at least one row the sum of whose elements is strictly less than 1.

The consumption matrices used in economic theory are very large, even as large as 400×400. For our purposes here, we must simplify the situation and consider only 2×2 and 3×3 consumption matrices. The only difference between our examples and the "real" case is that our computations are very much simpler.

Illustration 1 We consider the steel, coal, railroad example above. Let A represent its consumption matrix. Suppose that the daily production of these commodities is given by the *production vector:*

$$P = \begin{pmatrix} \$1{,}000 \\ \$\ 900 \\ \$1{,}100 \end{pmatrix} \begin{matrix} \text{steel} \\ \text{coal} \\ \text{R.R.} \end{matrix}$$

Then the product $AP = C_I$ is the *internal consumption vector*. Its components are the values of the three commodities consumed in the internal process of production. In this case

$$C_I = \begin{pmatrix} 0.20 & 0.30 & 0.10 \\ 0.30 & 0.10 & 0.30 \\ 0.20 & 0.40 & 0.10 \end{pmatrix} \begin{pmatrix} 1{,}000 \\ 900 \\ 1{,}100 \end{pmatrix} = \begin{pmatrix} \$580 \\ \$720 \\ \$670 \end{pmatrix} \begin{matrix} \text{steel} \\ \text{coal} \\ \text{R.R.} \end{matrix}$$

How much is left over for other consumers? This is the *external consumption vector* C_E, where

$$C_E = P - C_I$$

In this case

$$C_E = \begin{pmatrix} 1{,}000 \\ 900 \\ 1{,}100 \end{pmatrix} - \begin{pmatrix} 580 \\ 720 \\ 670 \end{pmatrix} = \begin{pmatrix} \$420 \\ \$180 \\ \$430 \end{pmatrix}$$

In the general situation we have the equations

$$AP = C_I \quad \text{and} \quad C_E + C_I = P$$

Hence

$$AP + C_E = P$$

By rearrangement this becomes the fundamental equation of this subject, namely,

Fundamental Equation:

$$(I - A)P = C_E$$

where A is a consumption matrix, I is the identity matrix, P is the production vector, and C_E is the external consumption vector. The elements of P and C_E are all nonnegative.

There are two types of problems which we can now attack:

Problem 1 Given A and P, find C_E. We can solve this by computing $(I - A)P$. There is no assurance that the resulting elements of C_E will be nonnegative. If one of them, say, does turn out to be negative, it follows that the assumed production cannot be achieved unless we are allowed to draw on an inventory of the corresponding commodity.

Problem 2 Given A and C_E, find P. Solving the fundamental equation, we find

$$P = (I - A)^{-1}C_E$$

Such a formula is meaningful only if the matrix $(I - A)$ is nonsingular and the resulting vector P has all of its components nonnegative. Although we cannot prove it here, our assumptions on the character of a consumption matrix A assure us that both of these conditions are satisfied.

Illustration 1 (continued)   In Illustration 1 we actually solved an example of Problem 1 in a somewhat roundabout fashion.   Let us repeat the solution using the formula $(I - A)P = C_E$.   The computation is

$$\begin{pmatrix} 0.8 & -0.3 & -0.1 \\ -0.3 & 0.9 & -0.3 \\ -0.2 & -0.4 & 0.9 \end{pmatrix} \begin{pmatrix} 1{,}000 \\ 900 \\ 1{,}100 \end{pmatrix} = \begin{pmatrix} \$420 \\ \$180 \\ \$430 \end{pmatrix}$$

Illustration 2   To give an example of the solution of Problem 2, let us suppose that A is the steel-coal-R.R. consumption matrix as above and that C_E is the vector $\begin{pmatrix} 420 \\ 180 \\ 430 \end{pmatrix}$.   Find P.   We already know the answer from Illustration 1, but let us find it without this knowledge.   We calculate

$$I - A = \begin{pmatrix} 0.8 & -0.3 & -0.1 \\ -0.3 & 0.9 & -0.3 \\ -0.2 & -0.4 & 0.9 \end{pmatrix}$$

$$(I - A)^{-1} = \frac{1}{0.423} \begin{pmatrix} 0.69 & 0.31 & 0.18 \\ 0.33 & 0.70 & 0.27 \\ 0.30 & 0.38 & 0.63 \end{pmatrix}$$

$$(I - A)^{-1} C_E = \frac{1}{0.423} \begin{pmatrix} 0.69 & 0.31 & 0.18 \\ 0.33 & 0.70 & 0.27 \\ 0.30 & 0.38 & 0.63 \end{pmatrix} \begin{pmatrix} 420 \\ 180 \\ 430 \end{pmatrix} = \begin{pmatrix} \$1{,}000 \\ \$ \ 900 \\ \$1{,}100 \end{pmatrix}$$

The same type of analysis can be applied in other situations which are illustrated in the problems.   For example, the consumption matrix A may refer to various sectors of the internal economy of the United States and the external consumption vector C_E to exports to foreign countries.

Problems 7.10

1   The consumption matrix for the interactions between oil and steel is assumed to be

	Steel	Oil
Steel	0.2	0.3
Oil	0.1	0.2

The desired external consumption is \$5,600 worth of steel and \$5,400 worth of oil.   Find the production vector P.

2   Given the steel-oil consumption matrix of Prob. 1 and the production vector whose components are steel $= \$15{,}000$, oil $= \$12{,}000$.   Find the external consumption vector C_E.

3   The consumption matrix for the interactions between oil and shipping is assumed to be

	Oil	Shipping
Oil	0.2	0.3
Shipping	0.1	0

If \$5,000 of oil and \$7,000 of shipping are produced, find the external consumption vector C_E.

4 Given the oil-shipping consumption matrix of Prob. 3 and the production vector whose components are oil = \$1,000, shipping = \$3,000. Find the external consumption C_E. Interpret your result.

5 Assume that the United States economy can be divided into three spending sectors. The interaction of these sectors is shown in the consumption matrix

	Industry	Government	Households
Industry	0.3	0.2	0.4
Government	0.2	0.2	0.2
Households	0.3	0.5	0.1

Suddenly the nation decides to assume a more active foreign-aid program, meaning increased industrial outlays, government gifts, and private investment abroad. The target figures for foreign aid spending are

Industry	$\$104 \times 10^9$
Government	4×10^9
Households	25×10^9

Find the production required for the three sectors when this new program goes into operation.

6 A model of the internal economy of the United States is given in the consumption matrix in Prob. 5. Assume that a sharp increase in international tension results in the following changes in this nation's foreign-aid spending program:

Industrial expenditures abroad decrease to $\$1 \times 10^9$.

Government gifts abroad increase to $\$66 \times 10^9$.

Private investments and expenditures abroad decrease to $\$5 \times 10^9$. Find the production required for each sector under these new circumstances.

7 Assume that a nation's economy can be divided into three producing sectors. The interaction of these sectors is shown in the consumption matrix

	Agriculture	Manufacturing	Services
Agriculture	0.4	0.2	0.1
Manufacturing	0.1	0.5	0.2
Services	0.1	0.2	0.4

In an effort to make a strong postwar recovery, the nation works toward the following external consumption goals:

Agricultural products	$\$\ 2 \times 10^9$
Manufacturing products	48×10^9
Services	23×10^9

What must be the *total* dollar output of each sector if the economy is to provide for these goals?

8 The economy represented in the matrix of Prob. 7 is stimulated by a sizable tax cut. External consumption is accelerated to the following levels:

Agricultural products 7×10^9
Manufactured products 56×10^9
Services 70×10^9

Find the production in each sector needed to provide for this consumption.

9 Suppose that the European Economic Community was joined by all those countries of Western Europe not already included in it and by the United States. The interaction of exports among members of this hypothetical common market is shown in the matrix

	United States	European Economic Community	Western Europe
United States	0	0.1	0.2
European Economic Community	0.1	0.4	0.4
Western Europe	0.1	0.2	0.3

What quantity of exports, in terms of dollars, must each member find outside the new common market to maintain its preunion export total of:

United States $\$19 \times 10^9$
European Economic Community 31×10^9
Western Europe 21×10^9

10 The matrix in Prob. 9 describes the interaction of exports within a hypothetical common market. Suppose that each member wishes to maintain the following dollar amounts of exports to countries outside the new union:

United States $\$15 \quad \times 10^9$
European Economic Community 9.5×10^9
Western Europe 10.5×10^9

Find the amounts that each member must export in *total*.

11 The economy of a nation under the guidance of socialist planners is divided into three producing sectors, the interaction of which is described in the consumption matrix

	Agriculture	Manufacturing	Services
Agriculture	0.4	0.3	0.1
Manufacturing	0.2	0.5	0.1
Services	0.1	0.1	0.3

A recent government plan predicts that the following external consumption targets will be reached within a year:

Agricultural products $\$6 \quad \times 10^9$
Manufacturing products $34 \quad \times 10^9$
Services 4.5×10^9

Physical limitations of the nation's resources make the following binding:

Value of agricultural production $< \$50 \times 10^9$

Value of manufacturing $\qquad < \$130 \times 10^9$

Value of services $\qquad < \$ 35 \times 10^9$

Find the value of the production necessary in each sector for the government's plan to be fulfilled, and determine whether the plan is feasible in light of the nation's physical limitations.

12 In the production of Gagets, the Brand X Corporation involves three interconnected, part production processes, A, B, and C, the interaction of which is shown in the matrix

$$
\begin{array}{c c c c}
 & A & B & C \\
\begin{matrix} A \\ B \\ C \end{matrix} & \begin{pmatrix} 0 & 0.4 & 0.5 \\ 0.6 & 0 & 0.3 \\ 0.2 & 0.3 & 0 \end{pmatrix}
\end{array}
$$

Physical limitations of the plant make the following conditions binding:

Total value of production of $A < \$50 \times 10^3$

Total value of production of $B < \$53 \times 10^3$

Total value of production of $C < \$48 \times 10^3$

Assume that the following values of A, B, C are consumed (externally) in the production of Gagets at its present level:

$$\$9 \times 10^3 \text{ of } A$$
$$\$8 \times 10^3 \text{ of } B$$
$$\$2 \times 10^3 \text{ of } C$$

The sales manager suggests that it would be wise to get into the Giget market. The production of the proposed number of Gigets would require an additional external consumption of

$$\$11 \times 10^3 \text{ of } A$$
$$0 \qquad \text{ of } B$$
$$\$ \ 2 \times 10^3 \text{ of } C$$

As production manager, find the total value of production of each part process necessary if both Gagets and Gigets are produced. Determine whether or not the existing plant could handle the expansion into the Giget market.

References

Beaumont, R. A.: "Linear Algebra", Harcourt, Brace & World, New York (1965).

Hohn, F. E.: "Elementary Matrix Algebra", Macmillan, New York (1964).

INEQUALITIES

8.1   Fundamental Properties

In Chap. 2 (Sec. 2.11) we introduced the concept of inequality between two real numbers.   In this chapter we shall treat this subject in more detail.

We noted (Sec. 2.2) that the set of real numbers is the union of the subsets:

1   The positive real numbers
2   The negative real numbers
3   Zero

Moreover, each real number belongs to just one of the subsets; the sum of two positive numbers is positive, the product of two positive numbers is positive, and the additive inverse of a positive number is negative.   In view of these remarks, we define inequality of real numbers as follows.

Definition   For two real numbers a and b, a is *greater* than b, written $a > b$, if and only if $a - b$ is positive.   Moreover, a is *less* than b, written $a < b$, if and only if $a - b$ is negative.

The symbol $a \geq b$, read *a is greater than or equal to b*, means $a > b$ or $a = b$.   Similarly, $a \leq b$, read *a is less than or equal to b*, means $a < b$ or $a = b$.

We next observe that we can prove that the real numbers have the following properties:

R12 For each pair of real numbers a and b, one and only one of the following relations is true:

$$a < b \qquad a = b \qquad a > b$$

This follows from the fact that $a - b$ is negative, zero, or positive.

Now consider the inequalities

$$a > b \qquad b > c$$

From these we can conclude that $a > c$. To see this we note that

$$a - b \text{ is positive} \qquad b - c \text{ is positive}$$

Hence the sum $(a - b) + (b - c) = a - c$ is positive, which shows that $a > c$. We call this the *transitive law* for inequalities and write it as property R13.

R13 If a, b, and c are real numbers, and if $a > b$ and $b > c$, then $a > c$.

Illustration 1 Since $2 > -3$ and $8 > 2$, it follows that $8 > -3$.

Next let us suppose that $a > b$, so that $a - b$ is positive. Then it follows that $a + c > b + c$, for $(a + c) - (b + c) = a - b$, which is positive. We write this as property R14.

R14 If a, b, and c are real numbers and if $a > b$, then

$$a + c > b + c$$

Illustration 2 Since $5 > 3$, it follows that $5 + 2 > 3 + 2$, or $7 > 5$.

Finally, we again suppose that $a > b$ and that c is positive. Then $ac - bc = (a - b)c$ is positive, for each factor is positive. Therefore $ac > bc$. We write this as property R15.

R15 If a, b, and c are real numbers, and if $a > b$, and $c > 0$, then $ac > bc$.

Illustration 3 Since $6 > 2$, it follows that $6 \times 3 > 2 \times 3$, or that $18 > 6$.

Exercises

A The conclusion of R15 is false if c is negative. Find a counterexample which illustrates this fact.

B Do the rational numbers have properties R12 to R15?

C Show that $a > 0$ if and only if a is positive.

D Show that $a < b$ if and only if $b > a$.

Properties R12 to R15 are thus established for real numbers. In terms of R1 to R15 we can now derive all the rules for operating with inequalities. In more advanced mathematics a system having properties R12 to R15 is said to be "linearly ordered". A system having properties R1 to R15 is called an "ordered field".

ordered
field

8.2   Theorems about Inequalities

From R1 to R15 we shall now derive the chief theorems on inequalities. These will enable us to manipulate our inequalities and to solve problems involving them.   All letters refer to real numbers.

Theorem 1   If $a > b$ and $c > d$, then $(a + c) > (b + d)$.

Proof

1	$a + c > b + c$	[R14]
2	$b + c > b + d$	[R14]
3	$a + c > b + d$	[R13]

Illustration 1   From $6 > -3$ and $8 > 4$ we conclude from Theorem 1 that $6 + 8 > -3 + 4$, or $14 > 1$.   Note that Theorem 1 says nothing about adding the corresponding sides of two inequalities, one of which contains a *less than* $(<)$ and the other a *greater than* $(>)$.

Exercise

A   Prove that, if $a < b$ and $c < d$, then $(a + c) < (b + d)$.

Theorem 2 $a > 0$ if and only if $-a < 0$.

Proof (Left to Right)

1	$a + (-a) > 0 + (-a)$	[R14]
2	$0 > -a$	[R4 and R3]
3	$-a < 0$	[Sec. 8.1, Exercise D]

Exercises

B   Complete the proof (right to left) of Theorem 2.
C   Prove: $a < 0$ if and only if $-a > 0$.

Theorem 3 $a > b$ if and only if $-a < -b$.

Proof (Left to Right)

1	$[(-a) + (-b)] + a > [(-a) + (-b)] + b$	[R14]
2	$-b > -a$	[R2, R3, and R4]
3	$-a < -b$	[Sec. 8.1, Exercise D]

Exercises

D   Complete the proof (right to left) of Theorem 3.
E   Prove: $a < b$ if and only if $-a > -b$.

Illustration 2

> From $7 > 2$, we conclude that $-7 < -2$.
> From $10 > -3$, we conclude that $-10 < 3$.
> From $-4 > -8$, we conclude that $4 < 8$.

This theorem is sometimes stated in the form:

> *If we change the signs of both sides of an inequality, we change its sense.* By changing the sense of an inequality we mean that we have replaced $>$ by $<$ or $<$ by $>$.

Theorem 4   If $a > b$ and $c < 0$, then $ac < bc$.

multiplica-
tion by
negative
number
reverses
sense of
inequality

Proof

1 Let $c = -d$, where $d > 0$.
2 $ad > bd$ [R15]
3 $-ad < -bd$ [Theorem 3]
4 $a(-d) < b(-d)$ [Sec. 2.4, Theorem 6]
5 $ac < bc$ [Substitution from (1)]

This means that if we multiply both sides of an inequality by a negative number, we change its sense.

Theorem 5   If $a \neq 0$, then $a^2 > 0$.

Proof If a is positive, then a^2 is positive; for the product of two positive numbers is positive. If a is negative, then $a < 0$. Hence, multiplying both sides by a and using Theorem 4, we find that $a^2 > 0$.

Theorem 6 $a > 0$ if and only if $(1/a) > 0$.

Proof (Left to Right)

1 Since $(1/a) \neq 0$, $(1/a)^2 > 0$ [Theorem 5]
2 $a(1/a)^2 > 0 \cdot (1/a)^2$ [R15]
3 $(1/a) > 0$ [Sec. 2.3]

Exercises

F Complete the proof (right to left) of Theorem 6.
G Prove: $a < 0$ if and only if $(1/a) < 0$.

Theorem 7   If $a > b$ and $c > 0$, then $(a/c) > (b/c)$.

Proof Multiply both sides by $(1/c)$, which is positive because of Theorem 6.

Exercise

H   Prove: If $a < b$ and $c > 0$, then $(a/c) < (b/c)$.

Theorem 8   If $a > b$ and $c < 0$, then $(a/c) < (b/c)$.
If $a < b$ and $c < 0$, then $(a/c) > (b/c)$.

Proof is left to the reader.

Exercise

I   Prove Theorem 8.

The net result of these theorems is that inequalities behave *almost* like equalities. We can add the corresponding sides of two inequalities having the same sense and obtain a true inequality. We can add (or subtract) equal quantities to (or from) both sides of an inequality. We can multiply or divide both sides of an inequality by a *positive* number. The only difference in the behavior of inequalities as compared to that of equations is that when we multiply or divide by a *negative* number, we must *change the sense of the inequality*.

8.3   Linear Inequalities

In many practical situations we encounter inequalities such as the following:

$$3x + 5 < x - 7 \qquad \text{or} \qquad 2x^2 - 4x + 9 < 0$$

We wish to "solve" each of these inequalities; i.e., we seek to identify those values of x which satisfy the inequalities. That is, we are interested in the sets $\{x \mid 3x + 5 < x - 7\}$ and $\{x \mid 2x^2 - 4x + 9 < 0\}$, and wish to rewrite the definitions of these sets in a simpler form from which we can read off their elements at once. These sets will be subsets of the real numbers and will generally consist of intervals or unions of intervals.

Let us first examine linear inequalities, i.e., those like $ax + b < 0$, $ax + b > 0$, $ax + b \le 0$, $ax + b \ge 0$. We can solve these at once by using the theorems introduced in Sec. 8.2. We proceed as in the illustration below.

Illustration 1

a   Solve $3x + 5 < x - 7$.
By R14 we may subtract $x + 5$ from each side. Doing this, we obtain $2x < -12$. By Theorem 7 we may divide both sides by 2, and we thus conclude that $x < -6$.

The graph of this solution set is given in Fig. 8.1*a*, where we use an open circle above -6 to indicate that this point is *not* part of the set.

Solution sets of linear inequalities

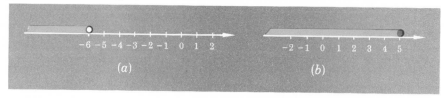

(a) (b)

Figure 8.1 (a) $3x + 5 < x - 7$; (b) $x + 8 \geq 5x - 12$.   The solution set is indicated by the colored strip.   The open circle indicates that the end point is *not* part of the solution set.   A filled circle indicates that the end point *is* included in the solution set.

b   Solve $x + 8 \geq 5x - 12$.

$$-4x + 8 \geq -12$$
$$-4x \geq -20$$
$$x \leq 5$$

The graph of this solution set is given in Fig. 8.1*b*, where we use a solid circle above 5 to indicate that this point *is* part of the set.

Thus we see that the method of solving inequalities is very similar to that for solving equations.   By using the operations which we justified in Sec. 8.2 we convert the given inequality into a series of equivalent inequalities, i.e., inequalities which define the same set as the given inequality.   From the last of these we can read off the answer.

solution of
$|x| > a$
and
$|x| < a$

As a variation on this type of problem, let us consider linear inequalities involving absolute values.   If we recall that $|-x| = |x|$, we observe that $|x| > a$, where $a > 0$ is satisfied by either $x > a$ or $-x > a$.   Since the second of these is equivalent to $x < -a$, we say that

$|x| > a$ is equivalent to $(x > a$ or $x < -a)$

Similarly

$|x + b| > a$ is equivalent to $(x + b > a$ or $x + b < -a)$
     or to $(x > a - b$ or $x < -a - b)$

By similar arguments we can show that

$|x| < a$ is equivalent to $-a < x < a$
$|x + b| < a$ is equivalent to $-a < x + b < a$
     or to $-a - b < x < a - b$
$|cx + b| < a$ is equivalent to $-a < cx + b < a$
     or to $-a - b < cx < a - b$

     or to $\dfrac{-a - b}{c} < x < \dfrac{a - b}{c}$

provided that c is positive.

Illustration 2 The inequalities stated in the left-hand column are equivalent to those in the right-hand column.

a $|x| > 4$ $(x > 4 \text{ or } x < -4)$
b $|x + 1| > 3$ $(x > 2 \text{ or } x < -4)$
c $|x| < 5$ $-5 < x < 5$
d $|x - 3| < 2$ $1 < x < 5$
e $|2x - 5| < 1$ $2 < x < 3$

Inequalities containing absolute values

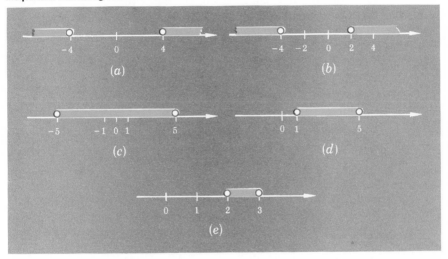

Figure 8.2 (a) $|x| > 4$; (b) $|x + 1| > 3$; (c) $|x| < 5$; (d) $|x - 3| < 2$; (e) $|2x - 5| < 1$. The graphs of inequalities can be finite intervals, infinite intervals, or unions of these.

The above theorems also help us to prove certain relationships by means of mathematical induction (Sec. 2.7).

use of
mathematical
induction

Illustration 3 Prove: For all integers $n \geq 1$, $2^n \geq 1 + n$. Using the notation of Sec. 2.7, we have

$$a = 1 \qquad A = \{n \geq 1\}$$
$$p_n: \quad 2^n \geq 1 + n$$

I p_1 is true, for $2 \geq 1 + 1$.
II For all integers $k \geq 1$: If $2^k \geq 1 + k$, then $2^{k+1} \geq 1 + (k + 1) = 2 + k$.

Proof of II By hypothesis $2^k \geq 1 + k$. Multiplying both sides of this inequality by 2, we obtain

$$2^{k+1} \geq 2 + 2k$$

Since $2 > 1$ and k is positive, it follows that $2k > k$. Hence

$$2^{k+1} \geq 2 + 2k > 2 + k \quad \text{or} \quad 2^{k+1} > 2 + k$$

which is the desired result.

Illustration 4 Prove: For all integers $n \geq 4$, $n! > 2^n$. We have

$$a = 4 \qquad A = \{n \mid n \geq 4\}$$
$$p_n: \quad n! > 2^n$$

I p_4 is true, for $4! = 24$, $2^4 = 16$, and $24 > 16$.
II For all integers $k \geq 4$: If $k! > 2^k$, then $(k + 1)! > 2^{k+1}$.

Proof of II By hypothesis $k! > 2^k$. Hence

$$(k + 1)! = k!(k + 1) > 2^k(k + 1)$$

Since $k \geq 4$,

$$k + 1 > 4 \qquad \text{and} \qquad 2^k(k + 1) > 2^k \cdot 4 > 2^k \cdot 2 = 2^{k+1}$$

Therefore $\qquad\qquad\qquad (k + 1)! > 2^{k+1}$

which is the desired result.

Illustration 5 Prove: For all integers $n \geq -2$, $2n^3 + 3n^2 + n + 6 \geq 0$.
We have

$$a = -2 \qquad A = \{n \mid n \geq -2\}$$
$$p_n: \quad 2n^3 + 3n^2 + n + 6 \geq 0$$

I p_{-2} is true, since $2(-2)^3 + 3(-2)^2 + (-2) + 6 \geq 0$.
II For all integers $k \geq -2$: If $2k^3 + 3k^2 + k + 6 \geq 0$, then

$$2(k + 1)^3 + 3(k + 1)^2 + (k + 1) + 6 \geq 0$$

Proof of II We write

$$2(k + 1)^3 + 3(k + 1)^2 + (k + 1) + 6$$
$$= 2k^3 + 9k^2 + 13k + 12$$
$$= (2k^3 + 3k^2 + k + 6) + (6k^2 + 12k + 6)$$
$$= (2k^3 + 3k^2 + k + 6) + 6(k + 1)^2$$

In the last expression, $(2k^3 + 3k^2 + k + 6) \geq 0$ by hypothesis, and $(k + 1)^2 \geq 0$. Hence the sum of these is greater than or equal to zero.

Exercise

A Show that $2n^3 + 3n^2 + n + 6 = (n + 1)(2n^2 + n) + 6$. Hence construct a proof of the inequality in Illustration 5 without using induction.

Problems 8.3

In Probs. 1 to 20 solve the stated inequalities, and plot their solution set on the line.

1	$x - 3 > 0$	**2**	$2x + 6 < 0$
3	$-3x - 9 > 0$	**4**	$-2x + 4 \leq 0$
5	$5x + 3 > x + 11$	**6**	$-2x + 6 < 3x - 4$
7	$|x| > 4$	**8**	$|x + 3| > 5$
9	$|x - 2| < 3$	**10**	$|x - 1| > 6$
11	$|2x - 3| > 7$	**12**	$|3x + 5| > -4$
13	$|x| < 3$	**14**	$|x + 3| < 7$
15	$|2x + 5| < 9$	**16**	$|4x - 7| < 5$
17	$|3x - 2| < 7$	**18**	$|-2x + 4| < 3$
19	$|x - 2| < 0.1$	**20**	$|x - 4| < 0.01$

In Probs. 21 to 28 write an inequality which states that:

21 x is within 3 units of 4.   **22** x is within 2 units of -5.

23 $2x$ is within 5 units of 1.   **24** $3x$ is within 4 units of 2.

25 The distance from x to 2 is greater than 3.

26 The distance from x to -3 is greater than 1.

27 The distance from $4x$ to 1 is greater than 6.

28 The distance from $3x$ to -5 is greater than 2.

29 Prove that the linear transformation $x' = ax + b$ with $a > 0$ preserves the order relationship; i.e., if $x_1 > x_2$, then $x_1' > x_2'$.

30 Prove that the linear transformation $x' = ax + b$ with $a < 0$ reverses the order relationship; i.e., if $x_1 > x_2$, then $x_1' < x_2'$.

In Probs. 31 to 36 prove the given statement by mathematical induction.

31 For all integers $n \geq 1$, $5^n \geq 1 + 4n$.

32 For all integers $n \geq 1$, $3^n \geq 1 + 2n$.

33 For all integers $n \geq 1$ and for $h \geq -1$, $(1 + h)^n \geq 1 + nh$.

34 For all integers $n \geq 1$ and for $h \geq 0$,

$$(1 + h)^n \geq 1 + nh + \frac{n(n - 1)}{2} h^2$$

35 For all integers $n \geq -1$, $2n^3 - 9n^2 + 13n + 25 > 0$.

36 For all integers $n \geq -2$, $2n^3 + 9n^2 + 13n + 7 > 0$.

37 Prove: For all integers $n \geq 1$ and for $-1 < h \leq \sqrt{2}$, $(1 - h)^n \leq 1/(1 + h)^n$. HINT: $(1 - h)^n(1 + h)^n = (1 - h^2)^n \leq 1$. Now divide this inequality by $(1 + h)^n$.

38 Prove: If $a \geq b$ and $ab > 0$, then $(1/a) \leq (1/b)$.

39 Prove: For all integers $n \geq 1$ and for $0 \leq h \leq \sqrt{2}$, $(1 - h)^n \leq 1/(1 + nh)$. HINT: Use the results of Probs. 37, 33, and 38.

40 Prove: For all integers $n \geq 1$ and for $0 \leq h < \sqrt{2}$,

$$(1 - h)^n \leq 1 \bigg/ \left(1 + nh + \frac{n(n - 1)}{2} h^2\right)$$

41 Show that $2x^3 - 3x^2 - 5x + 3 \leq 2|x|^3 + 3|x|^2 + 5|x| + 3.$
Hence show that when $-3 < x < 1$, $2x^3 - 3x^2 - 5x + 3 < 99.$
HINT: From $-3 < x < 1$, we conclude that $|x| < 3.$

42 Using the method of Prob. 41, show that when $-1 < x < 2$, $x^3 - 4x^2 - 3x + 5 < 35.$

*43 Show that if $|x| < 1$, then $\left|\dfrac{x+1}{x-2}\right| < 2.$

HINT: Use the following steps:
a $-1 < x < 1$; so $0 < x + 1 < 2$ and $-3 < x - 2 < -1.$
b $|x + 1| < 2$ and $|x - 2| > 1$

c $\left|\dfrac{x+1}{x-2}\right| = \dfrac{|x+1|}{|x-2|} < \dfrac{2}{1} = 2$

In this step we obtain an upper bound for the given expression by making its numerator as large as it can be and its denominator as small as it can be.

*44 Show that if $|x| < 2$, then $\left|\dfrac{x-3}{x+4}\right| < \dfrac{5}{2}.$

*45 Show that if $|x - 3| < 1$, then $\left|\dfrac{x+5}{x+1}\right| < 3.$

*46 Show that if $|x| < 1$, then $\left|\dfrac{x-2}{x+3}\right| < \dfrac{3}{2}.$

*47 Show that if $|x - 2| < 1$, then $|x^2 - 4| < 5.$
*48 Show that if $|x - 2| < 0.1$, then $|x^2 - 4| < 0.41.$
*49 Show that if $|x - 2| < 0.01$, then $|x^2 - 4| < 0.0401.$
*50 Show that if $|x - 2| < a$, then $|x^2 - 4| < 4a + a^2.$

8.4 Quadratic Inequalities

The properties of quadratic inequalities follow easily from those of quadratic equations. We are concerned with the inequalities

$$ax^2 + bx + c > 0$$
$$ax^2 + bx + c < 0$$
$$ax^2 + bx + c \geq 0$$
$$ax^2 + bx + c \leq 0$$

where $a > 0$, and a, b, and c are real. Only real values of x will be permitted.

Let us suppose, first, that $b^2 - 4ac > 0$, so that the solutions of $ax^2 + bx + c = 0$ are real and unequal. Then we can write

$$ax^2 + bx + c = a(x - r_1)(x - r_2)$$

where r_1, r_2 are real and $r_1 \neq r_2$. Suppose $r_1 < r_2$. Let us imagine x moving from left to right along the real line, and consider how the sign of $ax^2 +$

$bx + c$ varies in the process. At the extreme left (where $x < r_1 < r_2$), $(x - r_1)$ is negative and $(x - r_2)$ is negative. Hence, in this region, $a(x - r_1)(x - r_2)$ is positive. When x reaches r_1, $a(x - r_1)(x - r_2)$ becomes zero. Between r_1 and r_2, $(x - r_1)$ is positive and $(x - r_2)$ is negative. So, in this region, $a(x - r_1)(x - r_2)$ is negative. At r_2, $a(x - r_1) \times (x - r_2)$ is zero. To the right of r_2, $(x - r_1)$ and $(x - r_2)$ are both positive, so that $a(x - r_1)(x - r_2)$ is positive. These results are summarized in Fig. 8.3. We state them formally in the theorem.

Sign pattern for $ax^2 + bx + c$

Figure 8.3
This figure assumes that the roots of $ax^2 + bx + c = 0$ are real and distinct and that $a > 0$. Observe that the value of $ax^2 + bx + c$ changes sign only at the roots of $ax^2 + bx + c = 0$.

Theorem 9 If $ax^2 + bx + c = 0$ $(a > 0)$ has distinct real solutions r_1 and r_2 with $r_1 < r_2$, then when

quadratic
has real
solutions

$$x < r_1,\ ax^2 + bx + c > 0$$
$$x = r_1,\ ax^2 + bx + c = 0$$
$$r_1 < x < r_2,\ ax^2 + bx + c < 0$$
$$x = r_2,\ ax^2 + bx + c = 0$$
$$x > r_2,\ ax^2 + bx + c > 0$$

A similar method gives the next result.

Theorem 10 If $ax^2 + bx + c = 0$ $(a > 0)$ has equal real solutions $r_1 = r_2 = r$, then when

$$x < r,\ ax^2 + bx + c > 0$$
$$x = r,\ ax^2 + bx + c = 0$$
$$x > r,\ ax^2 + bx + c > 0$$

quadratic
has nonreal
solutions

The final situation is that in which $ax^2 + bx + c = 0$ has nonreal solutions; that is, $b^2 - 4ac < 0$. As we have just seen, the only way for $ax^2 + bx + c$ to change sign as x moves from left to right is for x to pass through a point at which $ax^2 + bx + c = 0$. In this case, however, this equation has no real solutions; hence there is no such point. Therefore, $ax^2 + bx + c$ must have a constant sign. At the extreme right (or left), $ax^2 + bx + c$ is certainly positive. Hence it is positive for all values of x. This proves the theorem.

Theorem 11 If $ax^2 + bx + c = 0$ $(a > 0)$ has nonreal solutions, then for all x, $ax^2 + bx + c > 0$.

Exercise

A Prove the converse of Theorem 11.

Illustration 1

a Solve $3(x - 1)(x - 4) > 0$. *Ans.*: $\{x \mid (x < 1) \text{ or } (x > 4)\}$. (See Fig. 8.4*a*.)

b Solve $2(x + 1)(x - 2) < 0$. *Ans.*: $\{x \mid -1 < x < 2\}$. (See Fig. 8.4*b*.)

c Solve $4(x - 3)^2 \geq 0$. *Ans.*: All real x. (See Fig. 8.4*c*.)

d Solve $(x + 1)(x + 2) \geq 0$. *Ans.*: $\{x \mid (x \leq -2) \text{ or } (x \geq -1)\}$. (See Fig. 8.4*d*.)

e Solve $x^2 + x + 1 > 0$. *Ans.*: All real x. (See Fig. 8.4*c*.)

Solution sets of quadratic inequalities

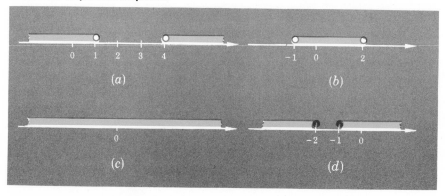

Figure 8.4 (*a*) $3(x - 1)(x - 4) > 0$; (*b*) $2(x + 1)(x - 2) < 0$; (*c*) $4(x - 3)^2 \geq 0$ and $x^2 + x + 1 > 0$; (*d*) $(x + 1)(x + 2) \geq 0$. To solve a quadratic inequality, first find the corresponding roots and then test each of the resulting intervals.

The method of this section is easily extended to inequalities of the form

$$a(x - r_1)(x - r_2)(x - r_3) \cdots (x - r_n) > 0, < 0$$

etc., where $a > 0$ and $r_1, \ldots, r_n$ are real.

Illustration 2 Solve $2(x - 1)(x - 2)(x - 3) > 0$.

When $x < 1$, each factor is negative; so the product is negative.

When $1 < x < 2$, $(x - 1)$ is positive, and $(x - 2)$ and $(x - 3)$ are negative. Hence the product is positive.

Continuing in this way we find the solution set: $\{x \mid (1 < x < 2) \text{ or } (3 < x)\}$, whose graph is given in Fig. 8.5.

A further extension of this method permits us to solve inequalities involving algebraic fractions. As in the solutions of equations involving such fractions (Sec. 6.6), the best procedure is to obtain a common denominator and collect all terms on one side of the equation. (Under no circumstances should you multiply both sides of the inequality by an expression containing x which is sometimes positive and sometimes negative. In such a case there is no way of deciding what the sense of the resulting inequality should be.) Next, find the values of x for which the numerator and denominator are each zero, arrange them in order along the line, and proceed as in Illustration 2.

Solution set of a cubic inequality

Figure 8.5
$2(x - 1)(x - 2)(x - 3) > 0$
The same method applies to products of three or more factors.

Illustration 3 Solve $\dfrac{7}{x - 1} - \dfrac{6}{x^2 - 1} < 5$

After simplification this inequality becomes

$$\frac{(x - 2)(5x + 3)}{x^2 - 1} > 0$$

The roots of the numerator are 2 and $-\frac{3}{5}$; those of the denominator are 1 and -1. Arranging these in order, we have -1, $-\frac{3}{5}$, 1, 2. We then examine each interval and plot the graph (Fig. 8.6).

Solution set of a fractional inequality

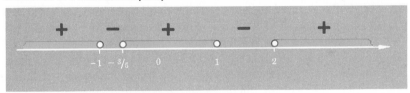

Figure 8.6 $\dfrac{7}{x - 1} - \dfrac{6}{x^2 - 1} < 5$ or $\dfrac{(x - 2)(5x + 3)}{x^2 - 1} > 0$

The method also applies to an inequality involving a fraction whose numerator and denominator are products of linear factors.

Problems 8.4

In Probs. 1 to 20 solve the given inequality, and plot the graph of its solution set.

1	$(x + 2)(x - 1) > 0$	**2**	$(x - 3)(x + 4) > 0$
3	$5(x - 2)^2 \geq 0$	**4**	$-3(x - 1)^2 \leq 0$
5	$2x^2 + 3x - 2 \leq 0$	**6**	$3x^2 + 10x + 3 \geq 0$
7	$4x^2 + 3x - 1 < 0$	**8**	$2x^2 - 3x - 2 > 0$
9	$x^2 + 3x + 3 > 0$	**10**	$-x^2 + 2x - 4 \leq 0$
11	$-x^2 - x - 1 \geq 0$	**12**	$3x^2 + 2x + 6 < 0$

13 $4(x - 3)(x + 2)(x + 5) > 0$ **14** $3(x + 1)(x + 2)(x - 3) < 0$

15 $-5(x + 2)^2(x - 3) \geq 0$ **16** $-2(2x - 1)(x + 4)(x) < 0$

17 $\dfrac{4}{x + 2} - \dfrac{1}{x - 1} > 0$ **18** $\dfrac{6}{x + 3} + \dfrac{1}{x - 2} > 2$

19 $\dfrac{3}{x + 4} + \dfrac{12}{x - 5} < -1$ **20** $\dfrac{5}{x + 1} - \dfrac{6}{x - 3} > 7$

*21 By mathematical induction show that: For all integers $n \geq 5$, $2^n > n^2$.

*22 By mathematical induction show that: For all integers $n \geq 5$, $4^n > n^4$.

In Probs. 23 to 30 try to establish the indicated theorem by mathematical induction. Point out why the method fails.

23 For $n \geq 1$, $2^n > n^2$ **24** For $n \geq 1$, $4^n > n^4$
25 For $n \geq 2$, $2^n > n^2$ **26** For $n \geq 2$, $4^n > n^4$
27 For $n \geq 3$, $2^n > n^2$ **28** For $n \geq 3$, $4^n > n^4$
29 For $n \geq 4$, $2^n > n^2$ **30** For $n \geq 4$, $4^n > n^4$

Applications

31 Prove: Let $a > 0$ and $b > 0$. Then, if $a^2 > b^2$, it follows that $a > b$. HINT: Consider $a^2 - b^2 > 0$ or $(a + b)(a - b) > 0$.

32 From the Pythagorean relation $a^2 + b^2 = c^2$, prove that the hypotenuse of a right triangle is longer than either leg.

33 Apply the result of Prob. 32 to the general triangle given in Fig. 8.7 to show that $a + b > c + d$. Hence the sum of two sides of a triangle is greater than the third side.

34 In Fig. 8.8 assume that $c > a$. Use the result of Prob. 33 to show that $c - a < b$. Hence any side of a triangle is greater than the absolute value of the difference of the other two sides.

Figure 8.7

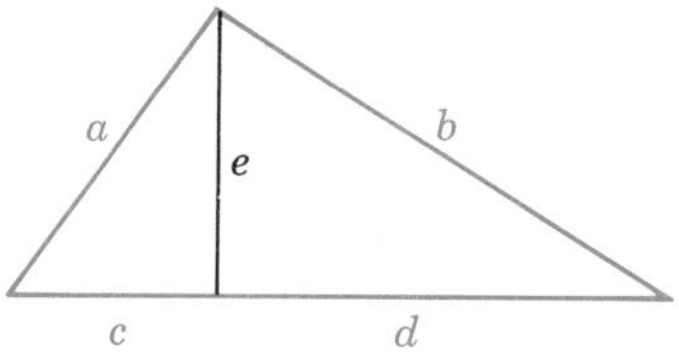

Figure 8.8

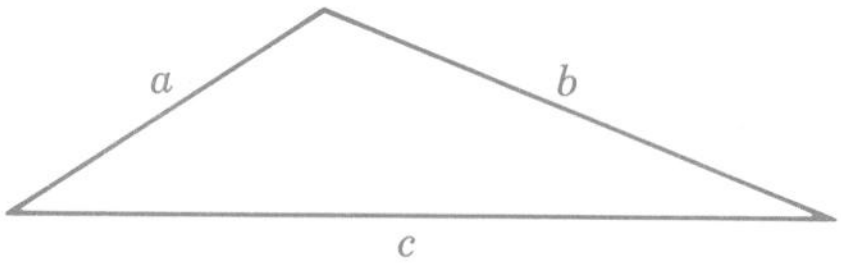

35 Prove that $|a + b| \leq |a| + |b|$. HINT: Since $|a|^2 = a^2$, etc., let us consider

$$\begin{aligned}
(a + b)^2 &= a^2 + 2ab + b^2 \\
&\leq a^2 + 2|a||b| + b^2 \\
&\leq [|a| + |b|]^2
\end{aligned}$$

When does equality hold in the above relation?

36   By mathematical induction prove that: For all integers $n \geq 2$, $|x_1 + x_2 + \cdots + x_n| \leq |x_1| + |x_2| + \cdots + |x_n|$.

37   Using the result of Prob. 35, prove that

$$|a - b| \geq |a| - |b|$$

38   Prove that $\sqrt{ab} \leq \dfrac{a + b}{2}$ if a and b are positive.   HINT: Show that this is equivalent to $0 \leq a^2 - 2ab + b^2$. $\sqrt{ab}$ is called the "geometric mean" of a and b; $(a + b)/2$ is called the "arithmetic mean".

The above result is a special case of the general theorem which states that, for a set of positive quantities $a_1, \ldots, a_n$, the geometric mean $\sqrt[n]{a_1 a_2 \cdots a_n}$ is less than or equal to the arithmetic mean $(1/n)(a_1 + \cdots + a_n)$.   When does equality hold in the above relation?

39   Prove that for any real numbers a_1, a_2, b_1, and b_2

$$(a_1 b_1 + a_2 b_2)^2 \leq (a_1{}^2 + a_2{}^2)(b_1{}^2 + b_2{}^2)$$

This is known as "Cauchy's Inequality".   HINT: Consider

$$(a_1 x + b_1)^2 = a_1{}^2 x^2 + 2a_1 b_1 x + b_1{}^2 \geq 0$$
$$(a_2 x + b_2)^2 = a_2{}^2 x^2 + 2a_2 b_2 x + b_2{}^2 \geq 0$$

Adding, we find that

$$(a_1{}^2 + a_2{}^2)x^2 + 2(a_1 b_1 + a_2 b_2)x + (b_1{}^2 + b_2{}^2) \geq 0 \qquad \text{for all } x$$

Now apply the result of Exercise A, Sec. 8.4.   When does equality hold in the above relation?

40   Generalize the result of Prob. 39 to two sets of numbers: $a_1, \ldots, a_n$ and $b_1, \ldots, b_n$.   The formula so obtained is of great importance in statistics and higher geometry.

41   Let $\mathbf{A}$ be the vector $(a_1, \ldots, a_n)$; $\mathbf{B}$ be the vector $(b_1, \ldots, b_n)$; $|\mathbf{A}|$ be the length of $\mathbf{A}$; and $|\mathbf{B}|$ be the length of $\mathbf{B}$.   Then from Cauchy's Inequality show that

$$\left(\frac{\mathbf{A} \cdot \mathbf{B}}{|\mathbf{A}||\mathbf{B}|}\right)^2 \leq 1$$

42   Show that of all rectangles with a given perimeter, the square has the largest area.   HINT: Let the dimensions of the rectangle be x and y.   Then the perimeter P is $P = 2x + 2y$.   The area A is xy. From Prob. 38

$$xy \leq \frac{(x + y)^2}{4} \qquad \text{or} \qquad A \leq \frac{P^2}{16}$$

Since P is fixed, A will be greatest when equality holds.   When is this?

8.5   The Graph of a Linear Inequality

Here we are concerned with the graph of the linear inequality $ax + by + c > 0$, or more properly with the graph of the set

$$\{(x,y) \mid ax + by + c > 0\}$$

We assume that at least one of the coefficients a and b is not zero.

First we plot the graph of the equation

$$ax + by + c = 0$$

which we know is a line.

This line divides the plane into two half planes A and B, neither of which contains the line. These half planes are each *connected*, in the sense that, if points P_1 and P_2 lie in one of them, then there is a smooth curve joining P_1 and P_2 which lies entirely in that half plane (Fig. 8.9). In particular, the curve does not have a point in common with the given line. Our procedure for graphing the solution set of $ax + by + c > 0$ depends strongly on the next theorem.

Separation of the plane by a line

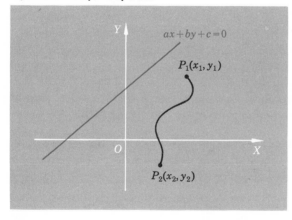

Figure 8.9
A line divides the plane into three disjoint subsets: (1) the line, (2) a half plane on one side of it, (3) a half plane on the other side of it.

Theorem 12   Let P lie in one of the two half planes into which the graph of $ax + by + c = 0$ divides the plane. If $ax + by + c > 0$ at P, then $ax + by + c > 0$ at every point of the half plane in which P lies.

Proof   Assume that P_1 and P_2 lie in the same half plane, that $ax + by + c > 0$ at P_1, and that $ax + by + c \leq 0$ at P_2.

We shall prove that this is absurd (indirect proof).

Let P_1 have coordinates (x_1,y_1) and P_2 have coordinates (x_2,y_2).

In the first place, we cannot have $ax_2 + by_2 + c = 0$, for in this case P_2 is on the given line, which is contrary to its assumed location. Hence our hypotheses reduce to: $ax_1 + by_1 + c > 0$; $ax_2 + by_2 + c < 0$; P_1 and P_2 in the same half plane.

Draw the curve P_1P_2, which lies entirely in the given half plane. Let $\bar{P}(\bar{x},\bar{y})$ move along this curve from P_1 to P_2. At P_1, $a\bar{x} + b\bar{y} + c > 0$, and at P_2, $a\bar{x} + b\bar{y} + c < 0$. Hence there must be a point on the curve P_1P_2 at which $a\bar{x} + b\bar{y} + c = 0$. This, however, is impossible, for such a point would have to be on the given line, and P_1P_2 has no points in common with this line.[1]

Exercise

test point in each half plane

A State similar theorems for the inequalities: $ax + by + c < 0$, $ax + by + c \geq 0$, and $ax + by + c \leq 0$.

As a result of this theorem, we have the following procedure for plotting the graph of the solution set of a linear inequality. (1) Plot the graph of the line $ax + by + c = 0$. (2) For a single point in each half plane determined by this line, test the truth of the given inequality. (3) Shade the half plane or half planes corresponding to points at which the given inequality is true.

Remark You will observe that if $ax + by + c > 0$ in one half plane, then $ax + by + c < 0$ in the other half plane. Nevertheless, it it wise to check both half planes. There are situations in which an inequality is true in both half planes. For instance, this is the case for the inequality

$$(ax + by + c)^2 > 0$$

Illustration 1 Plot the graph of the solution set of $3x + 2y - 6 > 0$. The points $(0,3)$ and $(2,0)$ satisfy the equation $3x + 2y - 6 = 0$, so we plot the line through them (Fig. 8.10). The point $(0,0)$ is in the lower half

Linear inequality in two variables

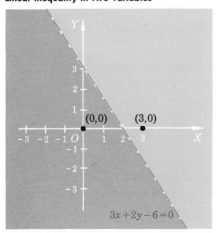

Figure 8.10
$3x + 2y - 6 > 0$
The solution set of a linear inequality is the colored half plane and may or may not also include the line. A *dashed* line is not part of the solution set; a *solid* line is included in the solution set (see Fig. 8.11).

---

[1] This proof assumes properties of continuity which you have not studied so far. Although precise details must be omitted here, the idea should be clear.

plane, and $0 + 0 - 6 > 0$ is false. On the other hand, $(3,0)$ is in the upper half plane and $9 + 0 - 6 > 0$ is true. Therefore we color the upper half plane. The line is dashed, since it is not part of the desired graph.

Illustration 2 Plot the graph of the solution set of $2x - 4y + 8 \leq 0$. We draw the line through $(-4,0)$ and $(0,2)$ (Fig. 8.11). The point $(0,0)$ does not satisfy the inequality. The point $(0,3)$ does satisfy the inequality, so we color the upper half plane, drawing the line solid, since it is part of the graph.

Linear inequality in two variables

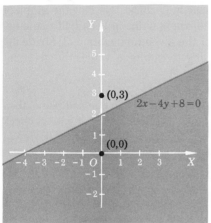

Figure 8.11
$2x - 4y + 8 \leq 0$
The solution set of a linear inequality is the colored half plane and may or may not also include the line. A *dashed* line is not part of the solution set (see Fig. 8.10); a *solid* line is included in the solution set.

8.6 Simultaneous Linear Inequalities

We now extend our treatment of linear inequalities (Sec. 8.5) to the case of systems of inequalities. A typical system of two linear inequalities is

$$a_1 x + b_1 y + c_1 > 0 \qquad a_1 \neq 0 \text{ or } b_1 \neq 0$$
$$a_2 x + b_2 y + c_2 > 0 \qquad a_2 \neq 0 \text{ or } b_2 \neq 0$$

where we assume that the coefficients are real numbers. The solution set of each inequality is a half plane, and the solution set of the pair of inequalities is the intersection of these two half planes. The best method of procedure is graphical.

Illustration 1 Graph the set determined by

$$2x + \ y - 3 > 0$$
$$x - 2y + 1 < 0$$

First draw the two lines, which intersect at $(1,1)$. We find that $2x + y - 3 > 0$ determines its right half plane and that $x - 2y + 1 < 0$ determines its left half plane. The region common to the two is colored in Fig. 8.12. It is the interior of an angle whose vertex is $(1,1)$.

This illustration is typical for two inequalities. But we can consider three or more simultaneous inequalities. The ideas and procedure are the same.

Two simultaneous linear inequalities

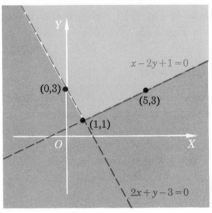

Three simultaneous linear inequalities

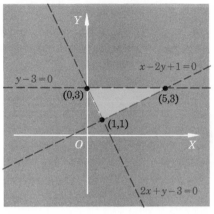

Figure 8.12
$$\begin{cases} 2x + y - 3 > 0 \\ x - 2y + 1 < 0 \end{cases}$$

Figure 8.13
$$\begin{cases} 2x + y - 3 > 0 \\ x - 2y + 1 < 0 \\ y - 3 < 0 \end{cases}$$

The solution set of a system of inequalities is the intersection of their individual solution sets.

Illustration 2 Graph the set determined by

$$2x + y - 3 > 0$$
$$x - 2y + 1 < 0$$
$$y - 3 < 0$$

The first two inequalities are the same as in Illustration 1; so we merely add the third line to Fig. 8.12. This gives a triangle with vertices $(1,1)$, $(0,3)$, and $(5,3)$. This line divides the colored region of Fig. 8.12 into two parts. We see that $y - 3 < 0$ determines its lower half plane. Hence the desired set is the interior of the colored triangle in Fig. 8.13.

Illustration 3 Let us add one more inequality to our picture, and consider the system

$$2x + y - 3 > 0$$
$$x - 2y + 1 < 0$$
$$y - 3 < 0$$
$$x + y - 5 < 0$$

Since $x + y - 5 < 0$ determines its left half plane, the desired set is the interior of the colored quadrilateral in Fig. 8.14.

Illustration 4   Instead of the system of Illustration 3, consider

$$2x + y - 3 > 0$$
$$x - 2y + 1 < 0$$
$$y - 3 < 0$$
$$x + y + 2 < 0$$

The inequality $x + y + 2 < 0$ determines its left half plane, which has no points in common with the triangle of Fig. 8.13.   Hence the desired set is empty, and nothing is colored in Fig. 8.15.

Four simultaneous linear inequalities

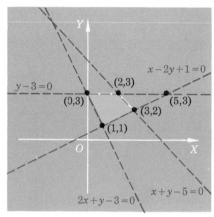

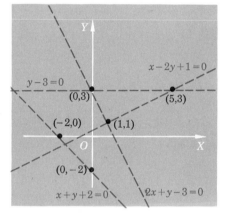

Figure 8.14 $\begin{cases} 2x + y - 3 > 0 \\ x - 2y + 1 < 0 \\ y - 3 < 0 \\ x + y - 5 < 0 \end{cases}$

The solution set of a system of inequalities may be the interior of a polygon.

Figure 8.15 $\begin{cases} 2x + y - 3 > 0 \\ x - 2y + 1 < 0 \\ y - 3 < 0 \\ x + y + 2 < 0 \end{cases}$

The solution set of a system of inequalities may be the empty set.

simultaneous inequalities determine a polygon

We see in this way that the graph of a system of linear inequalities can take many forms.   In the usual cases it is the interior of a convex polygon, which may be bounded as in Figs. 8.13 and 8.14 or unbounded as in Fig. 8.12.   In other cases it may be the empty set.   When the inequalities include $\geq$ or $\leq$, the possibilities of variation in form are even more numerous.

Simultaneous inequalities can arise in unexpected places in practical situations.   Let us consider two of these.

Illustration 5 The situation in a simplified version of a football game is as follows. There are just two plays, a running play and a pass play. We assume these facts to be true:

Play	Distance gained, yd	Time required, sec
Running	3	30
Pass	6	10

Also suppose that there are 60 yd to go for a touchdown and that 150 sec remain in the game. Ignore the requirement of having to make 10 yd in four downs and other considerations of score and strategy.

Problem What combinations of running and pass plays will secure a touchdown in the allotted time?

Solution Let r and p, respectively, represent the numbers of running and pass plays. Then the conditions of the problem may be written

$$3r + 6p \geq 60$$
$$30r + 10p \leq 150$$

These may be simplified to

$$r + 2p \geq 20$$
$$3r + p \leq 15$$

Solution of the football problem

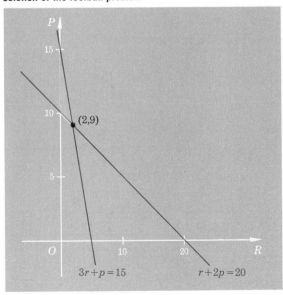

Figure 8.16
$$\begin{cases} r + 2p \geq 20 \\ 3r + p \leq 15 \end{cases}$$

The quarterback should use a combination of plays corresponding to a point in the interior or on the boundary of the colored triangle.

Plot the graph of this simultaneous system as in Fig. 8.16. The solution set is the colored triangle (including its sides). Since r and p must be integers, the solution consists of the following combinations:

r	0	0	$\cdots$	0	1	1	$\cdots$	1	2
p	10	11	$\cdots$	15	10	11	$\cdots$	12	9

The quarterback can then select his pattern of plays from this array according to his best judgment regarding strategy or other matters.

Illustration 6   The Minneapolis and Seattle Lumber Company can convert logs into either lumber or plywood. In a given week the mill can turn out 400 units of production, of which 100 units of lumber and 150 units of plywood are required by regular customers. The profit on a unit of lumber is $20 and on a unit of plywood is $30.

Problem   How many units of lumber and plywood should the mill produce (totaling 400) in order to maximize the total profit?

Solution   Let L and P, respectively, represent the number of units of lumber and plywood. Then the conditions of the problem are

$$L + P = 400 \qquad L \geq 100 \qquad P \geq 150$$
$$\text{Total profit} = 20L + 30P$$

Let us graph these in Fig. 8.17.

Solution of the lumber problem

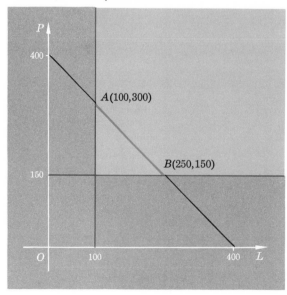

$A(100,300)$

$B(250,150)$

Figure 8.17
$L + P = 400$
$L \geq 100$
$P \geq 150$
Production must correspond to some point on the closed segment AB.

The possible solutions must lie on the line $L + P = 400$ and in the colored region of the plane. Hence they lie on the segment AB. The total profit at A is \$11,000 and at B is \$9,500. At other points of AB the profit lies between these values. Hence the mill should produce 100 units of lumber and 300 units of plywood.

Perhaps this solution is obvious to you without all this analysis, and it should be. The point of the illustration is that methods of this kind are extremely useful in more complicated problems where the answers are far from obvious. The mathematical subject which deals with such problems is called *Linear Programming* (Sec. 8.9).

8.7   The Graph of a Quadratic Inequality

In connection with the quadratic equation in two variables (Sec. 6.5) we can consider related inequalities of the form

$$y > ax^2 + bx + c \qquad y < ax^2 + bx + c$$

The parabola whose equation is

$$y = ax^2 + bx + c$$

divides the plane into two regions, in one of which $y > ax^2 + bx + c$, and in the other $y < ax^2 + bx + c$. Neither region contains the parabola itself. By an argument similar to that in the proof of Theorem 12, we can establish the following result.

Theorem 13   Let P lie in one of the regions into which a parabola divides the plane. If $y > ax^2 + bx + c$ at P, or if $y < ax^2 + bx + c$ at P, then the corresponding inequality is satisfied at every point of this region.

As a result of this theorem we can plot the graph of the solution set of inequalities of this form by (1) drawing the graph of $y = ax^2 + bx + c$, (2) testing the truth of the given inequality at a point in each of the two regions into which this parabola divides the plane, (3) coloring the region or regions in which the above test is affirmative.

Illustration 1   Plot the graph of $y > x^2 - 6x + 5$. First draw the graph (Fig. 8.18) of $y = x^2 - 6x + 5$. Then $(0,0)$ lies below the graph, and $0 > 0 - 0 + 5$ is false. On the other hand, $(2,1)$ lies above the graph, and $1 > 4 - 12 + 5$ is true. Hence we color the region above the graph.

This technique also permits us to solve simultaneous inequalities as suggested by the next illustration.

**Quadratic inequality
in two variables**

**Simultaneous linear and
quadratic inequalities**

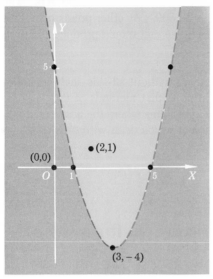

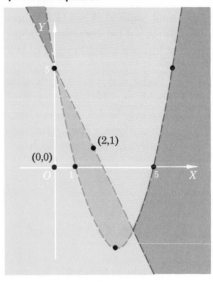

Figure 8.18 $y > x^2 - 6x + 5$
The dashed parabola is not included in the colored region
which represents the solution set.

Figure 8.19 $y > x^2 - 6x + 5$
$4x + 2y - 10 < 0$
The darker colored region is the
intersection of the individual solution sets of the two inequalities.

Illustration 2   Plot the graph of

$$y > x^2 - 6x + 5$$
$$4x + 2y - 10 < 0$$

The graph of the first is given by Illustration 1.   Now draw the line
(Fig. 8.19) $4x + 2y - 10 = 0$.   Since $(0,0)$ lies below the line and satisfies
the inequality, the graph of the second inequality is the half plane below
this line.   The solution of the pair of inequalities is the intersection of the
two colored regions.

8.8   Applications

Since you have probably not met inequalities and systems of inequalities
in your previous mathematical education, you may wonder how they arise
in practice and what they are good for.   In the first place the acceptable
solution of a problem may be limited by practical considerations such as
unavoidable restrictions on space, cost, use of materials, man power, or
time.   Each of these would be expressed analytically as an inequality.   On
the other hand, the solution may have to satisfy certain minimum requirements such as load-carrying capacity, food intake necessary to maintain

health, serviceable life of a product, or volume of sales necessary to justify the production of a certain item.

Our objectives are often stated in terms of inequalities. We do not try to manufacture a bolt whose diameter is exactly ½ in.; we are quite content if it is ½ in. plus or minus certain tolerances, which are expressed as inequalities. It is not necessary for our missiles to hit an exact spot in enemy territory; we are happy if they land within, say, half a mile of their objective. But to handle the analysis of their behavior we need inequalities.

Inequalities are of importance in modern mathematical economics and in recent theories such as game theory and linear programming. For elementary treatments of these topics we refer you to Secs. 8.9 and 12.17 of this book.

Problems 8.8

In Probs. 1 to 16 plot the graph of the solution set of the given inequality.

1 $3x - 4y + 12 > 0$ **2** $2x + 5y - 10 > 0$
3 $x + y + 1 \le 0$ **4** $x - y - 1 \ge 0$
5 $(2x + 3y - 6)^2 \ge 0$ **6** $(3x - 2y + 6)^2 < 0$
7 $y > x^2 + 9x + 8$ **8** $y > x^2 - 8x + 12$
9 $y \le -x^2 + 9x - 18$ **10** $y \ge -x^2 + 10x - 24$
11 $y < 4x^2 + 4x + 1$ **12** $y \le 9x^2 + 6x + 1$
13 $y - 2x^2 + x - 8 \ge 0$ **14** $y + 3x^2 - 2x + 2 < 0$
15 $(y + x^2 + 2x + 1)^2 \ge 0$ **16** $(y + 16x^2 + 8x + 1)^2 < 0$

In Probs. 17 to 26 plot the graph of the solution set of the simultaneous inequalities.

17 $4x + 3y - 12 > 0$ **18** $3x - 2y + 12 > 0$
  $4x - 5y + 20 > 0$ $x + 6y + 4 > 0$

19 $4x + 3y - 12 > 0$ **20** $3x - 2y + 12 > 0$
  $4x - 5y + 20 > 0$ $x - 4y + 4 > 0$
  $4x - y - 12 < 0$ $x + y - 6 < 0$

21 $4x + 3y - 12 > 0$ **22** $3x - 2y + 12 > 0$
  $4x - 5y + 20 > 0$ $x - 4y + 4 > 0$
  $4x - y - 12 > 0$ $x + y - 6 > 0$

23 $y > x^2 + 9x + 8$ **24** $y > x^2 - 10x + 24$
  $x + 2y - 4 < 0$ $x - y > 0$

25 $y > x^2 - 8x + 12$ **26** $y < -x^2 + 2x - 1$
  $y + 5 > 0$ $x + 3y + 3 < 0$

27 In the football problem (Sec. 8.6, Illustration 5), what combinations of plays will meet the required conditions if there are 180 sec left to play? If there are 90 sec left to play?

28 How should the Minneapolis and Seattle Lumber Company (Sec. 8.6, Illustration 6) arrange its production if the profit on a unit of plywood is $10 and the profit on a unit of lumber is $15?

8.9 Linear Programming

This subject was invented during World War II by George Dantzig as a method of solving certain logistic problems for the U.S. Air Force. In the subsequent years it has developed as an important tool in business and economic analysis. The basic problem in two dimensions is the following:

Basic Problem We are given a convex polygonal set R in the plane (defined below) and also a linear polynomial $f(x,y) = px + qy + r$ (p, q, and r are real and either $p \neq 0$ or $q \neq 0$), where the ordered pairs (x,y) correspond to points of R. We are required to find those points of R at which $f(x,y)$ has its greatest (or least) value.

In this section we shall show you how to solve this problem by graphical means. The analogous problem occurs with more variables in higher dimensions; but since graphing is impractical in such cases, more sophisticated methods are needed for their analysis. You can find a discussion of these methods in the many books devoted to this subject.

A typical two-dimensional problem is the following simplified form of the Diet Problem. This was one of the original problems which led to the development of this subject.

Illustration 1 Let us suppose that we have two types of synthetic foods, which we call A and B. You may think of K rations and C rations if you like, but the numbers given below are purely artificial and apply to no foods on the market. Let us suppose that the two foods contain the following nutritional components:

Food	Calories per ounce	Protein per ounce	Fat per ounce
A	100	50	0
B	200	10	30

The units for protein and fat are arbitrary and need not be specified. Let us also suppose that the minimum daily requirements for an active man are calories, 2,500; protein, 350; fat, 150.

Problem Which food or combination of foods should be employed in order to (1) fulfill the minimum daily nutritional requirements and (2) minimize the total weight?

Such a problem would be of importance to a mountain climber or to a military expedition. In order to attack the problem, let us introduce some notation.

Let a represent the number of ounces of food A that are required.
Let b represent the number of ounces of food B that are required.
Then the minimum daily requirements will be met if:

Calories	$100a + 200b \geq 2{,}500$
Protein	$50a + 10b \geq 350$
Fat	$30b \geq 150$

These three inequalities can be simplified to the following simultaneous system:

$$a + 2b \geq 25$$
$$5a + b \geq 35$$
$$b \geq 5$$

To these we should add the practical requirements that $a \geq 0$ and $b \geq 0$, for one cannot eat a negative amount of food. The graph of this simultaneous system is given in Fig. 8.20. The colored region is the region R referred to above.

Diet problem

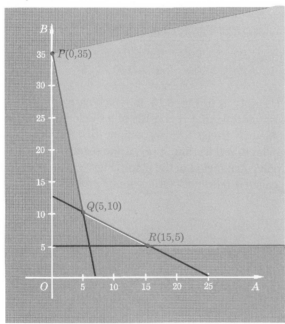

Figure 8.20
The colored region represents the points (a,b) which meet the minimum daily nutritional requirements.

The quantity to be minimized is the total weight $W = a + b$. So the problem is to find the point (or points) of R at which W is a minimum. Theorem 14 (below) tells us that this minimum will occur at one or more of the *vertices* labeled P, Q, R in the figure. In order to determine which vertex to use, let us do some arithmetic.

At P: $a = 0$, $b = 35$, $W = a + b = 35$
At Q: $a = 5$, $b = 10$, $W = a + b = 15$
At R: $a = 15$, $b = 5$, $W = a + b = 20$

Therefore, W is a minimum at Q, and the diet should consist of 5 oz of food A and 10 oz of food B.

The solution of problems of this type depends on the following definitions and theorems.

Definition   A convex polygonal set R is the set of points in the plane whose coordinates satisfy the finite system of inequalities

$$a_1x + b_1y \geq c_1$$
$$a_2x + b_2y \geq c_2$$
$$\cdots\cdots\cdots\cdots\cdots$$
$$a_nx + b_ny \geq c_n$$

where the a's, b's, and c's are real.

When R is two-dimensional, it is called a *convex polygonal region*.

Examples of such sets are given in Figs. 8.12 to 8.14, except that in the present situation the bounding segments are to be included in the set. In Fig. 8.16 the colored triangle and in Fig. 8.17 the segment AB are sets of this type.

Exercise

A   Show that the segment AB of Fig. 8.17 is defined by

$$L + P \geq \quad 400 \qquad L \geq 100$$
$$-L - P \geq -400 \qquad P \geq 150$$

It is possible that R is empty as in Fig. 8.15, in which case our problem has no solution. It is also possible that R consists of a single point, so that the solution to our problem is trivial. So we assume that R contains at least two points; this implies that R is a line, a ray, a line segment (Fig. 8.17) or a nonempty convex polygonal region in the plane (Figs. 8.12 to 8.14 and 8.16). A point of R is called a *feasible* solution of our problem.

Definition   The point $P(\bar{x},\bar{y})$ of R is called a *maximum* for the given poly-

max, min   nomial $f(x,y) = px + qy + r$ if and only if $f(x,y) \leq f(\bar{x},\bar{y})$ for all (x,y) in R. Similarly, P is a minimum if and only if $f(x,y) \geq f(\bar{x},\bar{y})$ for all (x,y) in R. A maximum or minimum point P is called an *optimal* solution of the problem.

Theorem 14   If $P(\bar{x},\bar{y})$ is a maximum point of $f(x,y) = px + qy + r$ (where $p \neq 0$ or $q \neq 0$) in R, then:

1   If R is a convex polygonal region in the plane,
   a P is a vertex of R and, at all other points of R, $f(x,y) < f(\bar{x},\bar{y})$, or
   b P is a point of a side S of R such that, on S, $f(x,y) = f(\bar{x},\bar{y})$ and, at all other points of R, $f(x,y) < f(\bar{x},\bar{y})$.
2   If R is a line segment or a ray,
   a P is an end-point of R and, at all other points of R, $f(x,y) < f(\bar{x},\bar{y})$, or
   b $f(x,y) = f(\bar{x},\bar{y})$ at all points of R.

Proof   Consider the line

$$L: f(x,y) - f(\bar{x},\bar{y}) = 0$$

or

$$(px + qy + r) - (p\bar{x} + q\bar{y} + r) = 0$$

As discussed in Sec. 8.5, this line divides the plane into two half planes in one of which, say I,

$$f(x,y) - f(\bar{x},\bar{y}) < 0 \quad \text{or} \quad f(x,y) < f(\bar{x},\bar{y})$$

Moreover, since $P(x,y)$ is a maximum,

$$f(x,y) \le f(\bar{x},\bar{y})$$

for every point (x,y) of R. It follows, therefore, that the points of R are either on L or on half plane I. This means that (Fig. 8.21)

1 If R is a convex polygonal region in the plane, the intersection of L and R (which contains P) is either

    a A vertex of R which is the only point of R where $f(x,y) = f(\bar{x},\bar{y})$ and, at all other points of R, $f(x,y) < f(\bar{x},\bar{y})$ or

    b A side S of R along which $f(x,y) = f(\bar{x},\bar{y})$ and, at all other points of R, $f(x,y) < f(\bar{x},\bar{y})$.

2 If R is a line segment or a ray, the intersection of L and R (which contains P) is either

    a An end-point of R which is the only point of R where $f(x,y) = f(\bar{x},\bar{y})$ and, at all other points of R, $f(x,y) < f(\bar{x},\bar{y})$ or

    b The whole of R, so that at all points of R $f(x,y) = f(\bar{x},\bar{y})$.

A similar result is true if P is a minimum.

Possible locations of maximum point P

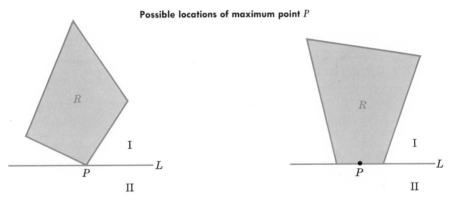

Figure 8.21 If P is a maximum point of $f(x,y)$ for (x,y) in R, it must lie on the boundary of R. It is, therefore, a vertex of R or a point of a side of R along which $f(x,y)$ is constant.

From this theorem our working procedure for finding a maximum point involves the following steps:

1 Check the situation to see whether a maximum exists. It always exists if R is bounded.

2 Find all vertices (or end-points) of R and the value of $f(x,y)$ at each of these.

3 If a maximum exists and if there is one vertex (or end-point) at which $f(x,y)$ is greater than at all the other vertices, this is the maximum point.

4 If a maximum exists and if there are two adjacent vertices (or end-points) at which $f(x,y)$ has equal values which are greater than its values at the other vertices, then all points of the side joining these vertices are maximum points.

No other possibilities can occur. A similar procedure holds for finding a minimum point.

Problems 8.9

1 In the Diet Problem (Illustration 1 above), assume all the information given there plus the fact that food A costs 10 cents/oz and food B costs 20 cents/oz. What combination of A and B should be used to provide an adequate diet at least cost?

2 In the Diet Problem (Illustration 1 above), suppose that 1 oz of food A occupies 2 cu in. and that 1 oz of food B occupies 3 cu in. Then what combination of A and B will provide an adequate diet and minimize the total volume?

3 In the Football Problem (Sec. 8.6, Illustration 5), suppose that on the average there is one injury in each five running plays and one injury in each ten pass plays. Then what combination of plays should the quarterback call to secure the touchdown in the allotted time with the minimum risk of injuries?

4 To provide sufficient nitrogen (N), phosphorus (P), and lime (L) for his 400-acre farm, Mr. Hill buys x bags of A and y bags of B per acre. A and B are two kinds of commercial fertilizer. The necessary data are shown in tabular form:

	N	P	L	First year cost/bag	Second year cost/bag
A	6	2	8	$ 5	$ 6
B	6	9	15	$10	$10
Minimum need/acre	108	85	235		

Find the values of x and y which will meet at least the minimum need for fertilizer and for which the cost per acre is a minimum, each year.

5 A diet contains carbohydrates (H), protein (P), and fats (F). Two standard rations are available, with the data in tabular form:

	H	P	F	Cost	Time	Weight
Ration A	100	50	30	$10	k	1
Ration B	200	10	0	$30	$2k$	1
ANPM	2,500	350	150			

ANPM means "minimum annual need per man".

x is the number of Ration A's used and y the number of Ration B's. Find the value of x and y which will provide at least the ANPM, where:

a Cost is a minimum.

b Time (of preparation) is a minimum.

c Weight is a minimum.

6 A ship arrives in port with 2,800 tons of grain. It must be unloaded in at most 35 hr. Available machinery is a conveyor belt, which runs with a crew of three men, costs \$80 per hr, and unloads 120 tons/hr and also a power shovel, which costs \$10 per hr, runs with a crew of ten men, and unloads 50 tons/hr. In each case, \$4 per hr per man of crew must be added. The conveyor belt and the shovel cannot operate at the same time.

The union contract stipulates that 1 hr of work must be provided some union man for each 20 tons of cargo on each ship. That means that 140 hr of labor must be provided. The conveyor belt runs for x hr and the power shovel for y hr. Find the minimum cost of unloading the ship. Also find the values of x and y which produce the minimum.

7 Accommodations must be provided for at least 435 persons at a remote mine. Buildings must be of two types, A and B. The pertinent data are:

House	Units of wood	Units of concrete	Capacity (persons)	Hours to build	Cost	No. of houses
A	10	7	12	40	\$ 5,000	x
B	5	25	25	50	\$12,000	y

At least 125 units of wood and 285 units of concrete must be used. Find the values of x and y for which:

a The cost of construction is a minimum.

b The time of construction is a minimum.

8 A certain machine tool company has found that it must have a core of at least 700 reliable customers to remain competitive in the local market. Furthermore, to earn the desired rate of profit, total sales of at least \$160,000 must be made yearly for a total selling cost of no more than \$4,600.

To meet these requirements, the sales manager has suggested contacting some customers directly by traveling salesmen and contacting others indirectly by written correspondence. The problem is determining how many to contact each way. When the traveling-salesman method of contact is used, it has been found that the company makes \$400 of sales per customer for a selling cost of \$7 per customer. When the written correspondence method of contact is used, it has been found that the company makes \$100 of sales per customer for a selling cost of \$4 per customer.

The company has a shortage of experienced employees and must attempt to minimize the use of labor hours per customer. Salesman contact requires 6 hr per customer, correspondence contact requires only 0.5 hr per customer. How many customers should be contacted by each method?

9 The personnel director of a corporation must find the most suitable method of testing the qualifications of job applicants. At least 35 men must be tested; the total cost of the test administration cannot exceed $6.25; and the total time for test administration cannot exceed 8.75 hr. Two types of qualification tests are available: a written aptitude test which indicates the applicant's potential in various job situations and a performance test which indicates the applicant's actual capabilities. The administration cost is $20 per 100 aptitude tests and $5 per 100 performance tests. The administration time is 10 hr per 100 aptitude tests and 25 hr per 100 performance tests.

The personnel director must stay within the three restrictions given above when deciding how many of each type of test to give, but he also must minimize the total number of inaccurate appraisals. If he has found that the appraisals of 18 per cent of the aptitude tests and 22 per cent of the performance tests are inaccurate, how many of each should he administer?   Let x be the number of aptitude tests and y be the number of performance tests.

10 A certain large winery is considering automating the wine-testing process. Below are the data comparing the old manual testers (a man with a glass) with the new electronic testers (a computer with a tube).

Electronic testers   Initial cost, $50 per tester; hourly cost of operation, $15 per tester; number of vats tested per hour, 23 per tester; daily man-hours of labor needed, 5 per tester, cups of wine wastage per day, 3 per tester.

Manual testers   Initial costs, $0 per tester, hourly cost of operation, $20 per tester; number of vats tested per hour, 4 per tester; daily man-hours of labor needed, 20 per tester; cups of wine wastage per day, 12 per tester.

In an effort to fight unemployment, the local union representing the winery workers objects to the introduction of electronic testers. The following compromise is finally agreed to by the union and the winery.

Demands by the union   (1) No more than $2,500 will be spent on electronic wine testers.   (2) A total of at least 850 man-hours of labor will be employed by the winery daily.

Demands by the winery   (1) The total hourly costs of operating the testers cannot be greater than $1,550.   (2) A total of at least 830 vats must be tested per hour.

Find a number of electronic testers which will combine with a number of manual testers to satisfy the above compromise and will minimize the total cups of wine wastage per vat.

11 In the Diet Problem (Illustration 1 above) does $W = a + b$ have a maximum in R?

12 In the Diet Problem (Illustration 1 above) does the cost $C = 10a + 20b$ have a maximum in R?

13 Given that R is defined by the inequalities (see Sec. 8.8, Prob. 19)

$$4x + 3y - 12 \geq 0$$
$$4x - 5y + 20 \geq 0$$
$$4x - y - 12 \leq 0$$

Find the maximum and minimum values in R of $f(x,y) = 3x - 4y + 5$.

14 Given that R is defined by the inequalities (see Sec. 8.8, Prob. 20)

$$3x - 2y + 12 \geq 0$$
$$x - 4y + 4 \geq 0$$
$$x + y - 6 \leq 0$$

Find the maximum and minimum values in R of $f(x,y) = 2x + y + 4$.

15 Given that R is defined by

$$x + 2y - 12 = 0$$
$$5x - 2y \geq 0$$
$$x - 2y \leq 0$$

Find the maximum and minimum values in R of $f(x,y) = 3x - y + 5$.

16 Given that R is defined by

$$3x - 2y = 0$$
$$3x + y - 9 \geq 0$$
$$x + 2y - 8 \leq 0$$

Find the maximum and minimum values in R of

$$f(x,y) = x + y + 5$$

17 What is the situation if both $p = 0$ and $q = 0$ in

$$f(x,y) = px + qy + r?$$

18 What is the situation if R is the whole of a line?

19 What conclusion can you draw if $f(x,y)$ is a maximum at an interior point of R?

20 What is the situation if R is the entire plane?

FUNCTIONS AND RELATIONS

Chapter **9**

9.1 Relations

In this chapter we shall be dealing with subsets of the set $X \times Y$, the set of all ordered pairs of real numbers (x,y). We have already met two important subsets of this type:

$$\{(x,y)\,|\,ax + by + c = 0\} \qquad [\text{Sec. 6.5}]$$
$$\{(x,y)\,|\,ax + by + c > 0\} \qquad [\text{Sec. 8.4}]$$

Here we are interested in other examples, and in the general situation as a whole. In order to have a name for these subsets, we define a *relation* as follows.

Definition A *relation* is a subset of the set of ordered pairs of real numbers, relation $X \times Y$.

There are a good many ways of defining specific relations, and so we give the most important of these below.

List of Ordered Pairs When the subset consists of a reasonably small (finite) number of ordered pairs, we just write them down. This list of pairs defines the relation.

Illustration 1 An example of such a relation is the set

$$\{(0,3),\ (2,-1),\ (2,4),\ (1,3),\ (0,0)\}$$

A relation consisting of a single ordered pair is the set $\{(1,3)\}$.

Illustration 2 As a physical illustration of such a set consider the pairs (T,D), where T represents the time of revolution in years of a planet and D represents its mean distance from the sun in units such that D for the earth is 1. Then we have the table:

	Mer- cury	Venus	Earth	Mars	Jupiter	Saturn	Uranus	Nep- tune	Pluto
T	0.241	0.615	1.00	1.88	11.9	29.5	84.0	165	265
D	0.387	0.723	1.00	1.52	5.20	9.54	19.2	30.1	41.3

The relation so defined is the set

$$\{(0.241, 0.387), (0.615, 0.723), \ldots, (265, 41.3)\}$$

Illustration 3 Relations of this type frequently appear as tables. You have probably met tables of logarithms which are good examples of this. In these the ordered pairs are of the form $(x, \log x)$. The set of all such pairs of entries is the relation so defined.

Equations The subset is often defined to consist of those ordered pairs which satisfy a certain equation. Then the relation is

$$\{(x,y) \mid (x,y) \text{ satisfy the given equation}\}$$

Illustration 4 Examples of such relations are

$$\{(x,y) \mid 3x + 2y = 5\} \qquad \{(x,y) \mid 4x^2 - 6xy + y^2 = 0\}$$
$$\{(x,y) \mid 2x^4 + y^2 - xy^2 + 7 = 0\}$$

Inequalities Here the subset is defined to consist of those ordered pairs which satisfy a certain inequality. An example is the relation

$$\{(x,y) \mid x + 3y > 2\}$$

Simultaneous Systems Relations may also be defined (as in Secs. 6.8, 6.10, 6.11, and 8.6) by a simultaneous system of equalities, or inequalities, or mixtures of these.

Illustration 5 The following simultaneous systems define relations:

$$\begin{cases} 2x + y = 3 \\ x > 1 \end{cases} \qquad \begin{cases} 3x + 2y = 9 \\ x + y = 1 \end{cases} \qquad \begin{cases} 2x - y < 0 \\ x + y \geq 1 \end{cases}$$

Exercise

A Graph the relations of Illustration 5.

Graphs In science relations sometimes are defined by a graph. This amounts to a curve in the plane, to a shaded region, or to a discrete set of points. The coordinates (x,y) of all points in the graph give us the desired subset of $X \times Y$ and define the relation.

A relation R defines a subset of X called its *domain* and a subset of Y called its *range*.

domain
and range

Definitions The *domain* of a relation is the subset of $X: \{x \mid x$ is the first element of at least one of the pairs (x,y) of $R\}$. The *range* of a relation is the subset of $Y: \{y \mid y$ is the second element of at least one of the pairs (x,y) of $R\}$.

Illustration 6 Consider the relation $\{(x,y) \mid 2x + 3y + 1 = 0\}$. For every x there is a pair (x,y) which satisfies this equation (solve for y in terms of x). Hence the domain is X. Similarly the range is Y.

For the relation $\{(x,y) \mid y = 3\}$, the domain is X, and the range is $\{3\} \subset Y$.

For the relation $\{(x,y) \mid x^2 + y^2 = 9\}$, x must lie in the interval $-3 \le x \le 3$ if y is to be real. Hence the domain is $\{x \mid -3 \le x \le 3\}$. Similarly the range is $\{y \mid -3 \le y \le 3\}$.

The methods of defining relations are often referred to as *rules*. Thus the relation $\{(x,y) \mid x + 2y - 1 > 0\}$ is defined by the rule which, in this case, is the inequality $x + 2y - 1 > 0$.

Problems 9.1

In Probs. 1 to 20 discuss the domain and range of each relation whose defining rule is:

1 $y = 4x + 3$ **2** $y = 6x - 5$

3 $y = x^2 + 1$ **4** $y = -x^2 + 2$

5 $x^2 + y^2 = 25$ **6** $x^2 + y^2 = 100$

7 $y = x^3$ **8** $y = 3x^4$

9 $x^2 + y^2 \le 1$ **10** $x^2 + y^2 \ge 1$

11 $x - y > 0$ **12** $x + y < 0$

13 $\begin{cases} x + y > 1 \\ x > 2 \end{cases}$ **14** $\begin{cases} x - y = 1 \\ x \le 2 \end{cases}$

15 $\begin{cases} 2x + y = 3 \\ y \ge 5 \end{cases}$ **16** $\begin{cases} 2x + 3y = 8 \\ 4x - y = 2 \end{cases}$

17 $\{(1,4), (4,6)\}$ **18** $\{(1,2), (1,3)\}$

19 $\{(3,5), (7,5)\}$ **20** $\{(-2,1), (-2,3), (0,3)\}$

9.2 Functions

A special kind of relation, called a function, is of great importance.

Definition A *function* f is a relation in which no two ordered pairs have the same first element.

This means that, given an x in the domain of f, there is a unique pair (x,y) belonging to the function. This determines a value y uniquely associated with the given x. We denote this value by the symbol $f(x)$, read "f of x", called the value of the function f at x.

Definition The *value $f(x)$ of the function f at x* (in the domain of f) is the second element of that unique ordered pair (x,y) belonging to f which has x as its first element.

$f(x)$ is value of a function at x

In a relation there may be many pairs (x,y_1), (x,y_2), ... with x as first element. Hence we do not speak of the "value of a relation".

Illustration 1 The relation $\{(x,y)\,|\,3x - 2y + 4 = 0\}$ is also a function since there is a unique $y = f(x) = (3x + 4)/2$ associated with each x. The domain is X, and the range is Y.

Illustration 2 The relation $\{(x,y)\,|\,y = x^2\}$ is also a function with $f(x) = x^2$. The domain is X, and the range is $\{y\,|\,y \geq 0\}$.

Illustration 3 The relation $\{(x,y)\,|\,x^2 + y^2 = 4\}$ is *not* a function. For to each x in the open interval $-2 < x < 2$ there are associated two values of y, namely, $y = \pm\sqrt{4 - x^2}$.

Illustration 4 The relation $\{(x,y)\,|\,x^2 + y^2 < 4\}$ is *not* a function. For to each x in the open interval $-2 < x < 2$ there are associated infinitely many values of y, namely, those such that $y^2 < 4 - x^2$.

Illustration 5 We may define a function by giving its values for each x in its domain. For example,

$$f(x) = \begin{cases} 1 & \text{if } x \geq 0 \\ -1 & \text{if } x < 0 \end{cases}$$

defines the function f whose ordered pairs are $(x,f(x))$. The domain is X; the range is $\{-1,1\} \subset Y$.

As in Illustration 5, we regularly define functions by stating what their values $f(x)$, or y, are to be. Thus we speak of "the function f whose values are $f(x) = 4x^2 - 9$" or "whose values are $y = 4x^2 - 9$". It is common practice to abbreviate this and to speak of the "function $f(x) = 4x^2 - 9$" or the "function $y = 4x^2 - 9$" or even the "function $4x^2 - 9$". These abbreviations can be misleading, for they confuse the function itself with its set of values. However, once you have understood the true meaning of a function, there is then no real harm in your using a convenient abbreviation. Consequently, we shall use the full, correct language while you are learning about functions in this chapter, but later we shall not hesitate to use abbreviations and to speak of the functions x, $3x^2 - 4$, $\sin x$, $\tan x$, e^x, $\log x$, etc., when we mean the functions whose values are given by these expressions.

In connection with functions, the terms variable and constant (Sec. 1.4) are frequently used. Some important definitions relating to functions are as follows.

Definitions

variable and constant

1 A *variable* is a symbol which may be replaced by any element of a given universal set.

2 A *constant* is a fixed element of a given set.

3 A *real variable* is a variable whose universal set is a subset of the real numbers (as it is in this chapter).

4 A *complex variable* is a variable whose universal set is a subset of the complex numbers.

5 When we are given a function with its domain and range, the variable whose universal set is the domain is called the *independent variable*. When x is the independent variable for a function f, we say that "f is a function of x".

6 When we are given a function and a particular value of its independent variable, a unique element of the range is determined. Hence the variable whose universal set is the range of this function is called the *dependent variable*.

So far we have considered functions f whose ordered pairs are (x,y). It is often necessary or desirable to use other letters to represent the function and the independent and dependent variables. To say what we are doing in such cases, we shall use notations such as $f:(x,y)$; $g:(w,z)$; $\phi:(r,T)$; etc. For instance, $f:(x,y)$ means "the function f whose ordered pairs are (x,y)"; $g:(w,z)$ means "the function g whose pairs are (w,z)"; and so forth.

Finally, we define two important special functions as follows.

Definitions The function $f:(x,k)$, k a constant, is called the *constant function*.

constant function and absolute-value function

The function $f:(x,|x|)$ is called the *absolute-value function*. We remind you that this function is the same as the function $g:(x, \sqrt{x^2})$.

Problems 9.2

In Probs. 1 to 4 which sets of ordered pairs are functions?

1 $\{(3,2), (4,6), (5,-1)\}$ 2 $\{(1,4), (3,4), (7,3)\}$
3 $\{(1,2), (1,3), (1,-2)\}$ 4 $\{(3,6), (3,7), (4,7)\}$

In Probs. 5 to 24 find the domain and the range of the indicated functions.

5 $f(x) = -x$ 6 $f(x) = 2x$
7 $f(x) = x^2 + 1$ 8 $f(x) = x^3$

9
$$\begin{array}{c|ccc} x & 0 & 3 & 5 \\ \hline f(x) & 1 & 2 & 1 \end{array}$$

10
$$\begin{array}{c|ccc} x & 5 & 7 & 8 \\ \hline f(x) & 1 & 1 & 2 \end{array}$$

11
$$\begin{array}{c|cc} x & 1 & 2 \\ \hline f(x) & 5 & 6 \end{array}$$

12
$$\begin{array}{c|cc} x & 1 & 2 \\ \hline f(x) & 7 & 7 \end{array}$$

13
$$\begin{array}{c|cc} x & 2 & 3 \\ \hline f(x) & 1 & 1 \end{array}$$

14
$$\begin{array}{c|cc} x & 3 & 2 \\ \hline f(x) & 7 & 7 \end{array}$$

15 $f(x) = \begin{cases} 1, & x \text{ rational} \\ 2, & x \text{ irrational} \end{cases}$

16 $f(x) = \begin{cases} 0, & 1 \le x \le 2 \\ 1, & 3 \le x \le 4 \\ 10, & \text{otherwise} \end{cases}$

17 $f(x) = \begin{cases} 0, & x \text{ an integer} \\ \text{undefined}, & \text{otherwise} \end{cases}$

18 $f(x) = \begin{cases} \sqrt{x}, & x \ge 0 \\ 0, & x < 0 \end{cases}$

19 $y = \dfrac{1}{x(x-1)}$

20 $y = \dfrac{1}{(x-1)(x-2)}$

21 $y = \dfrac{x-1}{x(x-2)}$

22 $y = \dfrac{x+2}{x^2-4}$

23 $y = \dfrac{x-2}{x^2-4}$

24 $y = x^2 - 2$

25 For $y = (x-1)(x-3)$ state the range if the domain is considered to be:

   a $\{x \mid 1 \le x \le 3\}$

   b $\{x \mid x \le 1\} \cup \{x \mid x \ge 3\}$

26 The dimensions of a rectangle are 10 and L. Show that the area A is a function of L, and state domain and range.

27 Given $f(x) = \dfrac{3x+1}{6}$; compute $f(x)$ for $x = 0, 1, 2, 3, -1, -2$.

28 Given $f(x) = x^2 + 1$; compute $f(x)$ for $x = 0, 1, \sqrt{2}, t, a+h$. Also compute $f(1) - f(3)$.

29 The equation $x^2 + y^2 = 25$ defines a relation. Graph the relation. Indicate three functions which are special cases (i.e., subsets) of this relation. State the domain, range, and defining rule of each.

30 The equation $x^2 - y^2 = 25$ defines a relation. Graph the relation. Indicate three functions which are special cases (i.e., subsets) of this relation. State the domain, range, and defining rule of each.

9.3 Algebra of Functions

We have studied the four elementary operations of arithmetic $+$, $-$, $\times$, $\div$ in connection with numbers (Chap. 2). These ideas can also be applied to functions according to the following definitions.

Definitions Consider the functions $f:(x,y)$ and $g:(x,z)$ whose domains are, respectively, indicated by d_f and d_g. The *sum* $f + g$, the *difference* $f - g$, the *product* fg, and the *quotient* f/g are defined as follows:

1 $f:(x,y) + g:(x,z) = (f + g):(x, y + z)$
2 $f:(x,y) - g:(x,z) = (f - g):(x, y - z)$
3 $f:(x,y) \times g:(x,z) = (fg):(x,yz)$

4 $\dfrac{f:(x,y)}{g:(x,z)} = \left(\dfrac{f}{g}\right):\left(x, \dfrac{y}{z}\right)$

1′ In the addition $f + g$ of two functions, the functional values are added.

2′ In the subtraction $f - g$, the functional values are subtracted (in the proper order).

3′ In the multiplication fg, the functional values are multiplied.

4′ In the division f/g, the functional values are divided (in the proper order).

The functional values of f and g are y and z, respectively. Thus, when f and g are defined by $y = f(x)$ and $z = g(x)$, the above definitions become:

1″ $y + z = f(x) + g(x)$
2″ $y - z = f(x) - g(x)$
3″ $yz = f(x) \times g(x)$

4″ $\dfrac{y}{z} = \dfrac{f(x)}{g(x)}$

The domain of each of $f + g$, $f - g$, and fg is the set of all elements x common to the domains of f and g; that is, it is the intersection of the sets d_f and d_g. Thus, in symbols $d_{f+g} = d_f \cap d_g$, $d_{f-g} = d_f \cap d_g$, and $d_{fg} = d_f \cap d_g$. The domain $d_{f/g} = d_f \cap d_g$ except for those x's for which $g(x) = 0$. (Division by zero is impossible.)

Illustration 1   Given the two functions f and g whose values are $y = x^2$ and $z = x^3$, the domain of each is a set of real numbers. Then

$$y + z = x^2 + x^3 \qquad\qquad d_{f+g} = X$$
$$y - z = x^2 - x^3 \qquad\qquad d_{f-g} = X$$
$$yz = x^2 \times x^3 = x^5 \qquad\quad d_{fg} = X$$

$$\frac{y}{z} = \frac{x^2}{x^3} = \frac{1}{x} \qquad\qquad d_{f/g} = X \text{ except } 0$$

Note that $x = 0$ is not in the domain of f/g since $g(0) = 0$.

Illustration 2   Let f and g have the values $y = 1 + 1/x$, $z = \sqrt{1 - x^2}$. The domain d_f is the set of all real numbers excluding 0; the domain d_g is the set of all real numbers between -1 and 1 inclusive. Then

$$y + z = 1 + \frac{1}{x} + \sqrt{1 - x^2} \qquad d_{f+g} = \{x \,|\, (-1 \le x < 0) \text{ or } (0 < x \le 1)\}$$

$$y - z = 1 + \frac{1}{x} - \sqrt{1 - x^2} \qquad d_{f-g} = d_{f+g}$$

$$yz = \left(1 + \frac{1}{x}\right)\sqrt{1 - x^2} \qquad d_{fg} = d_{f+g}$$

$$\frac{y}{z} = \frac{1 + (1/x)}{\sqrt{1 - x^2}} \qquad d_{f/g} = \{x \mid (-1 < x < 0) \text{ or } (0 < x < 1)\}$$

In f/g we must exclude $x = \pm 1$, since $g(-1) = g(1) = 0$.

One further operation in the algebra of functions is of great importance; it is the operation of forming the *composite* of two functions.

We illustrate with a special example. Consider first the two functions $f:(x,y)$ and $g:(y,z)$, whose values are given respectively by

$$y = 2x \qquad \begin{cases} \text{domain } X \\ \text{range } Y \end{cases}$$

$$z = +\sqrt{y^3 - 1} \qquad \begin{cases} \text{domain } \{y \mid y^3 - 1 \geq 0\}; \text{ that is, } \{y \mid y \geq 1\} \\ \text{range } \{z \mid z \geq 0\} \end{cases}$$

If there is no connection between these two functions f and g, then there is nothing but a notational difference intended in using ordered pairs of the form (x,y) for f and ordered pairs of the form (y,z) for g.

However, mathematics is filled with situations in which there *is* a connection—situations in which range values of one function f must serve as the domain values of another function g. In the above example, only those values $y \geq 1$ of the range of f can be used in the domain of g. This process leads to a *third* function whose ordered pairs are (x,z), where $z = +\sqrt{y^3 - 1} = +\sqrt{(2x)^3 - 1}$. This third and new function is called the *composite of g and f*, and we choose the symbol $g \circ f$ to represent it. The symbol $g(f)$ is also used.

Now let us describe the general situation where we are given two functions $f:(x,y)$ and $g:(y,z)$. Choose an x such that the y which f assigns to it is in the domain of g. Then g assigns a z to this y. This gives us the pair (x,z). The set of all pairs (x,z) which can be constructed in this fashion is the composite function $g \circ f$. The domain of $g \circ f$ is the set of all x's for which this process can be defined; if there are no such x's, the function $g \circ f$ is not defined.

Definition For two given functions $f:(x,y)$ and $g:(y,z)$, the set of ordered pairs (x,z) described above defines a function called the *composite* of g and f and written $g \circ f:(x,z)$. The value of the composite of g and f is written $(g \circ f)(x)$ or $g(f(x))$.

Illustration 3 Let $z = g(y) = 3y^2 - 2y + 1$ and $y = f(x) = 4x + 7$. The composite $g \circ f$ is given by

$$z = g(y) = 3y^2 - 2y + 1 = 3(4x + 7)^2 - 2(4x + 7) + 1$$

This can be simplified to yield

$$z = 48x^2 + 160x + 134$$

which defines $g \circ f:(x, 48x^2 + 160x + 134)$.

Illustration 4   A stone is dropped into a liquid, forming circles which increase in radius with time according to the formula $r = 4t$. How does the area of a given circle depend upon time?

Solution   The area A of a circle is $A = \pi r^2$, and we are given that $r = 4t$. These define two functions $g:(r,A)$ and $f:(t,r)$; and we seek the composite $g \circ f:(t,A)$.

$$A = \pi r^2 = \pi(4t)^2$$

or, reduced,
$$A = 16\pi t^2$$

Hence $g \circ f:(t,16\pi t^2)$ is the composite of g and f. Here we are not interested in negative values of t, although, mathematically, the maximal domain is the set of all real numbers. For the physical problem, a subset such as $0 \le t \le t_1$, where t_1 is sufficiently large, would suffice.

Since the letters used to represent the independent and dependent variables of a function can be replaced by other letters without change of meaning, we can speak of the composite $g \circ f$, where the functions f and g are given in the usual form $f:(x,y)$ and $g:(x,y)$. For we can rewrite g in the form $g:(y,z)$ if we wish. If we were interested in the composite $f \circ g$ we would rewrite f in the form $f:(y,z)$.

Illustration 5   Given f and g whose values are $f(x) = x^2 + 2$ and $g(x) = 1 - (1/x)$, form the composite functions $g \circ f$ and $f \circ g$.

Solution

For $g \circ f$	*For $f \circ g$*
Rewrite the defining equations in the form	Rewrite the defining equations in the form

$$y = f(x) = x^2 + 2$$
$$z = g(y) = 1 - (1/y)$$

$$z = f(y) = y^2 + 2$$
$$y = g(x) = 1 - (1/x)$$

Then

$$z = 1 - (1/y) = 1 - [1/(x^2 + 2)]$$

and the composite $g \circ f$ is

$$g \circ f:(x,z)$$

or $g \circ f:[x, \, 1 - [1/(x^2 + 2)]]$

Then

$$z = y^2 + 2 = [1 - (1/x)]^2 + 2$$

and the composite $f \circ g$ is

$$f \circ g:(x,z)$$

or $f \circ g:[x, \, [1 - (1/x)]^2 + 2]$

Illustration 6   Find $g \circ f$ when g and f have the values $g(x) = |x|$ and $f(x) = x^2 - 3x + 1$.

Solution   Write $z = g(y) = |y|$ and $y = f(x) = x^2 - 3x + 1$. Then $z = |y| = |x^2 - 3x + 1|$. Thus $g \circ f:(x, \, |x^2 - 3x + 1|)$. To evaluate $|x^2 - 3x + 1|$ for a given x, say $x = 1$, we first find $x^2 - 3x + 1$, which equals -1. Then we take its absolute value, which is $+1$.

Problems 9.3

In Probs. 1 to 7 find the values $(f + g)(x)$, $(f - g)(x)$, $(fg)(x)$, and $(f/g)(x)$. In each case state domain.

1 $f(x) = \dfrac{1}{x}$, $g(x) = \dfrac{1}{x + 1}$ **2** $f(x) = \dfrac{1}{\sqrt{x + 1}}$, $g(x) = \dfrac{1}{\sqrt{x + 2}}$

3 $f(x) = \dfrac{1}{x + 1}$, $g(x) = x + 2$ **4** $f(x) = x + 1$, $g(x) = x - 1$

5 $f(x) = x + 2$, $g(x) = (x + 2)^2$ **6** $f(x) = x + 1$, $g(x) = \dfrac{1}{x + 1}$

7 $f(x) = \dfrac{1}{x + 1}$, $g(x) = \dfrac{1}{(x + 1)(x - 1)}$

In Probs. 8 to 10 form the composite $(g \circ f)(x)$. State domain.

8 $z = y^2$, $y = x^2 + 3x + 5$ **9** $z = y^2 - 8y + 5$, $y = x^3$
10 $z = 3y^2 - 2y$, $y = 1/x$

In Probs. 11 to 16 form $(g \circ f)(x)$ and also $(f \circ g)(x)$. State domain.

11 $f(x) = x + 4$, $g(x) = x - 3$ **12** $f(x) = x - 1$, $g(x) = x^2$
13 $f(x) = |x|$, $g(x) = \sqrt{x}$ **14** $f(x) = x$, $g(x) = 1/x$
15 $f(x) = x^2$, $g(x) = 1/x^3$ **16** $f(x) = |x| - 1$, $g(x) = |x - 2|$
17 Evaluate $|x^2 - 3x + 2|$ at: **18** Simplify:
   a $x = 0$ **a** $\sqrt{x^2 - 2x + 1}$
   b $x = 3$ **b** $\sqrt{1/x^2}$
   c $x = \frac{3}{2}$

9.4   Graphs

In Chaps. 6 and 8 we plotted the graphs of certain special relations and in Sec. 6.5 we discussed the general concept of the graph of a relation. Here we shall study general methods for graphing relations and functions. By way of review, we define graph as follows:

Definition The *graph* of a relation whose ordered pairs are (x,y) is the set of points in the XY-plane whose coordinates are the given pairs.

When the relation is defined by an equation, the basic method of plotting its graph is to find a reasonable number of points (x,y) whose coordinates satisfy the equation. Then we join these points by a smooth curve. There are two disadvantages to this method: (1) We may need to compute a rather large number of points in order to see just how the graph should look. (2) Even then, we may overlook some abnormal features of the graph which occur in the gaps between the plotted points. It is, therefore, desirable for us to develop some general aids to graphing which will cut down on your work and improve your accuracy. These apply only to relations (or functions) which are defined by equations.

Intercepts The x-intercepts are the x-coordinates of the points at which the graph crosses (or meets) the X-axis, and the y-intercepts are the y-coordinates of the corresponding points on the Y-axis. To find the x-intercepts, put $y = 0$ in the equation which defines the relation, and solve for x. To find the y-intercepts, put $x = 0$, and solve for y.

When the relation is a function, the x-intercepts correspond to those x's for which the value of the function is zero. These x's are called the *zeros of the function*. We shall devote considerable attention in Chap. 10 to methods for finding the zeros of polynomial functions.

Illustration 1 Find the intercepts of the graph of $y = x^2 - 3x + 2$.

Solution Setting $y = 0$, we find that the solutions of $x^2 - 3x + 2 = 0$ are $x = 1$ and $x = 2$. The x-intercepts are 1 and 2. By setting $x = 0$ we find $y = 2$, which is the y-intercept (Fig. 9.1).

Intercepts

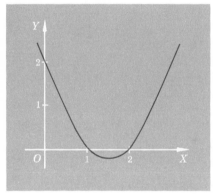

Figure 9.1
The zeros of a function are the x-intercepts of its graph. Here $y = x^2 - 3x + 2$.

domain
and range

Domain and Range It is very useful to know the domain and range of a relation, for this knowledge tells us about regions of the plane to which the graph is confined or from which it is excluded. It is useless to try to plot points in excluded regions. There are two common situations in which the domain or range is restricted to a subset of the whole axis.

The first of these is based upon the principle that y cannot take on values which require x to be complex and similarly that x cannot take on values which require y to be complex.

Illustration 2 Discuss the domain and range of $x^2 + y^2 = 4$.

Solution First solve for x, and obtain

$$x = \sqrt{4 - y^2}$$

The right-hand side is real if and only if $4 - y^2 \geq 0$, or $4 \geq y^2$. Hence y must lie in the interval $-2 \leq y \leq 2$. Solving for y, we obtain

$y = \sqrt{4 - x^2}$ and arrive at the similar conclusion that x is in the interval $-2 \leq x \leq 2$.

Illustration 3 Discuss the domain and range of $y = x^2 - 3x + 2$.

Solution Since all values of x give real values of y, there is no restriction on x and the domain is the real line X. To find any possible restrictions on the range, solve for x. We have

$$x^2 - 3x + (2 - y) = 0$$

which yields

$$x = \frac{3 \pm \sqrt{9 - 4(2 - y)}}{2}$$

$$= \tfrac{3}{2} \pm \tfrac{1}{2}\sqrt{1 + 4y}$$

Since x must be real, this requires that y satisfy the inequality $1 + 4y \geq 0$, or $y \geq -\tfrac{1}{4}$. No part of the graph can therefore lie below the horizontal line $y = -\tfrac{1}{4}$ (Fig. 9.1).

The second principle is that expressions equal to a perfect square can never be negative. The application of this may give us inequalities which x or y must satisfy.

Illustration 4 Discuss the domain and range of $y^2 = (x - 1)(x + 3)$.

Solution Since $y^2 \geq 0$, we must have $(x - 1)(x + 3) \geq 0$. This is a quadratic inequality of the type discussed in Sec. 8.4. Using the methods developed there, we find that x cannot lie in the interval $-3 < x < 1$.
Solving for x, we find

$$x = -1 \pm \sqrt{4 + y^2}$$

Since $4 + y^2$ can never be negative, there are no restrictions on y (Fig. 9.2).

Restrictions on domain

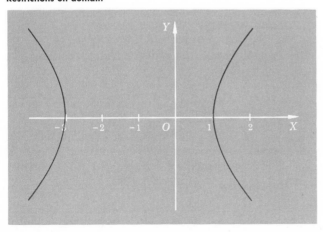

Figure 9.2 The domain of a real-valued function cannot include values of x which correspond to nonreal values of y. Here $y^2 = (x - 1)(x + 3)$.

symmetry

Symmetry   The points (x,y) and $(x,-y)$ are symmetric with respect to the X-axis, the one being the mirror image of the other.   Either point is called a *reflection* of the other about the X-axis.   The graph will be symmetric about the X-axis if for every point (x,y) on the graph the corresponding point $(x,-y)$ also lies on it.   To test for symmetry, we therefore replace y in the equation of our relation by $-y$.   If the resulting equation is the same as the given one, the graph is symmetric about the X-axis.   In particular the graph is symmetric about the X-axis when y appears in the given equation to an *even* power only, for $y^{2k} = (-y)^{2k}$.

In a similar manner, a graph is symmetric about the Y-axis when replacement of x by $-x$ leaves the equation unchanged, e.g., when x occurs to an even power only.

Further, since a line joining (x,y) and $(-x,-y)$ passes through the origin and the distance from (x,y) to the origin is the same as the distance from $(-x,-y)$ to the origin, the graph will be symmetric about the origin if replacement of (x,y) with $(-x,-y)$ leaves the given equation unchanged.

Exercises

A   Examine $|y| - x = 0$, $y - |x| = 0$, $|x| + |y| - 1 = 0$ for symmetry.

B   Show that if there is symmetry with respect to both axes there is, necessarily, symmetry with respect to the origin, but not conversely.

Illustration 5

a   The graph of $x^2 - x + y^4 - 2y^2 - 6 = 0$ is symmetric about the X-axis, but not about the Y-axis or the origin.

b   The graph of $x^2 - x^4 + y - 5 = 0$ is symmetric about the Y-axis, but not about the X-axis or the origin.

c   The graph of $x^4 + 2x^2y^2 + y^4 - 10 = 0$ is symmetric about both axes and the origin.

d   The graph of $xy = 1$ is symmetric about the origin, but not about either axis.

asymptotes

Asymptotes   When we solve the given equation for x or y, we may get an expression which contains a variable in the denominator.   For example, we may have

$$y = \frac{x}{x - 1}$$

We have seen before that we cannot substitute $x = 1$ on the right, for this would make the denominator zero.   We can, however, let x take values nearer and nearer to 1 and see how the graph behaves.   Construct the table of values:

x	1.1	1.01	1.001	1.0001
y	11	101	1,001	10,001

It is clear that, as x approaches 1 from the right, y is becoming very large in the positive direction. Similarly, as x approaches 1 from the left, y becomes very large in the negative direction (Fig. 9.3).[1] The line $x = 1$ is now called a *vertical asymptote*.

Vertical asymptote

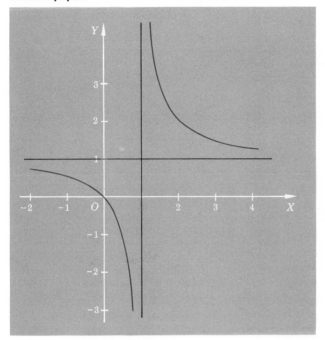

Figure 9.3 Values of x for which a denominator is zero but for which the corresponding numerator is not zero usually give vertical asymptotes. Similar values of y give horizontal asymptotes. Here $y = x/(x - 1)$.

If we solve the above equation for x, we obtain

$$x = \frac{y}{y - 1}$$

The same argument can now be applied to show that $y = 1$ is a *horizontal asymptote*.

To find asymptotes, the procedure is as follows: Solve for y and x if possible. Values of x or y which make the corresponding denominator zero correspond to vertical or horizontal asymptotes, provided that for these values the numerator is not also zero. The behavior of the graph near an asymptote must be determined by examining points near it, as was done above.

---

[1] The language here is very imprecise, but is the best that can be presented to you at this stage. More advanced textbooks write

$$\lim_{x \to 1^+} \frac{x}{x - 1} = +\infty \quad \text{and} \quad \lim_{x \to 1^-} \frac{x}{x - 1} = -\infty$$

and define these terms more precisely.

There is a more general definition of asymptote which applies to lines in other directions, but we shall not give it here.

Illustration 6   Find the horizontal and vertical asymptotes, if any, of

$$y = \frac{x(x - 1)}{x + 2}$$

Solution   Since the denominator is zero for $x = -2$, there is a vertical asymptote at $x = -2$. Solving for x, we find

$$x = \frac{1 + y \pm \sqrt{y^2 + 10y + 1}}{2}$$

Since y does not occur in the denominator, there are no horizontal asymptotes.

In the illustrations below we shall use these methods as needed to plot a number of graphs.

Illustration 7   Plot the graph of the function whose values are $y = 4x^2 - 3$.

Solution   To determine the x-intercepts, or the zeros of the function, we set $4x^2 - 3 = 0$ and compute $x = \pm \sqrt{3}/2$. The y-intercept is $y = -3$.

Since any x gives a real value of y, the domain is the entire X-axis. We solve for x to determine the range. We find that $x = \pm \frac{1}{2}\sqrt{y + 3}$. Therefore, the graph does not lie below the line $y = -3$. There is symmetry about the Y-axis.

Graph of $y = 4x^2 - 3$

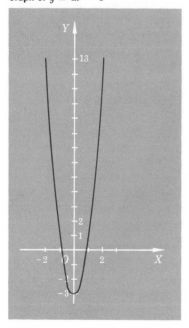

Figure 9.4
Before plotting points, determine the intercepts, symmetry, domain, range, and asymptotes of the graph.

We construct a short table of values:

x	0	± 1	± 2
y	-3	1	13

The graph is plotted in Fig. 9.4.

Illustration 8   Plot the graph of the function whose values are

$$y = \frac{x}{(x - 1)(x + 2)}$$

Solution   The origin is both the x-intercept and the y-intercept. The domain is the entire X-axis with the exception of $x = 1$ and $x = -2$, where the function is not defined. In order to find the range, we solve for x and find

$$x = \frac{1 - y \pm \sqrt{9y^2 - 2y + 1}}{2y}$$

Hence we must choose y so that $9y^2 - 2y + 1 \geq 0$. This is a quadratic inequality of the type discussed in Sec. 8.4. Since $b^2 - 4ac = -32$, the equation $9y^2 - 2y + 1 = 0$ has nonreal solutions, and so from Theorem 11, Sec. 8.4, the inequality is satisfied for all y, and there is no restriction on y. The appearance of y in the denominator suggests, however, that we must exclude $y = 0$. But when we put $x = 0$ in the original equation, we get $y = 0$; so it is not to be excluded from the range.

There is no symmetry, but we find a vertical asymptote at $x = 1$ and at $x = -2$ and a horizontal asymptote at $y = 0$. We construct the following table of values:

x	0	1	2	3	-1	-2	-3	-4	$\tfrac{1}{2}$	$\tfrac{3}{2}$	$-\tfrac{1}{2}$	$-\tfrac{3}{2}$	$-\tfrac{5}{2}$
y	0	—	$\tfrac{1}{2}$	$\tfrac{3}{10}$	$\tfrac{1}{2}$	—	$-\tfrac{3}{4}$	$-\tfrac{2}{5}$	$-\tfrac{2}{5}$	$\tfrac{6}{7}$	$\tfrac{2}{9}$	$\tfrac{6}{5}$	$-\tfrac{10}{7}$

The graph is plotted in Fig. 9.5.

Graph of $y = x/[(x - 1)(x + 2)]$

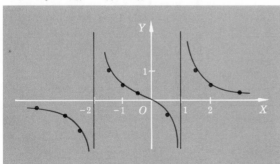

Figure 9.5
A curve may cross its own asymptote. Here the curve crosses its horizontal asymptote at (0,0).

Illustration 9 Plot the graph of each of the functions defined by the equations (a) $y = \sqrt{x^3 - x^2}$ and (b) $y = -\sqrt{x^3 - x^2}$.

Solution The domain of definition is restricted to those values of x for which $x^3 - x^2 \geq 0$. This requires that $x^2(x - 1) \geq 0$. This is satisfied if $x \geq 1$ or $x = 0$. The domain of each is therefore $1 \leq x < \infty$ and also $x = 0$. When $x = 0$, $y = 0$. The range of (a) is $0 \leq y < \infty$; the range of (b) is $-\infty < y \leq 0$.

For the zeros, we set $x^3 - x^2 = 0$; that is, $x^2(x - 1) = 0$ or $x = 0,1$. Construct tables.

For $y = \sqrt{x^3 - x^2}$

x	0	1	2	3
y	0	0	2	$3\sqrt{2}$

For $y = -\sqrt{x^3 - x^2}$

x	0	1	2	3
y	0	0	-2	$-3\sqrt{2}$

Each graph has an isolated point $(0,0)$. Each graph is shown in Fig. 9.6.

Graph of $y = \pm\sqrt{x^3 - x^2}$

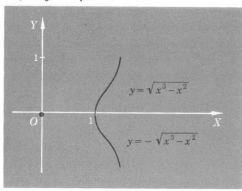

Figure 9.6
Watch for isolated points of graphs, such as $(0,0)$ in this figure.

Problems 9.4

In Probs. 1 to 30 plot the graphs of the relations.

1	$y = 2x - 1$	**2**	$y = 3 - x$
3	$y = 3$	**4**	$y = x^2 - 1$
5	$y = x^2 + 1$	**6**	$y = 2\sqrt{x}$
7	$y = -1/x$	**8**	$y = \sqrt{2 - x}$
9	$y = (x + 3)(1 - x)$	**10**	$y = 2 - x - x^2$
11	$y = -8x^2 + 12x + 8$	**12**	$y = x^2 + x + 1$
13	$y = -\sqrt{1 + x^2}$	**14**	$y = \sqrt{1 + x^2}$
15	$y^2 = x(x^2 - 9)$	**16**	$y^2 = x(x^2 - 4)$
17	$x^2 + y^2 = -x^3$	**18**	$x^2 + y^2 = x^3$
19	$y^3 = x^2$	**20**	$y^2 = x^3$
21	$x = -y^4$	**22**	$y = x^4$

23 $y = \frac{3}{2}\sqrt{x}$　　　　　　　　　**24** $x^4 + y^4 = 1$

25 $y = \dfrac{(x+1)(x-1)}{x-1}$　　　　**26** $y = 2/\sqrt{x+1}$

27 $y = x(x-1)(x+1)$　　　　**28** $y = 1/x^3$

29 $y = 2\sqrt{4-x^2}$　　　　　　**30** $y = \dfrac{x-1}{x(x+1)}$

In Probs. 31 to 42 analyze and sketch the graphs of the relations.

31 $4x^2 + 4y^2 = 1$　　　　　　**32** $x^2 + y^2 = 1$
33 $x^2 - y^2 = 1$　　　　　　　**34** $x^2 - y^2 = 0$
35 $x^2 - 4y^2 = 1$　　　　　　**36** $4x^2 - y^2 = 1$
37 $y^2 = 9x$　　　　　　　　　**38** $y^2 = 4x$
39 $y^2 = (x-1)(x+2)(x+3)$　　**40** $y = (x-1)(x+2)(x+3)$

41 $y^2 = \dfrac{x(x-2)}{x+3}$　　　　　**42** $y = \dfrac{x(x-2)}{x+3}$

43 Can the graph of a function be symmetric with respect to the Y-axis?
44 Can the graph of a function be symmetric with respect to the X-axis?

9.5　Graphs (Continued)

When the given relation is defined by an inequality or in some other way, additional ideas may be needed to plot its graph. We have seen (Secs. 8.5 and 8.7) how to plot the graphs of linear and quadratic inequalities. The following illustration extends this idea.

Illustration 1　Plot the graph of the relation defined by the inequality $x^2 + y^2 < 9$.
We first plot the graph of $x^2 + y^2 = 9$, which is a circle (Fig. 9.7).

Graph of $x^2 + y^2 < 9$

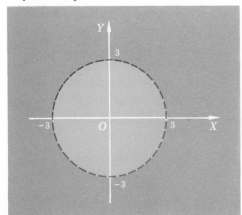

Figure 9.7
The graph of a relation may be a region in the plane.

This divides the plane into two regions, in one of which $x^2 + y^2 > 9$ and in the other $x^2 + y^2 < 9$. To see which region we want, test a typical point, say $(0,1)$. At this point the inequality becomes $0^2 + 1^2 < 9$, which is true. Hence this is the desired region, and we color the graph as shown.

When the relation is defined in a different way, we must use methods suitable to the case in hand.

Illustration 2   Plot the graph of the function f defined to be the set

$$\{(x,y) \mid x \text{ is a positive integer} > 1, \text{ and } y \text{ is the least prime not less than } x\}$$

The domain is the set of positive integers > 1, and the range is the set of primes. The graph is a discrete set of points which must not be joined by a continuous curve (Fig. 9.8).

Discrete set of points

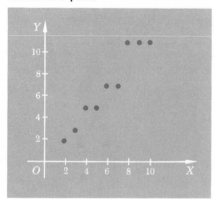

Figure 9.8
The graph of a relation may be a discrete set of points. This is the graph of the set $\{(x,y) \mid x$ is a positive integer > 1 and y is the least prime not less than $x\}$.

Problems 9.5

In Probs. 1 to 26 plot the graphs of the relations.

1	$x + y^2 < 0$	**2**	$x + y^2 = 0$
3	$x - y^2 > 0$	**4**	$x - y^2 = 0$
5	$x^2 - y^2 = 0$	**6**	$x^2 + y^2 = 0$
7	$x + y^4 \geq 0$	**8**	$x - y^4 \leq 0$
9	$y + x > 0$	**10**	$y \geq x$
11	$3x + 2y - 12 < 0$	**12**	$3x + 2y - 12 > 0$
13	$|y| = |x|$	**14**	$|y| = x$
15	$x^2 + y^2 \geq 1$	**16**	$x^2 + y^2 < 1$
17	$|x| + |y| \leq 1$	**18**	$|x| + |y| = 1$

19 $y = \begin{cases} 2, & x \text{ an integer} \\ -1, & \text{otherwise} \end{cases}$

20 $y = \begin{cases} 1 + x, & x > 0 \\ x, & x \leq 0 \end{cases}$

21 $y = \begin{cases} 1, & x \text{ rational} \\ -1, & x \text{ irrational} \end{cases}$

22 $y = \begin{cases} 0, & x \text{ an even integer} \\ 1, & x \text{ an odd integer} \\ \text{otherwise undefined} \end{cases}$

23 $y = x - |x|$

24 $y = |x|$

25 Let $[x]$ stand for the greatest integer not exceeding x. Plot $y = [x]$.
26 $y = |x - 1|$
27 The rate of postage on first-class letters is 5 cents per ounce or fraction thereof. This defines a function. Plot it.

9.6 Inverse Functions

A function $f:(x,y)$ is a set of ordered pairs such that no two of the ordered pairs have the same first element x. Several ordered pairs could have the same second element y, however. If a function f is of such character that no two pairs have the same second element, then there exists a function f^{-1} called the *inverse function* of f defined below.

Definition Given the function f such that no two of its ordered pairs have the same second element, the inverse function f^{-1} is the set of ordered pairs obtained from f by interchanging in each ordered pair the first and second elements.

given
$f:X \to Y$
then
$f^{-1}:Y \to X$

Thus the function f has elements (ordered pairs) of the form (x_1,y_1), (x_2,y_2), ..., while the inverse function has elements (ordered pairs) of the form (y_1,x_1), (y_2,x_2), ... We may write $f:X \to Y$ and $f^{-1}:Y \to X$. The range of f is the domain of f^{-1}, and the domain of f is the range of f^{-1}.

This notion is well illustrated by Fig. 9.9. We begin with point $R(0,y)$ on the Y-axis and draw a horizontal line which meets the graph in one or more points Q. If there is just one such point (as in Fig. 9.9a), draw QP perpendicular to the X-axis and determine $P(x,0)$. Thus we have a mapping $Y \to X$, which is the inverse function f^{-1}. If, however, there are several intersections Q (Fig. 9.9b), these determine several points P. Two or more values of x are then assigned to the given y, and the relationship

Examples of existence and nonexistence of inverse functions

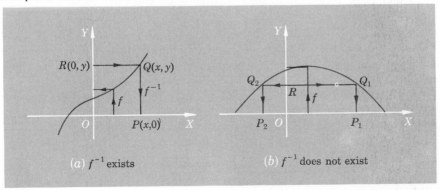

(a) f^{-1} exists

(b) f^{-1} does not exist

Figure 9.9 A function has an inverse if and only if no horizontal line meets its graph in more than one point.

is not a function. In this case f^{-1} is not defined. Note that, when f^{-1} exists, the graphs of f and f^{-1} are identical, for the X- and Y-axes are not to be interchanged when we graph f^{-1}.

If f is given by a simple formula $y = f(x)$, we can often obtain f^{-1} by solving this for x so that $x = f^{-1}(y)$. There are a number of difficulties with this procedure, which will be clarified by the following illustrations.

Illustration 1 Let f be given by $y = 3x + 1$ and have the domain $\{x \,|\, 0 \leq x \leq 1\}$. The range of f is then $\{y \,|\, 1 \leq y \leq 4\}$. Find f^{-1} and its domain and range.

Solution We solve the given equation for x and find

$$x = \frac{y - 1}{3}$$

This is the inverse function $f^{-1}:(y,x)$. Its domain is $\{y \,|\, 1 \leq y \leq 4\}$; its range is $\{x \,|\, 0 \leq x \leq 1\}$.

Illustration 2 Let f be defined by $y = \frac{1}{2}\sqrt{4 - x^2}$ with the given domain $\{x \,|\, -2 \leq x \leq 0\}$ and range $\{y \,|\, 0 \leq y \leq 1\}$. Find f^{-1}, its domain and range.

Solution In order to solve for y we square both sides and obtain

$$4y^2 = 4 - x^2$$

This process is risky, for $4y^2 = 4 - x^2$ is also obtained by squaring $y = -\frac{1}{2}\sqrt{4 - x^2}$, which is not the given function. With our fingers crossed, we now solve for x. The result is

$$x = \pm 2\sqrt{1 - y^2}$$

This is not exactly what we want, for we can use only one sign. We recall that the domain of f is $\{x \,|\, -2 \leq x \leq 0\}$, and this tells us to use the minus sign. The inverse function f^{-1} is therefore given by

$$x = -2\sqrt{1 - y^2}$$

Its domain is $\{y \,|\, 0 \leq y \leq 1\}$, and its range is $\{x \,|\, -2 \leq x \leq 0\}$. The common graph of f and f^{-1} is given in Fig. 9.10.

Graph of function and its inverse

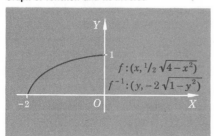

$f:(x, \frac{1}{2}\sqrt{4 - x^2})$

$f^{-1}:(y, -2\sqrt{1 - y^2})$

Figure 9.10

$f:X \to Y$ and $f^{-1}:Y \to X$ have identical graphs.

Notice that if we had chosen the maximum domain for f, namely, $\{x|-2 \leq x \leq 2\}$, then f^{-1} would not have existed.

restriction
of domain

This illustration emphasizes an important point which will come up later on: *If f does not have an inverse, we may be able to restrict its domain so that the restricted function does have an inverse.*

Exercise

A The function $y = x^2(-\infty < x < \infty)$ does not have an inverse. Find a domain for x such that the restricted function does have an inverse. How many domains can you find?

Illustration 3 Let f be defined by $y = 2^x$, where the domain is $\{x|-\infty < x < \infty\}$ and the corresponding range is $\{y|0 < y < \infty\}$. Find f^{-1}.

Graph of $y = 2^x$

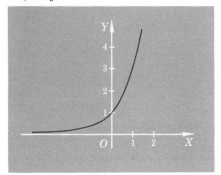

Figure 9.11
The inverse of a familiar function may be a new, unfamiliar function.

Solution Since the graph of $y = 2^x$ rises steadily as x increases (Fig. 9.11), no two values of x give the same value of y. So the inverse f^{-1} clearly exists. The trouble is that we do not know how to solve for x in terms of y by means of any formula. The function f^{-1} is therefore a new function unnamed at present. If we have frequent occasion to refer to this function, we shall find it convenient to give it a name and to investigate its properties. This is indeed a common method of obtaining new functions from known functions. You may already know the name of f^{-1} when $f(x) = 2^x$. It is called the *logarithm of y to the base 2*, written $\log_2 y$. We shall study this function in detail in Chap. 11.

This illustration emphasizes another point of importance: *If a function f has an inverse which cannot be calculated by elementary means, we shall often give this inverse a name and add it to our list of useful functions.*

Exercises

B Give a definition of $\sqrt[3]{y}(y > 0)$ as the inverse of some function.
C If you have studied trigonometry, show that

$$y = \sin x, \left(-\frac{\pi}{2} \le x \le \frac{\pi}{2}\right)$$

has an inverse. What is the name of this inverse?

The example of $y = 2^x$ in Illustration 3 suggests the following new term.

Definition   A function f whose domain and range are subsets of the reals is *strictly monotone increasing* if and only if, for every pair x_1 and x_2 such that $x_2 > x_1$, we have $f(x_2) > f(x_1)$. Similarly, f is *strictly monotone decreasing* if and only if, for every pair x_1 and x_2 such that $x_2 > x_1$, we have $f(x_2) < f(x_1)$.

monotone function

This enables us to state the following theorem.

Theorem 1   A function f whose domain and range are subsets of the reals has an inverse if it is either strictly monotone increasing or strictly monotone decreasing.

Proof   We must show that there are no two pairs (x_1,y) and (x_2,y) with the same second element and different first elements. This, however, is obviously true for strictly monotone functions.

Remark   The converse of this theorem is true if we add the hypothesis that f is defined in an interval in which it is continuous in a sense defined in books on calculus.

The process of defining new functions as the inverses of known functions introduces a further complication. In Illustration 3, we defined $x = \log_2 y$ as the inverse of $y = 2^x$. Now we should like to forget about the origin of this new function and treat it like any other function. Although it is not logically necessary, we generally use x to represent the independent variable and y to represent the dependent variable, but these roles are reversed in the case of $x = \log_2 y$. In order to conform to the usual practice, we therefore interchange the names of our variables and write $y = \log_2 x$. In general, the inverse of $y = f(x)$ is $x = f^{-1}(y)$, which we write as $y = f^{-1}(x)$ after interchanging variables. The graph of $y = \log_2 x$ is obtained from the common graph of $x = \log_2 y$ and $y = 2^x$ by reflection in the line $y = x$ (Fig. 9.12). The same procedure is valid in connection with other inverse functions, $y = f(x)$ and $y = f^{-1}(x)$, and we shall employ it in Chap. 11.

interchange of variables

In order to introduce a final important property of inverse functions, let us define the identity function whose domain and range are the same set X.

Graph of function and its inverse

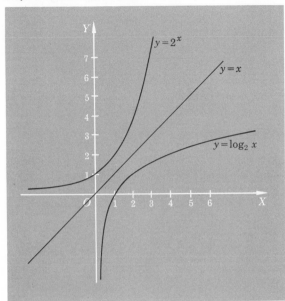

Figure 9.12
By interchanging x and y we may regard f^{-1} as a function $f^{-1}:X \to Y$. Then the graphs of $f:X \to Y$ and $f^{-1}:X \to Y$ are symmetrical about the line $y = x$.

Definition The function $E:X \to X$ whose elements are the ordered pairs (x,x) is called the identity function. This function maps each x into itself. (See Fig. 9.13.)

identity
function

The identity function

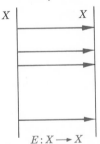

Figure 9.13
The identity function $E:X \to X$ consists of the ordered pairs (x,x).

Now let us consider a function $f:X \to Y$ and its inverse $f^{-1}:Y \to X$. The composite function $f^{-1}(f)$ maps each x into some y (under f) and then back into itself (under f^{-1}). We see then that

$$f^{-1}(f) = E \quad \text{and similarly} \quad f(f^{-1}) = E$$

The composite of f and f^{-1}

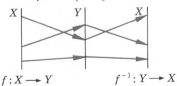

Figure 9.14
$f^{-1}(f) = E$.

Illustration 4   In Illustration 1, $f(x) = 3x + 1$ and $f^{-1}(y) = (y - 1)/3$. Hence

$$f^{-1}(f(x)) = \frac{(3x + 1) - 1}{3} = x$$

Also

$$f(f^{-1}(x)) = 3\left(\frac{x - 1}{3}\right) + 1 = x$$

Exercise

D   In Illustration 2, verify that $f(f^{-1}) = f^{-1}(f) = E$.

Problems 9.6

In Probs. 1 to 12 find f^{-1} if it exists, its domain, and range, and sketch $y = f(x)$, $y = f^{-1}(x)$.

1	$y = x + 1$	**2**	$y = 3x + 7$
3	$y = ax + b$	**4**	$y = ax - b$
5	$y = x$	**6**	$y = -x$
7	$y = x^2$, $x =$ all reals	**8**	$y = x^2$, $x \geq 0$
9	$y = x^3$, $x =$ all reals	**10**	$y = x^3$, $x \leq 0$
11	$y = \sqrt{1 - x^2}$, $0 \leq x \leq 1$	**12**	$y = \sqrt{x^2 - 1}$, $x \geq 1$

13   Show that the graphs of $y = f(x)$ and $y = f^{-1}(x)$ are symmetric with respect to the line $y = x$; that is, one is the mirror image of the other, the mirror being the line $y = x$.

9.7   Functions Derived from Equations

At first sight it may seem that any equation can be used to define a function by solving it for one of the variables. This process, however, often has a number of difficulties which are suggested by the illustrations below.

Illustration 1   The equation $2x - 3y + 1 = 0$ is called a "linear equation" because the pairs (x,y) which satisfy it lie on a straight line. From this equation we can derive two functions:

$$y = f(x) = \frac{2x + 1}{3} \qquad x = g(y) = \frac{-1 + 3y}{2}$$

Exercise

A   Show that $f^{-1} = g$.

Illustration 2   The equation $s = 16t^2$ gives the distance s in feet through which a body falls from rest under the influence of gravity in t sec. As such it defines a function. We may ask, however: "How long does it take for the body to fall 64 ft?" To answer this, we solve for t:

$$t^2 = \frac{s}{16} \qquad t = \pm \frac{1}{4}\sqrt{s}$$

This gives two functions defined by: $t = \frac{1}{4}\sqrt{s}$; $t = -\frac{1}{4}\sqrt{s}$. In terms of the physical situation only the first makes *practical* sense, but *both* make *mathematical* sense. Therefore we choose $t = \frac{1}{4}\sqrt{s}$, put $s = 64$, and find $t = 2$.

Exercise

B Are there any physical situations in which $t = -\frac{1}{4}\sqrt{s}$ makes *practical* sense?

This illustration makes the point that, although an equation may lead to several functions, not all of these necessarily have meaning in a practical situation. You will have to use your head and discard those which are nonsense.

Illustration 3 Consider the equation $x^2 + y^2 = 4$, which represents a circle of radius 2. If we solve for y, we obtain $y = \pm\sqrt{4 - x^2}$. Of the

Graph of $x^2 + y^2 = 4$

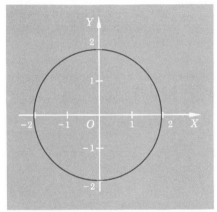

Figure 9.15
Two or more functions may be derived from an equation.

many functions which can be obtained from this, two have outstanding importance:

$$y = f(x) = \sqrt{4 - x^2} \qquad -2 \le x \le 2$$
$$y = g(x) = -\sqrt{4 - x^2} \qquad -2 \le x \le 2$$

The graph of f is the upper semicircle, and the graph of g is the lower semicircle.

Exercises

C Solve $x^2 + y^2 = 4$ in Illustration 3 for x, and describe the graphs of the two functions so obtained: Call these F and G.

D Does F^{-1} equal f or g?

Illustration 4   The equation $x^2 + xy + 4 = 0$ is quadratic in x and thus has the solution

$$x = \frac{-y \pm \sqrt{y^2 - 16}}{2}$$

This yields two functions defined by

$$x = f(y) = \frac{-y + \sqrt{y^2 - 16}}{2} \qquad |y| \geq 4$$

$$x = g(y) = \frac{-y - \sqrt{y^2 - 16}}{2} \qquad |y| \geq 4$$

When we solve for y, we obtain

$$y = -\frac{x^2 + 4}{x}$$

which gives the function h, where

$$y = h(x) = -\frac{x^2 + 4}{x} \qquad x \neq 0$$

Exercise

E   What is the domain of $f \circ h$? Show that in this domain $f \circ h = E$. Answer the same questions for $g \circ h$.

All these functions derived from equations have a common property: their elements (x,y) satisfy the equation. This suggests the more general definition.

Definition   If the elements (x,y) of a function f satisfy an equation in x and y, the function f is said to be *derived* from this equation.

Graph of $x^5y + xy^5 = 2$

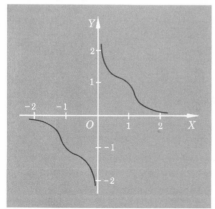

Figure 9.16
Functions may be derived from equations even though we cannot solve directly for x or y.

implicit
function

In many textbooks and older works a function thus derived from an equation is said to be given "implicitly" by the equation. The functions themselves are called "implicit functions".

In the examples given above the functions derived from an equation were obtained by solving for one of the variables. It is important to note that derived functions may exist even when we are unable to carry through such a solution. We shall sometimes wish to consider such functions, an example of which is given below.

Illustration 5 The equation $x^5y + xy^5 = 2$ has a graph given by Fig. 9.16. There is no simple way of solving this equation for x or y, but the graph indicates that functions $y = f(x)$ and $x = g(y)$ exist which are derived from this equation.

Problems 9.7

In Probs. 1 to 14 find some function derived from the given equation, and state its domain and range.

1	$4x - 3y = 4$	**2**	$3x - 4y = 5$
3	$2x^2 - y^2 = 1$	**4**	$x^2 - 2y^2 = 1$
5	$x^2y + y = 1$	**6**	$x^2y + x = 1$
7	$|x| - |y| = 0$	**8**	$|x| + |y| = 1$

9 $s = -16t^2 + v_0t$, where s is the distance at time t above ground of a particle fired at $t = 0$ vertically upward with initial velocity v_0.

10 $v^2 - 2gs = 0$, where v is the velocity of a body falling from rest, s is the distance fallen, and g is a given positive constant.

11	$x + y \geq 1, x \geq 0, y \geq 0$	**12**	$2^x = 2^y$
13	$x < y$	**14**	$|x| < y$

9.8 Supply and Demand Functions

In this section and the next one we shall apply the ideas of this chapter to the solution of some simple problems in economics dealing with supply and demand functions.

The demand for a particular commodity depends upon many factors in the market situation which influence the consumers to buy the quantity of this commodity that they do. In order to keep our discussion simple, we shall assume that the demand depends solely on the price of the commodity so that the other variables remain constant. We shall further assume that each individual consumer has complete discretion over the quantity of the product which he demands, but has no influential power over the producers, the prices at which they offer goods, or other consumers. Under these circumstances the demand for the product is a function of its price, called the *demand function*. We write

$$q = D(p)$$

demand
function

where q is the quantity of the product which is sold at the price p.

Although prices are expressed as an integral number of cents or dollars, for convenience we assume that the price can be any nonnegative real number. Hence the domain of D is the set $\{p \mid p \geq 0\}$. Similarly, we take the range of D to be the set $\{q \mid q \geq 0\}$. For practical purposes, however, the domain and range of D are usually limited to some subsets of these. In most cases the demand decreases as the price increases. Thus D is a strictly decreasing monotone function. We shall ignore the minor exception to this conclusion which holds for products with snob appeal, for which the demand increases as the price rises. Since we have assumed that D is strictly monotone, it has an inverse, namely, $p = D^{-1}(q)$ which tells us the price at which a quantity q can be sold.

In summary, we have the following definition:

Definition   The demand for a product in terms of its price is given by the *demand function* $q = D(p)$. The domain and range of D are subsets of the set of nonnegative real numbers. D is strictly decreasing, and hence has an inverse $p = D^{-1}(q)$.

demand curve   The graph of the demand function is called the *demand curve*. In economic theory it is customary to reverse our previous convention and to plot the independent variable p on the vertical axis and the dependent variable q on the horizontal axis. Thus the demand curve is really the graph of the inverse function $p = D^{-1}(q)$. Because of the restrictions on the domain and range of D the demand curve lies entirely in the first quadrant.

Illustration 1   The demand function for bread has been found to be

$$D: q = \frac{75 - 2p}{5}$$

where p is in units of cents/lb and q is in billions of lb/yr. The demand curve is plotted in Fig. 9.17. From this curve we can see that, if the price ever rose to 37.5 cents/lb, the demand would be zero and that, if bread were free, the demand would be 15 billion lb/yr. These extreme positions are hardly useful in practical economics.

Although actual demand functions must be determined empirically, it is convenient to approximate them by relatively simple mathematical formulas. We shall use functions whose graphs are straight lines, parabolas, and hyperbolas. For convenience, we represent a demand curve by an equation such as

$$D: 2p + 5q = 75$$

The demand function is obtained from the equation by solving for q:

$$q = \frac{75 - 2p}{5}$$

as in Illustration 1. The inverse D^{-1} is similarly found by solving for p.

Demand curve for bread

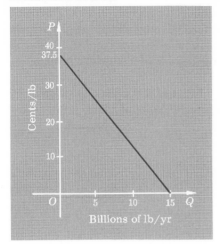

Figure 9.17
The demand function is $D: q = (75 - 2p)/5$.

Illustration 2   The demand curve for cigarettes is given by the equation

$$D: p + 16 = \frac{96}{q}$$

where p is in units of dollars per 1,000 cigarettes and q is in billions of cigarettes/yr.   The demand curve is plotted in Fig. 9.18.   In plotting the hyperbola in Fig. 9.18 it is convenient to forget the economic restriction that $p \geq 0$ and $q \geq 0$ and to plot the entire geometric curve.   We see that there is a horizontal asymptote at $p = -16$.   The portion of the curve which lies outside the first quadrant is dashed to indicate that it has no economic meaning.

Demand curve for cigarettes

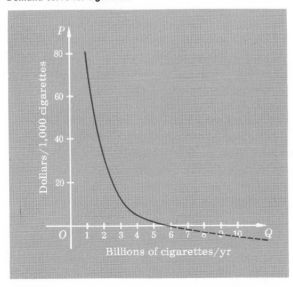

Figure 9.18
The equation of the demand curve is $D: p + 16 = 96/q$.

Demand functions and curves can represent either one consumer's demand for a certain product or an aggregation of all consumers' demand for that product throughout the economy. The aggregation process relies heavily on the assumption that only one, strictly homogeneous, product is being considered and that it is being both produced and consumed by many economic units. It is very important for an industry to understand the nature of the aggregate consumer demand curve which it faces.

supply
function
A supply function, $q = S(p)$, expresses the number of units q of a product which the producer is willing to supply at a given price p. Again we assume that all other variables are constant and that there is pure competition among the sellers. The domain and range of $S(p)$ are subsets of the nonnegative real numbers. As p increases, more units of the product will be placed on the market, and so $S(p)$ is strictly monotonic increasing. Here we are ignoring the special situation in which production is at full capacity so that q cannot increase no matter what the price. The graph of $q = S(p)$, or more properly of $p = S^{-1}(q)$, is called the *supply curve*. As in the case of demand functions, we shall limit ourselves here to supply curves which are lines, parabolas, or hyperbolas. In summary, we have the definition:

Definition The supply of a product in terms of its price is given by the *supply function $q = S(p)$*. The domain and range of S are subsets of the set of nonnegative real numbers. S is strictly increasing and hence has an inverse $p = S^{-1}(q)$.

Illustration 3 The supply function for bread has been found to be

$$q = \frac{9p - 90}{5}$$

where p is in units of cents/lb and q is in billions of lb/yr. Plot the supply curve (Fig. 9.19).

Supply curve for bread

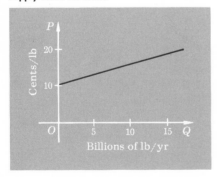

Figure 9.19
The supply function is S: $q = (9p - 90)/5$.
Note that if $p < 10$, no bread will be offered for sale.

market
equilibrium
When we examine a purely competitive market from both the buyers' and the sellers' points of view simultaneously, we find a condition of market equilibrium at the price at which the supply is equal to the demand.

This point is, then, the point of intersection of the supply and demand curves. It may be obtained either algebraically or graphically.

Illustration 4 Find the point of market equilibrium for bread assuming the data of illustrations 1 and 3.

We can rewrite the equations of the demand and supply curves as the linear equations

$$D: 2p + 5q = 75$$
$$S: 9p - 5q = 90$$

The solution of this system of equations is $q = 9$, $p = 15$. The graph is given in Fig. 9.20.

Market equilibrium for bread

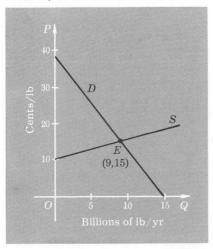

Figure 9.20
The point of market equilibrium is $q = 9$,
$p = 15$. D: $2p + 5q = 75$ S: $9p - 5q = 90$

In Illustration 4 suppose the actual market price of bread was 20 cents/lb. At this price level the quantity supplied (18 billion lb) would be greater than the quantity demanded (7 billion lb). The market would be flooded with bread, and sellers would be forced to settle for lower prices in order to unload their surplus. If the price dropped below 15 cents/lb, the quantity demanded would exceed the quantity supplied, consumers would be willing to pay more for the scarce commodity, and the price would be forced up. The situation is illustrated by Table 1.

Table 1   Comparison of demand and supply for bread at various prices

p	$D(p)$	$S(p)$	Pressure on price
37.5	0	49.5	Falling
20	7	18	Falling
15	9	9	Neutral
10	11	0	Rising
0	15	0	Rising

Until market equilibrium is reached, quantities will be adjusted and prices will fluctuate in a continuous process. At equilibrium there is no internal impetus for change, and the market comes to rest. In markets such as the stock exchange and the Chicago Board of Trade, the fluctuations settle at equilibrium almost instantaneously.

Illustration 5 Find the point of market equilibrium for cigarettes given that the demand and supply curves are obtained from the equations

$$D: p + 16 = \frac{96}{q} \qquad S: 2p - 3q - 4 = 0$$

Units: $p = $ dollars/1,000 cigarettes; $q = $ billions of cigarettes/yr. The solution of this system of equations is $q = 4$, $p = 8$. The graph is given in Fig. 9.21.

Market equilibrium for cigarettes

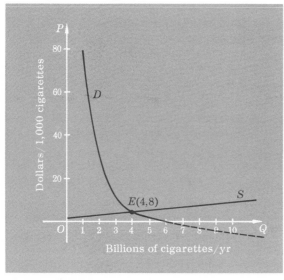

Figure 9.21
The point of market equilibrium is $q = 4$, $p = 8$.

$$D: p + 16 = \frac{96}{q}$$

$$S: 2p - 3q - 4 = 0$$

Problems 9.8

Problems 1 to 20 give equations from which demand and supply functions can be derived. It is understood that $p \geq 0$ and $q \geq 0$. Other restrictions on p and q are stated where they are necessary.

a Find the point of market equilibrium by algebraic means.

b Plot the demand and supply curves and find the point of market equilibrium graphically.

1 Crude oil

$$D: 2p + 3q = 27 \qquad S: 4p - 3q + 9 = 0$$

Units: $p = $ dollars/barrel; $q = $ millions of barrels/day.

2 Pies

$$D: 7p + 100q = 245 \qquad S: 7p - 150q = 70$$

Units: p = cents/lb; q = billions of lb/yr.

3 Raw cotton

$$D: p = \frac{264}{q} \qquad S: 3p - 22q + 198 = 0$$

Units: p = cents/lb; q = 10^3 bales/yr.

4 A particular movie at a particular theater

$$D: p = \frac{900}{q} \qquad S: q = 15$$

Units: p = cents/ticket; q = hundreds of tickets/day.

5 British draught beer

$$D: q^2 = -40(p - 20) \qquad S: 5p - 2q - 10 = 0$$

Units: p = £/per barrel; q = millions of barrels/yr.

6 Steel (pig iron)

$$D: p + 40 = \frac{6,000}{q} \qquad S: 5p - 2q + 100 = 0$$

Units: p = dollars/gross ton; q = millions of tons/yr.

7 Butter

$$D: p + 3q^2 - 70 = 0 \qquad S: p - 12q^2 - 10 = 0$$

Units: p = cents/lb; q = billions of lb/yr.

8 No. 2 cans, soda pop

$$D: 16p + 3q^2 - 720 = 0 \qquad S: p - 10 = \frac{23q^2}{64}$$

Units: p = dollars/1,000 cans; q = billions of cans/yr.

9 Sulfuric acid

$$D: p + 20 = \frac{1,000}{q} \qquad S: 5p - 2q - 50 = 0$$

Units: p = dollars/ton; q = millions of short tons/yr.

10 Benzene

$$D: p + q^2 - 4q - 36 = 0, \text{ where } q \geq 2 \qquad S: p - 8q + 24 = 0$$

Units: p = cents/gallon; q = hundreds of millions of gal/yr.

11 Donuts

$$D: p = \frac{150}{5q + 3} \qquad S: p - 8 = \frac{22}{11 - 25q}, \text{ where } q < 0.44$$

Units: p = cents/lb; q = billions of lb/yr.

12 Milk

$$D: pq + 9q - 540 = 0 \qquad S: p + 2 + \frac{250}{3q - 185} = 0, \quad \text{where} \quad q \le 61$$

Units: $p = $ dollars/100 lb; $q = $ billions of lb/yr.

13 Natural gas

$$D: pq + 6p - 840 = 0 \qquad S: p - 35q + 140 = 0$$

Units: $p = $ cents/thousand cu ft; $q = $ trillion cu ft/yr.

14 Brazilian coffee

$$D: p + 10q - 270 = 0 \qquad S: p^2 - 900q + 8{,}100 = 0, \text{ where } q \ge 9$$

Units: $p = $ cents/lb; $q = $ millions of bags/yr.

15 Airlines (passenger service)

$$D: pq + 12q - 432 = 0 \qquad S: 4p - 3q + 48 = 0$$

Units: $p = $ passenger revenue in cents/mile; $q = $ billions of passenger miles/yr.

16 Cake

$$D: p = \frac{350}{3q + 7} \qquad S: p = 20 + 15q$$

Units: $p = $ cents/lb; $q = $ billions of lb/yr.

17 Rayon (staple fiber)

$$D: 8p + q^2 + 12q = 400$$
$$S: p + q^2 - 23q + 90 = 0 \qquad \text{where } q \le 11.5$$

Units: $p = $ cents/lb; $q = $ millions of lb/yr.

18 Bituminous coal

$$D: pq + 17p - 374 = 0$$
$$S: p + 10 = \frac{810}{115 - 17q} \qquad \text{where } q < 6.8$$

Units: $p = $ dollars/ton; $q = $ hundreds of millions of tons/yr.

19 Aluminum (pig)

$$D: p^2 - 80p - 45q + 1{,}600 = 0 \qquad \text{where } p \le 40$$
$$S: p = -q^2 + 34q - 120 \qquad \text{where } q \le 17$$

Units: $p = $ dollars/lb; $q = $ millions of tons/yr.

20 Raw sugar

$$D: p + 20 = \frac{200}{q} \qquad S: p + 8 = \frac{208}{56 - 5q}, \text{ where } q < 11.2$$

Units: $p = $ cents/lb; $q = 10^6$ short tons/yr.

9.9 Shifts of Supply and Demand Curves

Once a market reaches equilibrium under a given set of conditions, there is no internal impetus for further movements along the supply and demand curves. If, however, we relax the limitation that everything else remains constant, new forces may act on the market, which will shift the demand and supply curves and thus disturb the point of market equilibrium. An example of such a force is the imposition of a tax on the product.

Illustration 1 Suppose that we start from the demand and supply curves for bread which were used in Sec. 9.8. These were given by the equations

$$D: 2p + 5q = 75$$
$$S: 9p - 5q = 90$$

Units: $p =$ cents/lb; $q =$ billions of lb/yr.

Now the government imposes a tax on the bakers of 6⅑ cents/lb on bread. What is the effect of this on the demand and supply curves and on the point of market equilibrium?

effect of a tax

There is no reason for the demand curve to change, for the consumers know nothing of the tax which the bakers are now paying. There will be, however, a major change in the supply curve. The inverse of the supply function, $p = S^{-1}(q) = (5q + 90)/9$, gives us the original price at which q units of bread will be offered for sale. The tax is now added to this price, so we have for $S^{-1}(q)$ the new expression

$$p = \frac{5q + 90}{9} + 6\frac{1}{9} = \frac{5q + 145}{9}$$

The equations for the demand and supply curves after the tax are therefore

$$D: 2p + 5q = 75$$
$$S: 9p - 5q = 145$$

The new point of market equilibrium is $q = 7, p = 20$.

Market equilibrium for bread after tax

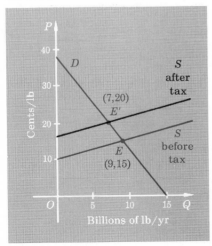

Figure 9.22
The point of market equilibrium after the tax is $q = 7, p = 20$. After the tax the equations are
D: $2p + 5q = 75$ S: $9p - 5q = 145$
The tax of 6⅑ cents/lb raises the supply curve by 6⅑ units and shifts the point of market equilibrium from E to E'.

Note that the price rise of bread (5 cents) did not reflect the entire tax (6⅑ cents). Part of the tax burden was absorbed by the bakers, for the bakers now receive only 13⁸⁄₉ cents/lb for their bread (13⁸⁄₉ = 20 − 6⅑) and also sell a smaller quantity.

total revenue

The total revenue received by the bakers is calculated by multiplying the price per unit times the number of units sold at the equilibrium price. Before taxes, the total revenue to the bakers is

$$9 \times 15 \times 10^9 \text{ cents} = \$1{,}350{,}000{,}000$$

After the tax the total revenue to the bakers is

$$7 \times (20 - 6\tfrac{1}{9}) \times 10^9 \text{ cents} \approx \$972{,}000{,}000$$

The income from the tax to the government is

$$7 \times 6\tfrac{1}{9} \times 10^9 \text{ cents} \approx \$428{,}000{,}000$$

A government subsidy of a product causes the supply curve to move down. A change in consumers' tastes for or against a product causes a similar shift in the demand curve to the right or to the left. In the next illustration we analyze the effect on the market for cigarettes of a disclosure of new evidence that cigarette smoking contributes to cancer.

Market equilibrium for cigarettes after disclosure

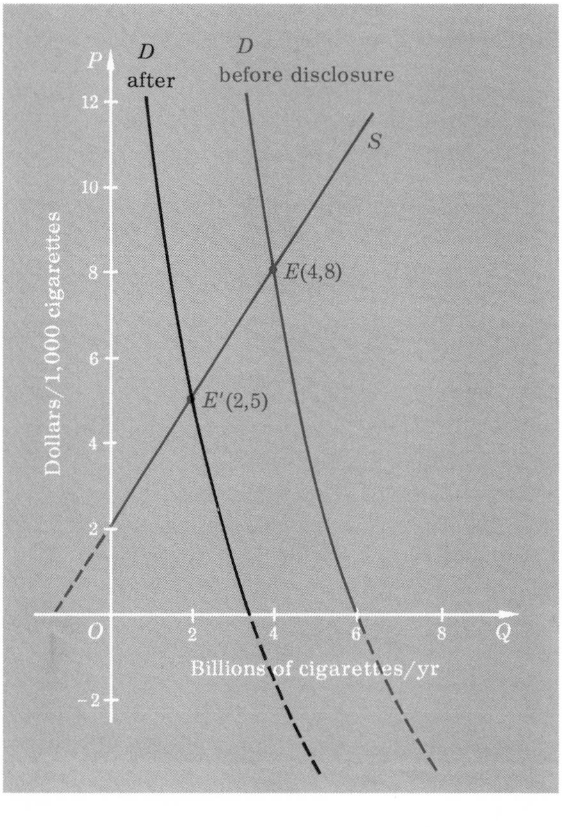

Figure 9.23
The point of market equilibrium after the disclosure is $q = 2$, $p = 5$. After the disclosure the equations are

$$D\text{: } q = \frac{384 - 18p}{7(p + 16)}$$

$$S\text{: } 2p - 3q - 4 = 0$$

The possible effect of cigarette smoking on cancer moves the demand curve by 2⁴⁄₇ units to the left and shifts the point of market equilibrium from E to E'.

Illustration 2   We recall the equations for the demand and supply curves for cigarettes given in the preceding section.

$$D: p + 16 = \frac{96}{q} \qquad S: 2p - 3q - 4 = 0$$

change in demand

Units: $p = $ dollars/1,000 cigarettes; $q = $ billions of cigarettes/yr.   The equilibrium point is $q = 4, p = 8$.   Because of the evidence of the effect of cigarette smoking on cancer, the demand drops by $2\frac{4}{7}$ billions of cigarettes per year.   To find the effect of this on the demand curve, we write $q = D(p)$ as

$$q = \frac{96}{p + 16}$$

The equation of the new demand curve is then

$$D: q = \frac{96}{p + 16} - 2\frac{4}{7} = \frac{384 - 18p}{7(p + 16)}$$

The supply curve remains unchanged.   The new equilibrium point is $q = 2, p = 5$ (Fig. 9.23).   The total revenue before the disclosure is

$$8 \times 10^{-3} \times 4 \times 10^9 \text{ dollars} = \$32,000,000$$

The total revenue after the disclosure is $10,000,000.

Problems 9.9

In Probs. 1 to 12 use the equations of the supply and demand curves of the various products as given in the corresponding problems in Sec. 9.8.

1   Crude oil.   Cause of shift: $0.75 per barrel additive tax.   Find the new equilibrium point and the total revenue received by the producers before and after the shift.

2   Pies.   Cause of shift: 5 cents/lb additive tax.   Find the new equilibrium point and the total revenue received by the bakers before and after the shift.

3   Raw cotton.   Cause of shift: A poor cotton crop results in 2.6×10^3 fewer bales being offered per year on the market at every possible price.   For revenue calculation, assume 1 bale = 478 lb.   Find the new equilibrium point and the total revenue before and after the shift.

4   Movie theater.   Cause of shift: Local authorities condemn the balcony of the theater, resulting in a loss of one-fifth of the seats.   Find the new equilibrium point and the total revenue before and after the shift.

5   British draught beer.   Cause of shift: During the particular year that the British Empire Games were held in Britain, the demand for draught beer was 13×10^6 barrels higher than normal at every possible price level.   Find the new equilibrium point and the total revenue before and after the shift.

6   Steel.   Cause of shift: Concentrated preparation for a World War would increase the demand for steel by $24\frac{1}{11}$ million tons per year. Find the new equilibrium point and the total revenue before and after the shift.

7   Butter
- **a**   What additive tax will increase the equilibrium price by 9 cents?
- **b**   What subsidy will decrease the original equilibrium price by 3 cents?
- **c**   Find the total revenue received by the producer before the shifts and after each shift.

8   No. 2 soda pop cans
- **a**   How much must the demand for canned pop shift to change the equilibrium point of cans to $q = 9.14$, $p = 40$?
- **b**   How much must it shift to change the equilibrium point of cans to $q = 7.46$, $p = 30$?
- **c**   Find the total revenue before the shifts and after each shift.
  Assume a constant supply function.

9   Sulfuric acid.   Cause of shift: Cleaning fluids using sulfuric acid are found to be dangerous in homes.   As a result, at every possible price $12\%_{19}$ million fewer short tons per year are used by the fluid producers as they substitute safer chemicals.   Find the new equilibrium point and the total revenue before and after the shift.

10   Benzene.   Cause of shift: A backward shift in the demand for refined oil eventually is reflected by the closing of many small refineries. Benzene, being a by-product of the refining process, suffers a $1\frac{7}{8}$ hundreds of millions gallons decrease in the amount supplied per year at every possible price.   Find the new equilibrium point and the total revenue before and after the shift.

11   Donuts.   Cause of shifts: The demand for donuts would increase x billion lb/yr if they were produced with hexagonal holes, but the high expense of production would force bakers to offer y billion lb/yr fewer at every possible price.   The new equilibrium point would be where 0.3 billion lb/yr are sold at 40 cents/lb.   Calculate x and y. Find the total revenue before and after the shifts.

12   Milk.   Cause of shifts: The repeal of all state liquor laws turns many young milk drinkers to other beverages.   The demand for milk decreased by 14 billion lb/yr.   Dairy farmers are faced with declining revenue.   They appeal for and receive a subsidy of $\$1\frac{1}{13}$ per 100 lb. Find the new point of equilibrium and the total revenue to the producers before and after these shifts.

The taxes considered so far have been given as a fixed amount for each unit of the product.   Now suppose that a tax is levied as a percent of the sales price of the product.   This will again alter the supply curve and result in a new equilibrium point.   In Probs. 13 to 16 write the equation of the new supply curve and find the new point of equilibrium.

13 Bread (see Illustration 1).

A tax of 20% of the sales price is added.

[HINT: The new supply curve is given by:

$$p = \frac{5q + 90}{9} + \frac{1}{5}\left(\frac{5q + 90}{9}\right) = \frac{2q + 36}{3}\Bigg]$$

14 Cigarettes (see Illustration 2).

A sales tax of 146⅔% ($= {}^{19}\!/_{13}$) of the sales price is added.

15 Crude oil (See Prob. 1, Sec. 9.8).

A tax of 300% of the sales price is added.

16 Pies (See Prob. 2, Sec. 9.8).

To encourage the production of pies, a subsidy of 25% of the sales price is granted to the producer by the government.

ALGEBRAIC FUNCTIONS

Chapter **10**

10.1　Introduction

In Chap. 2 we discussed some of the properties of real and complex numbers. The four operations of addition, subtraction (inverse of addition), multiplication, and division (inverse of multiplication) are called the *rational operations* of arithmetic. The *algebraic operations* of arithmetic include these and also those of taking roots and raising to powers. In this chapter the universal set will be the field of complex numbers which is closed under each of these processes.

10.2　Polynomial Functions

We now wish to apply three of the rational operations to a *variable x* in order to generate a very important special type of function. We restrict ourselves to addition, subtraction, and multiplication, and with these we build up such functions as those given by

1	2	**2**	$(3x - 5)(x + 6)$
3	$\sqrt{3}x^5 - \pi x^2 - 1$	**4**	$(2^{1/3}x - x^2)(x + 2)^3 + (7 - 3i)x - i$

But we do not obtain such functions as those defined by

5	$(1/x) - 3i$	**6**	$\sqrt{x}$
7	2^x	**8**	$|x|$

The functions illustrated by 1 through 4 are special cases of what are called polynomial functions according to the following definition.

Definition A function P is called a *polynomial function* if it is given by

polynomial
function

(1) $$y = P(x) = a_0 x^n + a_1 x^{n-1} + \cdots + a_{n-1} x + a_n$$

where n is a positive integer or zero and the coefficients $a_0, a_1, \ldots, a_n$ are complex numbers. Its domain and range are each the set of complex numbers.

Exercise

A Show that function 4 above can be written in the form (1).

We say that P is of *degree n*, provided that $a_0 \neq 0$. In the theory of polynomial functions it is customary to call the right-hand side of (1), namely, the expression

(2) $$a_0 x^n + a_1 x^{n-1} + \cdots + a_{n-1} x + a_n$$

a *polynomial*, which we designate by the symbol $P(x)$, read "polynomial in the variable x". $P(x)$ also stands for the value of P at x, but this should not lead to any confusion. Note that, since n may be zero, a constant is to be considered a polynomial (and a polynomial function). If $a_0 = n = 0$, then $P(x) = 0$ is called the zero polynomial.

We have stated that the domain of P is the set of complex numbers and that its range is the set of complex numbers. It is possible to discuss polynomial functions with other sets as domain and range and with other types of coefficients as well. For example, we may consider real polynomials in which x, y, and the a's are all real numbers. Or we may require that the a's be rational numbers or even integers and then let x and y be real or complex. All of these are special cases of the general definition given above. In discussing these special cases, we shall have to take great pains in stating the types of coefficients and variables which are under consideration.

Exercises

B Prove that the sum of two polynomials is a polynomial.

C Prove that the product of the two polynomials $(ax^2 + bx + c)$ and $(Ax^2 + Bx + C)$ is a polynomial. (As a matter of fact, the product of any two polynomials is a polynomial.)

D Show by an example that the quotient of two polynomials may be a polynomial. Find another example in which the quotient is not a polynomial.

E Prove that the composite $g \circ f$ of the two polynomial functions given by $f(x) = ax^2 + bx + c$, $g(x) = Ax^2 + Bx + C$ is a polynomial function. (As a matter of fact, the composites of any two polynomial functions are polynomial functions.)

10.3   Rational Functions

$R(x) =$
$P(x)/Q(x)$
      The next simplest type of function is a *rational function* which is so called because we now permit the use of division along with the other rational operations.

Definition   A function R defined by $y = R(x) = P(x)/Q(x)$, where $P(x)$ and $Q(x)$ are polynomials, is called a *rational function*.

The remarks made above about the domain and range of a polynomial function apply equally well to a rational function but here we must be a little careful: The function R is not defined at points where $Q(x) = 0$. This is made clear by the following illustrations.

Illustration 1

a $y = 1/x$ is not defined at $x = 0$.
b $y = (x - 1)/(x + 2)$ is not defined at $x = -2$.
c $y = 3x^3/[(x - 1)(x^2 + 1)]$ is not defined at either $x = 1$ or at $x = \pm i$.

Illustration 2

a $y = x/x$ is not defined at $x = 0$.   For other values of x, however, $x/x = 1$.   The two functions x/x and 1 are consequently not identical. This illustration brings up an important point: The cancellation of a common factor in the numerator and denominator may change the function involved.

b   As a similar example, consider the two rational functions

$$y = \frac{x(x - 1)}{x - 1} \qquad \text{and} \qquad y = x$$

These have the same values for $x \neq 1$, but at $x = 1$ the first is undefined, whereas the second has the value 1.   Hence they are different functions.

Illustration 3   Some functions which are not written in the explicit form $y = P(x)/Q(x)$ are nevertheless equivalent to rational functions.   Consider

$$y = \left(\frac{1}{x} - \frac{4}{x + 1}\right)\left(1 + \frac{1}{x - 1}\right)$$

$$= \frac{x + 1 - 4x}{x(x + 1)} \cdot \frac{x - 1 + 1}{x - 1}$$

$$= \frac{(1 - 3x)(x)}{x(x + 1)(x - 1)}$$

$$= \frac{-3x^2 + x}{x^3 - x}$$

In this bit of algebra we did not cancel out the x in numerator and denominator.   For the function

$$y = \frac{-3x + 1}{x^2 - 1}$$

is not equivalent to the given function. Why?

In Exercises A to D below assume that the sum, product, and composites of two polynomials are each polynomials.

Exercises

A Prove that the product of two rational functions is a rational function.
B Prove that the sum of two rational functions is a rational function.
C Prove that the quotient of two rational functions is a rational function.
D Prove that the composites of two rational functions are rational functions.

10.4 Explicit Algebraic Functions

Explicit algebraic functions constitute the next important class of functions. These include the polynomial and rational functions as special cases, and they are generated by a finite number of the algebraic operations. Thus the function whose values are given by

$$\frac{\sqrt{1 + x} - \sqrt[3]{x^5}}{\sqrt[6]{(2 + x - x^2)^3} - 8}$$

is an example of an explicit algebraic function. Because of the possible appearance of (even) roots in the equation defining such a function, it may happen that the value y of the function is real only when x is restricted to a very limited subset of the real numbers. For the example above, it is seen first of all that x (real) must be greater than or equal to -1 if $\sqrt{1 + x}$ in the numerator is to be real. Similarly in $\sqrt[6]{(2 + x - x^2)^3} - 8$ it must be true that $(2 + x - x^2)^3 > 8$; that is, $2 + x - x^2 > 2$ or $x(1 - x) > 0$. This says that x must lie between 0 and 1 exclusively. The domain of definition is therefore $0 < x < 1$.

Of course, if x and y are not required to be real, then the only values of x for which the above function is not defined are $x = 0$ and $x = 1$.

Our interest in the most general explicit algebraic functions will be confined mainly to their graphs.

Problems 10.4

In Probs. 1 to 4 state which of the following define polynomial functions. (Assume x real.)

1 a $y = \dfrac{1}{x}$ **2 a** $y = x + 5 - \dfrac{2}{i}$

   b $y = x$ **b** $y = x^2 + \sqrt{2}x - 1$

   c $y = 3^x + 2$ **c** $y = x^2 + 4x^{3/2} - 2$

3 **a** $y = 1 + \dfrac{x^2}{6} - \dfrac{x^5}{11}$

  b $y = \dfrac{x + 3i}{1 + i}$

  c $y = (x + a)^7$

4 **a** $f:(x,x^2)$

  b $g:(y,y^3)$

  c $h:(z,z^{-2})$

In Probs. 5 to 8 state which of the following define rational functions. (Assume x real.)

5 **a** $y = 3/x^2$

  b $y = \dfrac{x^{1/2}}{(x + 1)^{2/3}}$

  c $y = |x| - 3$

6 **a** $y = 3^x + 4$

  b $\dfrac{x}{x - i}$

  c $y = \dfrac{x^2 + 3x - 7}{5}$

7 **a** $y = \frac{3}{2}$

  b $y = \dfrac{x^2 - 8x + 1}{2x - 3}$

  c $y = 2^x + 3$

8 **a** $y = \dfrac{\sqrt{x} + 2}{\pi - 1}$

  b $y = \dfrac{x + 2}{x^2 + \sqrt{2}}$

  c $y = x^3/3^x$

In Probs. 9 to 12 state which of the following define explicit algebraic functions. (Assume x real.)

9 **a** $y = x^{1/3} + x^{1/2}$

  b $y = \dfrac{1}{\sqrt{x}}$

  c $y = 1 + \dfrac{x^2}{2i} - \dfrac{x^4}{\sqrt{2}}$

10 **a** $y = |x| - \sqrt{x}$

  b $y = \sqrt{1 - \sqrt{x}}$

  c $y = 1 + x + \dfrac{x^2}{2!} + \cdots + \dfrac{x^n}{n!}$

11 **a** $y = \begin{cases} a, & x \text{ rational} \\ b, & x \text{ irrational} \end{cases}$

  b $y = \sqrt{x} + 4^x$

  c $y = |\sqrt{x}|, \; x \geq 0$

12 **a** $f:(x,\sqrt{a})$, a real

  b $g:(\theta,\theta^{1/3})$, θ real

  c $h:(u,|u^4|)$, u real

In Probs. 13 to 18 state kind of function and domain (we assume x, y, etc., real and n a positive integer).

13 **a** $y = x^2 + \sqrt{2}\,x + 1$

  b $y = x^2 + \sqrt{2x} + 1$

  c $y = \dfrac{x^2 + \sqrt{2}\,x}{x^3}$

14 **a** $y = x^{1/4}$

  b $y = \sqrt{x^2 - 3x + 2}$

  c $y = \sqrt{x - 2} + \sqrt{x - 1}$

15 **a** $y = \dfrac{x}{(x + 2)(x - 3)}$

   b $y = \sqrt{x^2}$

   c $y = \sqrt{1 - x^2}$

16 **a** $\left(\dfrac{1}{x}\right)^n$

   b $2 + \dfrac{3}{4 - x}$

   c $\dfrac{\sqrt{x} + x}{\sqrt{x}}$

17 **a** $f:(x, 1 - \sqrt{x})$

   b $g:(y,(y + 1/y^2)^n)$

   c $h:(\theta, \sqrt{\theta + 1})$

18 **a** $f:(y,a)$

   b $g:(y, \sqrt{a - y^2})$

   c $h:(x, \sqrt{-x} + 5)$

10.5 Graphs and Continuity

We have already considered methods of plotting the graphs of functions and relations, or, what amounts to the same thing, of plotting the graph of an equation (Chap. 9). We must still rely upon intuition when we speak about the *continuity* of a function or about a *continuous* graph, but we wish at this time to make some pertinent remarks on the continuity of an algebraic function. For this discussion we restrict ourselves to the field of real numbers, since we plot only the real elements of a function. An element (x,y) is real when and only when x and y are both real. We shall refer indifferently to a continuous function or a continuous graph, the latter being merely descriptive geometric language.

continuity

Polynomial Functions A polynomial function, defined by $y = P(x)$, where $P(x)$ is a polynomial, is continuous everywhere. The graph of a polynomial function is a continuous curve. The domain of definition is the set of real numbers; the range is a subset of the real numbers (which could be the whole set). As an example, see Sec. 9.4, Illustration 7.

Definition The *zeros* of P are the values of x for which $P(x) = 0$.

zeros of a polynomial

Illustration 1 Sketch the graph of the polynomial function given by $y = x^4 - 2x^2$.

Solution For purposes of graphing we now consider the domain as the set of real numbers. Since

$$x^4 - 2x^2 = x^2(x^2 - 2)$$

the zeros are seen to be $x = 0, 0, \pm\sqrt{2}$. (For the factor x^2, we write $x = 0, 0$; see Theorem 4, and following remark.)

The graph is continuous everywhere. It is symmetric with respect to the Y-axis since x appears to even powers only. Some values of the function defined by this equation are given in the following table.

x	-2	$-\sqrt{2}$	-1	0	1	$\sqrt{2}$	2
y	8	0	-1	0	-1	0	8

The graph is shown in Fig. 10.1. The least value of y occurs when $x = \pm 1$, although we cannot prove this without the use of calculus. The range is $-1 \le y < \infty$.

Graph of polynomial function

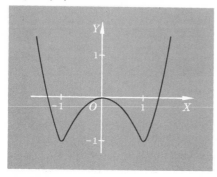

Figure 10.1
The graph of a polynomial function is continuous. Here $y = x^4 - 2x^2$.

Rational Functions   A rational function, defined by

$$y = \frac{P(x)}{Q(x)}$$

where $P(x)$ and $Q(x)$ are polynomials and $Q(x)$ is not the zero polynomial, is continuous everywhere with the exception of, at most, a finite number of isolated values of x, namely, those for which $Q(x) = 0$. These values must be excluded from the domain. The range is a subset of the reals. The graph of a rational function is a continuous curve with the exception of at most a finite number of points. It is not continuous at a point x_1 where $Q(x_1) = 0$. At such a point the function is undefined, as in $(x - 5)/(x - 5)$ at $x = 5$ or as in $1/x$ at $x = 0$.

Illustration 2   Sketch the graph of the rational function defined by $y = f(x) = (x + 2)/(x - 1)^3$.

Solution   The domain is the set of real numbers excluding $x = 1$. The following intuitive argument will tell us something about the range: If x is just a little larger than 1, y is positive and very large; if x is a very large number, y is positive and very small. For $x > 1$, the range is, therefore, the set of positive real numbers. There is no value of y for $x = 1$, and the function is not continuous there. Further, at $x = -2$, $y = 0$ and $x = -2$ is the only zero of the function. If x is negative and a little larger than -2, y is negative and in absolute value very small; if x is positive and a little less than 1, y is negative and very large in absolute value. Therefore, for $-2 \leq x < 1$, y ranges over all nonpositive real numbers. Hence the range is the set of real numbers.

We compute the following table of values.

x	-4	-3.5	-3	-2	-1	0	½	1	1.5	2	3	11
y	$\dfrac{1}{62.5}$	$\dfrac{1}{60.75}$	$\dfrac{1}{64}$	0	$-\dfrac{1}{8}$	-2	-20	—	28	4	$\dfrac{5}{8}$	$\dfrac{13}{1{,}000}$

We have included the value of y at $x = -3.5$ because it is the largest value of y for $x < 1$. (Methods of the calculus are needed to prove this.) The graph is plotted in Fig. 10.2.

Graph of rational function

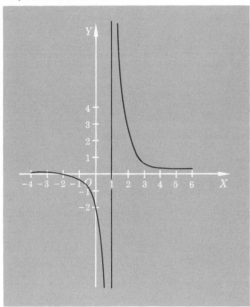

Figure 10.2
The graph of a rational function may not be continuous. Here $y = (x + 2)/(x - 1)^3$.

Explicit Algebraic Functions Explicit algebraic functions as defined include the rational functions as special cases. An explicit algebraic function may, therefore, fail to be continuous for the reasons given above. In addition it may have isolated points as indicated in the examples of Illustration 9, Sec. 9.4. It may also have end-points as indicated in Illustration 3 below. The domain and range are subsets of the reals.

Illustration 3 Sketch the graph of the explicit algebraic function defined by

$$y = \sqrt{x^2 - 1}$$

Since y must be real, we must have $x^2 - 1 \geq 0$, or $x^2 \geq 1$. Hence the domain of the function is $-\infty < x \leq -1$; $1 \leq x < +\infty$. The range is $0 \leq y < \infty$. The graph is symmetric with respect to the Y-axis, since substitution of $-x$ for x in $\sqrt{x^2 - 1}$ does not change this expression. Thus we compute a table of values.

x	± 1	± 2	± 3	± 4
y	0	1.73	2.83	3.87

The graph is shown in Fig. 10.3. It has end-points at $x = \pm 1$.

Graph of explicit algebraic function

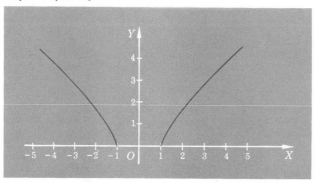

Figure 10.3   The graph of an explicit algebraic function may have end-points. Here $y = \sqrt{x^2 - 1}$.

It is interesting to compare Fig. 10.3 with the graph of $y = x^2 - 1$ (Fig. 10.4) plotted from the table of values.

x	0	± 1	± 2	± 3
y	-1	0	3	8

The ordinates of Fig. 10.3 are the square roots of those of Fig. 10.4 except in the interval $-1 < x < 1$ in which these square roots become imaginary. In cases like this considerable information about graphs such as Fig. 10.3 can be obtained by first plotting the related graph such as Fig. 10.4.

Graph related to Fig. 10.3

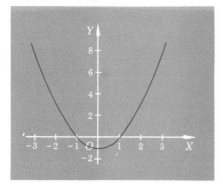

Figure 10.4
The graph of $y = \sqrt{f(x)}$ can be derived from the graph of $y = f(x)$. Here $y = x^2 - 1$. In this way we may obtain Fig. 10.3 from Fig. 10.4.

Illustration 4 Sketch the graph of

$$y = \frac{x}{\sqrt{x-1}}$$

Since $\sqrt{x-1}$ must be real and nonzero, we must have $x > 1$. The domain is therefore $1 < x < +\infty$, and $x = 0$. At $x = 1$ there is a vertical asymptote. The value of y diminishes as x increases; but later, as x gets very large, y increases again. (By the methods of calculus it can be shown that, for $x > 1$, y is least when $x = 2$.)

We plot the table of values.

x	0	1	1.5	2	3	4	5
y	0	$\cdots$	2.12	2	2.12	2.31	2.5

The graph is shown in Fig. 10.5. The point $(0,0)$ is an isolated point, and the graph is continuous for $x > 1$.

Graph of $y = x/\sqrt{x-1}$

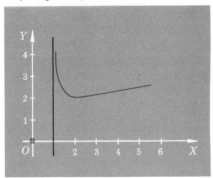

Figure 10.5
Be on the lookout for isolated points.

Problems 10.5

In Probs. 1 to 21 discuss type of algebraic function, domain, range (if possible), zeros, isolated points, end-points, and asymptotes. Sketch.

1 $y = x^3 - x$
2 $y = x^3 - 4x$
3 $y = \sqrt{x^3 - x}$
4 $y = \sqrt{x^3 - 4x}$
5 $y = x^2 - x^4$
6 $y = x^4 - x^2$
7 $y = \sqrt{x^2 - x^4}$
8 $y = \sqrt{x^4 - x^2}$
9 $y = x^3 + 1$
10 $y = 8 - x^3$
11 $y = \sqrt{x^3 + 1}$
12 $y = \sqrt{8 - x^3}$
13 $y = \sqrt[3]{x^3 + 1}$
14 $y = 1/(x^2 + 1)$
15 $y = (1 - x^{1/2})^2$ [portion of a parabola]
16 $y = 3x^4 - x^3 + 2$ [minimum value occurs at $x = \frac{1}{4}$ and is $^{511}\!/_{256}$]
17 $y = 8/(4 + x^2)$ [the Witch of Agnesi]
18 $y = \sqrt{(1 - x^{2/3})^3}$ [portion of a hypocycloid]

19 $y = \sqrt{x^3/(2-x)}$ [portion of the cissoid of Diocles]

20 $y = \sqrt{x(x-1)(x+3)}$

21 $y = -\sqrt{x(x-1)(x+3)}$

In Probs. 22 to 27 sketch the graph of each equation, and discuss.

22 $x^{1/2} + y^{1/2} = 1$ (see Prob. 15)

23 $x^{2/3} + y^{2/3} = 1$ (see Prob. 18)

24 $(2-x)y^2 - x^3 = 0$ (see Prob. 19)

25 $y^2 - 4x^2 = 0$

26 $y^2 - x^3 + x^2 = 0$

27 $(x+y)(xy+1) = 0$

28 Sketch the graph of the function defined jointly by $y = 4/(x^2 - 3x)$ and the condition $y > 1$.

29 Discuss the graph of $y = x^n$, n an even integer; again for n odd.

10.6 Properties of Polynomials

We have defined a polynomial as an expression of the form

$$P(x) = a_0 x^n + a_1 x^{n-1} + \cdots + a_{n-1}x + a_n$$

We now wish to discuss some of the more important properties of polynomials and polynomial functions.

Definitions A polynomial with complex coefficients we shall call a *complex* polynomial. A polynomial with real coefficients we shall call a *real* polynomial. A polynomial with rational coefficients we shall call a *rational* polynomial. The domain of definition in any case is the set of complex numbers. Associated with every polynomial $P(x)$ is the polynomial function P defined by $y = P(x)$, whose zeros are also called the *roots* of the polynomial equation $P(x) = 0$. A polynomial equation is called complex, real, or rational according as the coefficients are complex, real, or rational, respectively.

complex, real, rational polynomials

We have previously referred to the following theorem.

Theorem 1   The Fundamental Theorem of Algebra   Every complex polynomial equation $P(x) = 0$ of degree ≥ 1 has at least one root.

The proof of this theorem is beyond the scope of this text.

We pause here to tell you of a remarkable theorem first proved by the Norwegian mathematician Abel (1802–1829). The nature of the problem is easily described. You know that the general equations of the first and second degree

$$ax + b = 0 \qquad a \neq 0$$
$$ax^2 + bx + c = 0 \qquad a \neq 0$$

can be solved explicitly for x, giving respectively

$$x = -\frac{b}{a}$$

and
$$x = \frac{-b \pm \sqrt{b^2 - 4ac}}{2a}$$

So can the general equations of the third and fourth degree, namely,

$$ax^3 + bx^2 + cx + d = 0 \qquad a \neq 0$$
and
$$ax^4 + bx^3 + cx^2 + dx + e = 0 \qquad a \neq 0$$

be solved explicitly for x, using only a finite number of the algebraic operations. Abel proved that it is impossible to solve the general fifth-degree equation, using only a finite number of these operations. The proof is difficult. It is now known that no general equation of degree greater than four can be solved algebraically, i.e., by a finite number of the algebraic operations.

We seek properties of the roots of polynomial equations. First, we derive Theorem 2.

Theorem 2 Remainder Theorem If a complex polynomial $P(x)$ is divided by $x - b$ (where b is a complex number) until a remainder R free of x is obtained, then $P(b) = R$.

Proof Let us divide $P(x)$ by $x - b$ until we obtain a remainder not containing x. In accordance with the Division Algorithm (Sec. 3.5), this division may be written

$$(1) \qquad P(x) = (x - b)Q(x) + R$$

where $Q(x)$ is a complex polynomial of degree $n - 1$ called the quotient and R is a complex number.

Since Eq. (1) is true for all complex values of x, we may substitute b for x and obtain

$$P(b) = 0 + R \qquad \text{or} \qquad P(b) = R$$

Illustration 1 Let $P(x) = x^3 + 2x^2 - 3$. Find $P(2)$ by the remainder theorem. By division, we have

$$\frac{x^3 + 2x^2 - 3}{x - 2} = x^2 + 4x + 8 + \frac{13}{x - 2}$$

Hence $R = 13$ and $P(2) = 13$. This can be checked by noting that

$$P(2) = 8 + 8 - 3 = 13$$

Theorem 3 Factor Theorem If r is a root of a complex polynomial equation $P(x) = 0$, then $x - r$ is a factor of $P(x)$.

Proof   The statement "r is a root of $P(x) = 0$" is equivalent to the statement "$P(r) = 0$". Divide $P(x)$ by $x - r$ as in Eq. (1). By the Remainder Theorem, $R = P(r) = 0$. Hence $x - r$ is a factor of $P(x)$.

Exercise

A   Prove the converse of the Factor Theorem: If $(x - r)$ is a factor of $P(x)$, then r is a root of $P(x) = 0$.

Illustration 2   We may use the Factor Theorem to find a polynomial equation with given roots. Suppose we are given $r_1 = 1, r_2 = -2, r_3 = 0$ and are asked to find an equation with these roots. From the Factor Theorem $x - 1$, $x + 2$, and x are factors. Hence an equation with the desired property is

$$(x - 1)(x + 2)x = x^3 + x^2 - 2x = 0$$

Illustration 3   We use the converse of the Factor Theorem to help us solve polynomial equations which we can factor. Consider the problem: Solve

$$(x + 2)(x - 1)(x^2 + x + 1) = 0$$

Solution   Since $x + 2$ and $x - 1$ are factors, we know that two roots are $r_1 = -2, r_2 = 1$. The other roots are solutions of

$$x^2 + x + 1 = 0$$

or
$$x = \frac{-1 \pm i\sqrt{3}}{2}$$

Theorem 4   Number-of-roots Theorem   A complex polynomial equation $P(x) = 0$ of degree n $(n \geq 1)$ has exactly n roots.

multiple roots

It is possible, of course, for two or more of these roots to be equal. If k roots are all equal to r, say, the common language used is "r is said to be a root of multiplicity k". For Theorem 4 to be true, it is necessary to count a root of multiplicity k as k roots.

Proof   We have one root r_1 of $P(x) = 0$ from the Fundamental Theorem of Algebra. Therefore, from the Factor Theorem,

$$P(x) = (x - r_1)Q(x)$$

where $Q(x)$ is a polynomial of degree $n - 1$. Unless $Q(x)$ is a constant, the equation $Q(x) = 0$ also has a root r_2; thus $Q(x) = (x - r_2)S(x)$, where $S(x)$ is a polynomial of degree $n - 2$. Thus

$$P(x) = (x - r_1)(x - r_2)S(x)$$

Continue this process as long as possible. It must stop when n factors have been obtained; for the product of more factors would have a degree higher than n. Hence we have

$$P(x) = (x - r_1)(x - r_2) \cdots (x - r_n)a_0$$

This theorem tells us how many roots to look for. If we have an equation of fifth degree and have found three roots, we still have two more to find.

Problems 10.6

In Probs. 1 to 10 find a polynomial equation of lowest degree which has the given roots:

1	$2, 3$	**2**	$1, 2, 3$
3	i	**4**	$i, -i$
5	$0, 1, 2, 3$	**6**	$1 + i, 1 - i$
7	$0, 1, i$	**8**	$1 + i\sqrt{2}, 1 - i\sqrt{2}$
9	$1, 1, 2$	**10**	$1, i, -i$

In Probs. 11 to 18 find a polynomial of third degree which has the following zeros (if possible):

11	$1, -2, 0$	**12**	$i, 2i, 0$
13	$2, \sqrt{3}, -\sqrt{3}$	**14**	$2, 1 + 2i, 1 - 2i$
15	$a + ib, a - ib$	**16**	a, b
17	$a, 1, 1$	**18**	$1, 1, 1, 1$

In Probs. 19 to 23 use the Remainder Theorem to find:

19 $P(0)$ when $P(x) = x^4 - 6x^3 + 8x^2 - 9x + 3$
20 $P(1)$ when $P(x) = x^3 - 3x^2 + 3x - 4$
21 $P(-3)$ when $P(x) = x^2 - 8x + 15$
22 $P(i)$ when $P(x) = x^4 + x^2 + 2$
23 $P(2)$ when $P(x) = x^3 - 4x^2 + 8x - 1$

In Probs. 24 to 27 by using the converse of the Factor Theorem find all the roots of:

24 $x^3 - 1 = 0$ **25** $x^4 - 16 = 0$
26 $x^3 - 7x + 6 = 0$ **27** $x^3 - 6x^2 + 6x - 5 = 0$
28 How many roots does $x^4 + 3x^3 + 6x^2 = 0$ have?
29 How many roots does $x^5 - 3 = 0$ have?
30 How many roots does $(3 + i)x^2 - 7ix - 6 = 0$ have?
31 How many roots does $(10 + i)x^3 - 3i = 0$ have?
32 How many zeros does the polynomial $x(x - r_1)(x - r_2)(x - r_3) \cdots (x - r_k)$ have?
33 Show that a polynomial equation of degree n cannot have more than $n/2$ double roots. (A double root is a root of multiplicity 2.)
*34 As a consequence of Theorem 4 show that, if

$$P(x) = a_0 x^n + a_1 x^{n-1} + \cdots + a_{n-1} x + a_n = 0$$

has $n + 1$ roots, then each coefficient a_i is zero.
*35 Prove that $P(x) = 0$ and $Q(x) = 0$ have all their roots equal if and only if there is a constant c $(\neq 0)$ so that, for all x, $P(x) - cQ(x) = 0$.

10.7   Synthetic Division

The Remainder Theorem gives us a convenient short cut for finding the value $P(b)$, say, for it tells us that $P(b) = R$ and R is easy to compute. To perform the division called for in the Remainder Theorem, we use a short method, called synthetic division. To illustrate the method, we consider the case of the general cubic (complex) polynomial

$$P(x) = a_0x^3 + a_1x^2 + a_2x + a_3$$

which is to be divided by $x - b$. The work is exhibited in all detail below, where R is the remainder.

$$
\begin{array}{l}
x - b \,\big|\, a_0x^3 + a_1x^2 + a_2x + a_3 \,\big|\, a_0x^2 + (a_0b + a_1)x + (a_0b^2 + a_1b + a_2) \\
\quad\underline{a_0x^3 - a_0bx^2} \\
\qquad (a_0b + a_1)x^2 + a_2x \\
\qquad \underline{(a_0b + a_1)x^2 - (a_0b^2 + a_1b)x} \\
\qquad\qquad (a_0b^2 + a_1b + a_2)x + a_3 \\
\qquad\qquad \underline{(a_0b^2 + a_1b + a_2)x - (a_0b^3 + a_1b^2 + a_2b)} \\
\qquad\qquad\qquad a_0b^3 + a_1b^2 + a_2b + a_3 = R
\end{array}
$$

But, surely, we have written down more detail than we actually need; the following, where we have suppressed every x, is quite clear:

$$
\begin{array}{l}
-b \,\big|\, a_0 + a_1 + a_2 + a_3 \,\big|\, a_0 + (a_0b + a_1) + (a_0b^2 + a_1b + a_2) \\
\quad\underline{-a_0b} \\
\qquad (a_0b + a_1) + a_2 \\
\qquad \underline{-(a_0b^2 + a_1b)} \\
\qquad\qquad (a_0b^2 + a_1b + a_2) + a^3 \\
\qquad\qquad \underline{-(a_0b^3 + a_1b^2 + a_2b)} \\
\qquad\qquad\qquad a_0b^3 + a_1b^2 + a_2b + a_3 = R
\end{array}
$$

We have also omitted the second writing of a_0, $a_0b + a_1$, and $a_0b^2 + a_1b + a_2$ inasmuch as they are going to cancel by subtraction anyway. We will further simplify the process by changing the sign of $-b$ to $+b$ in the divisor, and hence the subtractive process to an additive one. Also there is no need of writing the quotient Q in the little box to the right since every term there is to be found in the work below, which is finally written on just three lines:

$$
\begin{array}{l}
\underline{b \,\big|} \quad a_0 \quad a_1 \qquad\quad a_2 \qquad\qquad\quad a_3 \\
\qquad\quad\; \underline{a_0b \qquad\quad a_0b^2 + a_1b \qquad a_0b^3 + a_1b^2 + a_2b} \\
\qquad a_0 \quad a_0b + a_1 \quad a_0b^2 + a_1b + a_2 \quad a_0b^3 + a_1b^2 + a_2b + a_3 = R
\end{array}
$$

Although we have skeletonized the work, the details can still be extracted: We are dividing $a_0x^3 + a_1x^2 + a_2x + a_3$ by $x - b$, and we get a quotient of $a_0x^2 + (a_0b + a_1)x + (a_0b^2 + a_1b + a_2)$ and a remainder of $a_0b^3 + a_1b^2 + a_2b + a_3$. Note that the remainder is $P(b)$ as it should be by the Remainder Theorem.

Illustration 1   Divide $x^4 - 3x^3 + x + 3$ by $x - 2$ synthetically.

Solution Form the array, noting that the coefficient of x^2 is zero. (We normally place the "2" associated with the divisor $x - 2$ on the right.)

$$
\begin{array}{rrrrr|r}
1 & -3 & 0 & 1 & 3 & \underline{2} \\
  & 2 & -2 & -4 & -6 & \\
\hline
1 & -1 & -2 & -3 & -3 = R &
\end{array}
$$

The quotient $Q(x) = x^3 - x^2 - 2x - 3$, and the remainder $R = -3$. By direct computation we also find that $P(2) = -3$.

Illustration 2 Given $P(x) = 3x^4 - 4x^3 - 2x^2 + 1$, compute $P(-1)$, $P(0)$, $P(1)$, $P(2)$, $P(3)$, $P(-0.3)$, and sketch $y = P(x)$.

Solution Directly from $P(x)$ we compute $P(0) = 1$, $P(1) = -2$, and $P(-1) = 6$. In the slightly more complicated cases of $P(2)$, $P(3)$, and $P(-0.3)$ we use synthetic division:

$$
\begin{array}{rrrrr|r}
3 & -4 & -2 & 0 & 1 & \underline{2} \\
  & 6 & 4 & 4 & 8 & \\
\hline
3 & 2 & 2 & 4 & 9 = P(2) &
\end{array}
$$

$$
\begin{array}{rrrrr|r}
3 & -4 & -2 & 0 & 1 & \underline{3} \\
  & 9 & 15 & 39 & 117 & \\
\hline
3 & 5 & 13 & 39 & 118 = P(3) &
\end{array}
$$

$$
\begin{array}{rrrrr|r}
3 & -4 & -2 & 0 & 1 & \underline{-0.3} \\
  & -0.9 & 1.47 & 0.159 & -0.0477 & \\
\hline
3 & -4.9 & -0.53 & 0.159 & 0.9523 = P(-0.3) &
\end{array}
$$

The preceding table is self-explanatory. Note especially the value $P(-0.3) = 0.9523$ and the corresponding dip in the graph (Fig. 10.6).

Graph of $y = 3x^4 - 4x^3 - 2x^2 + 1$

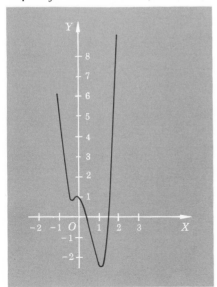

Figure 10.6
Some of the points are located by means of synthetic division.

This kind of variation cannot be discovered in general without the methods of the calculus.

Exercises

A Compute $P(-1)$ by synthetic division, and note the alternating signs in the last line of your work. Explain why this tells us there are no real zeros of P to the left of $x = -1$. Generalize for the case where $P(x)$ is a real polynomial.

B Examine the line where $P(2) = 9$ and state what follows about real zeros of P to the right of $x = 2$. Generalize for the case where $P(x)$ is a real polynomial.

Problems 10.7

In Probs. 1 to 6 use synthetic division to find:

1 $P(1)$ and $P(-1)$ when $P(x) = x^3 + 3x^2 - 2x + 6$
2 $P(-3)$ and $P(2)$ when $P(x) = 4x^3 - 3x + 8$
3 $P(1)$ and $P(-1)$ when $P(x) = x^4 + x^3 + x^2 + x + 1$
4 $P(1)$ and $P(2)$ when $P(x) = x^4 - x^3 + 2x^2 - 3x + 1$
5 $P(\frac{1}{3})$ and $P(\frac{1}{2})$ when $P(x) = 4x^2 - 3x + 2$
6 $P(\frac{2}{3})$ and $P(\frac{3}{4})$ when $P(x) = 3x^2 - x + 1$

In Probs. 7 to 16 use synthetic division to find quotient and remainder.

7 $\dfrac{4x^3 + 2x^2 - 3x + 1}{x - 1}$

8 $\dfrac{3x^3 - 6x^2 + 8x - 5}{x + 2}$

9 $\dfrac{x^3 + 3x^2 - 6x - 8}{x - 1}$

10 $\dfrac{5x^3 - 2x^2 + 8x - 6}{x - 3}$

11 $\dfrac{x^4 + 3x - 6}{x + 1}$

12 $\dfrac{3x^4 + 4x^3 + 9x - 2}{x - 2}$

13 $\dfrac{x^2 + 3x - 7}{x - \frac{1}{2}}$

14 $\dfrac{x^2 - 6x + 2}{x + \frac{1}{2}}$

15 $\dfrac{x^3 + 8x^2 - 3x + 6}{x - \frac{1}{3}}$

16 $\dfrac{x^3 + 8x^2 - 3x + 6}{x - \frac{3}{10}}$

In Probs. 17 to 22 use synthetic division to show that the first polynomial is a factor of the second.

17 $x + 2,\ 5x^3 + 8x^2 - x + 6$
18 $x - 4,\ 4x^3 - 13x^2 - 17x + 20$
19 $x + 2,\ 4x^4 + 6x^3 - x^2 + 2x - 8$
20 $x - 3,\ x^5 - 3x^4 + 6x^2 - 20x + 6$
21 $x - \frac{3}{2},\ 6x^4 - 5x^3 + 4x^2 - 13x - 3$
22 $x + \frac{5}{4},\ 4x^5 + 13x^4 + 10x^3 - 4x - 5$
23 Divide $ax^2 + bx + c$ by $x - r$ by long division and also by synthetic division. Compare the results.

24 Divide $ax^2 + bx + c$ by $x + r$ by long division and also by synthetic division. Compare the results.

25 For what value of k does $x^2 + kx + 4$ yield the same remainder when divided by either $x + 1$ or $x - 1$?

26 For what values of k is $x^2 + 3x + k$ divisible by $x - k$?

27 If the polynomial $P(x) = Ax^4 + Ax^3 - 33x^2 + 17x - 10$ is such that $P(5) = 0$, what is $P(4)$?

28 If the polynomial $P(x) = kx^3 - 4x + 2$ is such that $P(1) = -1$, what is $P(2)$?

29 When $x^2 + 2x - 4$ is divided by $x - r$, the remainder is 31. Find r.

30 When $3x^2 - 6x + 2$ is divided by $x + r$, the remainder is -1. Find r.

31 Use the factor theorem to prove that $x^n - a^n$ is divisible by $x - a$ when n is a positive integer.

32 Use the factor theorem to prove that $x^n + a^n$ is divisible by $x + a$ when n is an odd positive integer.

10.8 Roots of Polynomial Equations

Because of its practical importance, much effort has been spent on the question of how to calculate the roots of a polynomial equation. We have mentioned that formulas for the roots exist for $n = 1, 2, 3$, and 4; but there is no simple method of handling equations of higher degree. The general procedure consists of two steps:

I Find all roots which can be obtained by elementary means; then use the Factor Theorem or other methods to factor the polynomial into polynomials of lower degree.

II Find the zeros of the factors by known formulas or by approximate methods.

When the coefficients of $P(x)$ are general complex numbers, there is little that can be said here which will help you in these steps, for the known methods are too complicated to be treated in this book. We can make progress, however, if we consider only polynomials whose coefficients are real numbers. In this case we can prove the following theorem.

Theorem 5 A real polynomial $P(x)$, with real coefficients, can always be represented as a product of factors each of which is either of the form $ax + b$ or $cx^2 + dx + e$, where a, b, c, d, and e are real numbers.

Proof We know that the roots of $P(x) = 0$ are complex numbers, but some of them may actually be real. Corresponding to each real root r, the Factor Theorem tells us that there is a factor $(x - r)$. Therefore we can write

$$P(x) = (x - r_1)(x - r_2) \cdots (x - r_s)Q(x) = 0$$

where $r_1, r_2, \ldots, r_s$ are its real roots and $Q(x)$ is a polynomial of degree $n - s$ which has no real zeros. We must show that $Q(x)$ can be factored into quadratic factors of the form $cx^2 + dx + e$.

Suppose that $\alpha + i\beta$ with $\beta \neq 0$ is a root of $Q(x) = 0$. Construct the quadratic polynomial

$$(x - \alpha - i\beta)(x - \alpha + i\beta) = (x - \alpha)^2 + \beta^2 = S(x)$$

Note that $S(\alpha + i\beta) = 0$ and $S(\alpha - i\beta) = 0$. Now divide $Q(x)$ by $S(x)$, and obtain

$$Q(x) = S(x) \cdot R(x) + px + q$$

Substitute $x = \alpha + i\beta$ into this equation. Since $Q(\alpha + i\beta) = 0$ and $S(\alpha + i\beta) = 0$, we get

$$p(\alpha + i\beta) + q = 0$$

or
$$p\alpha + q = 0 \qquad p\beta = 0$$

Since $\beta \neq 0$, this shows that $p = 0$ and $q = 0$. Therefore $S(x)$ is a factor of $Q(x)$ and hence of $P(x)$. The same process can now be applied to $R(x)$, and we continue until we get

$$P(x) = (x - r_1) \cdots (x - r_s)S_1(x) \cdots S_t(x)a_0$$

where $s + 2t = n$. This is of the required form.

Corollary   If $\alpha + i\beta$ is a root of a real polynomial equation, then $\alpha - i\beta$ is also a root of this equation.

complex
roots come
in pairs

Exercises

A Construct an example which shows this corollary false when the coefficients of $P(x)$ are no longer real.
B Show that the degree of $Q(x)$ must be even.

This theorem tells us a lot about the nature of the roots of $P(x) = 0$, but it does not help us to find them. Special methods for finding the roots of certain simple types of equations are given in the next two sections.

Problems 10.8

Solve the following equations.

1 $2x^2 - 3x + 1 = 0$ **2** $4x^2 - 11x - 3 = 0$
3 $3x^2 - 20x + 12 = 0$ **4** $4x^2 + 7x - 2 = 0$
5 $16x^3 - 34x^2 - 15x = 0$ **6** $2x^3 + 3x^2 - 9x = 0$
7 $(x - 1)^2 - 7(x - 1) + 12 = 0$ **8** $(x - 2)^4 - 1 = 0$

9 $\dfrac{1}{x^2} - 4 = 0$ **10** $\dfrac{1}{(x - 1)^2} = 1$

11 $3x^3 + 7x^2 - 14x - 24 = 0$ **12** $2x^3 - 7x^2 - 7x + 12 = 0$

13 $x^4 + x^2 = 0$

14 $y^3 - y^2 = 0$

15 $\dfrac{1}{x^2} + \dfrac{1}{x} - 6 = 0$

16 $6 - \dfrac{1}{x} - \dfrac{12}{x^2} = 0$

17 $2t^3 - 3t^2 - 20t = 0$

18 $12t^4 - 41t^3 + 35t^2 = 0$

19 $z^6 - z^4 = 0$

20 $2w^4 + 7w^2 - 30 = 0$

21 $(x^2 + x - 1)^2 - 5(x^2 + x - 1) + 6 = 0$

22 $(x^2 - x - 3)^2 + (x^2 - x - 3) - 2 = 0$

10.9 Rational Roots of Rational Polynomial Equations

We now restrict ourselves to rational polynomial equations, i.e., to polynomial equations of the form $P(x) = 0$, where the coefficients in $P(x)$ are rational numbers.

Exercise

A Show that a rational polynomial can be written in the form $A \cdot P(x)$, where A is a rational number and where $P(x)$ has integer coefficients. Hence show that a given rational polynomial equation has the same roots as a certain polynomial equation in which the coefficients are integers. [Multiplying both sides of an equation by a constant ($\neq 0$) does not change the roots.]

There is a simple method in this case for obtaining quickly all those roots of $P(x) = 0$ which happen to be rational numbers. Of course, there is no necessity that any of these roots be rational; therefore this method may not produce any of the roots at all since it exhibits only the rational roots.

Theorem 6 Rational-root Theorem If

$$P(x) = a_0 x^n + a_1 x^{n-1} + \cdots + a_{n-1} x + a_n$$

has integers for coefficients, and if $r = p/q$ is a rational root (in lowest terms) of $P(x) = 0$, then p is a factor of a_n and q is a factor of a_0.

Proof We are given that

$$a_0 \frac{p^n}{q^n} + a_1 \frac{p^{n-1}}{q^{n-1}} + \cdots + a_{n-1} \frac{p}{q} + a_n = 0$$

Multiply through by q^n; the result is

(1) $a_0 p^n + a_1 p^{n-1} q + \cdots + a_{n-1} p q^{n-1} + a_n q^n = 0$

This may be written as

$$p(a_0 p^{n-1} + a_1 p^{n-2} q + \cdots + a_{n-1} q^{n-1}) = -a_n q^n$$

Now p is a factor of the left-hand side of this equation, and therefore p is a factor of the right-hand side, $-a_nq^n$. Since p/q is in lowest terms, p and q^n are relatively prime; and since p is a factor of a_nq^n, it follows (from Theorem 7, Chap. 2) that p *is a factor of* a_n.

Equation (1) can also be written

$$a_0p^n = -q(a_1p^{n-1} + \cdots + a_{n-1}pq^{n-2} + a_nq^{n-1})$$

By a similar argument q *is a factor of* a_0.

Illustration 1   Solve the equation $2x^4 + 5x^3 - x^2 + 5x - 3 = 0$.

Solution   The possible rational roots are $\pm 1, \pm 3, \pm \tfrac{1}{2}, \pm \tfrac{3}{2}$. Using synthetic division, we find that -3 is a root, for

$$
\begin{array}{rrrrr|r}
2 & 5 & -1 & 5 & -3 & \underline{-3} \\
  & -6 & 3 & -6 & 3 & \\
\hline
2 & -1 & 2 & -1 & 0 &
\end{array}
$$

Therefore

$$2x^4 + 5x^3 - x^2 + 5x - 3 = (x + 3)(2x^3 - x^2 + 2x - 1)$$

The remaining roots of the given equation are thus roots of the "reduced" equation

$$2x^3 - x^2 + 2x - 1 = 0$$

Its possible rational roots are $\pm 1, \pm \tfrac{1}{2}$. Using synthetic division, we find that $\tfrac{1}{2}$ is a root, for

$$
\begin{array}{rrrr|r}
2 & -1 & 2 & -1 & \underline{\tfrac{1}{2}} \\
  & 1 & 0 & 1 & \\
\hline
2 & 0 & 2 & 0 &
\end{array}
$$

The new reduced equation is

$$2x^2 + 2 = 0 \qquad \text{or} \qquad x^2 + 1 = 0$$

This is solved by the usual methods for quadratic equations and yields $x = \pm i$.

The roots of the original equation are therefore $\tfrac{1}{2}, -3, i, -i$. In this case each real root is a rational number.

Problems 10.9

In the following equations find the rational roots, and, where possible, solve completely.

1 $12x^3 + 7x^2 - 42x - 40 = 0$ **2** $3x^3 + 4x^2 - 13x + 6 = 0$
3 $x^3 - x^2 - 14x + 24 = 0$ **4** $2x^3 - 7x^2 - 27x - 18 = 0$
5 $x^3 + x^2 - 8x + 6 = 0$ **6** $x^3 + x^2 - x - 10 = 0$
7 $12x^4 - 67x^3 - 15x^2 + 484x - 480 = 0$

8 $6x^4 - 37x^3 + 79x^2 - 68x + 20 = 0$
9 $2x^4 + 5x^3 - 11x^2 - 20x + 12 = 0$
10 $x^4 - 4x^3 - 2x^2 + 21x - 18 = 0$
11 $2x^4 + x^3 + x^2 - 7x + 3 = 0$

10.10   Real Roots of Real Polynomial Equations

In Sec. 10.9 we discussed the general method of obtaining the roots of rational polynomial equations when those roots are rational numbers. There is no simple general way in which a root can be determined exactly when it is not rational and when the degree of the polynomial exceeds 4. Indeed, about the only method available to us is an approximation method which is best described as a graphical one.   This method will yield those roots of $P(x) = 0$ which are real but gives no information concerning other roots.   The method applies equally to other types of equations as well, provided that the graphs of these equations are continuous.

A real root of $f(x) = 0$ or a zero of $f:(x,y)$ corresponds to a value of x at which the graph of $y = f(x)$ crosses or touches the X-axis.   Hence the procedure is to construct an accurate graph from which the zeros may be read off (approximately).

Most graphs will only be accurate enough to locate the desired zero between successive integers, and a refined technique is needed to obtain more decimal places.   To be definite, suppose that we have located a single root between 2 and 3, so that $f(2)$ and $f(3)$ have opposite signs.   We may calculate $f(2)$, $f(2.1)$, $f(2.2)$, . . . , $f(2.9)$, $f(3)$ in turn and thus locate the root between the adjacent pair of these which have opposite signs. Since this process is tedious, we try to speed it up graphically by a procedure which suggests which of these tenths to try first.

Suppose the situation is as in Fig. 10.7.   Draw a straight line between the points $(2,f(2))$ and $(3,f(3))$, and observe where this crosses the axis.

Linear approximation to a root

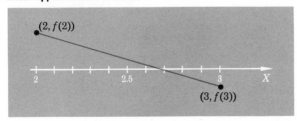

Figure 10.7
This graph helps us to estimate the first decimal place of the root.

Now try tenths in the neighborhood of this crossing.   When the root is located between successive tenths, the process may be repeated for hundredths, etc., as far as desired.   Usually, however, the graphic method is abandoned after the tenths have been obtained, and refined numerical techniques (beyond the scope of this book) are employed.

We should say a final word about the use of a straight line with which to approximate a curve whose equation is $y = f(x)$ within some interval $[a,b]$. We shall assume that within $[a,b]$ the curve is continuous and strictly monotone (increasing or decreasing). Then we draw the line segment joining $(a,f(a))$ and $(b,f(b))$ and treat it as an approximation to the graph of $y = f(x)$ within $[a,b]$. The accuracy of the approximation depends upon the length of $[a,b]$ and the behavior of $y = f(x)$ within it; but for the functions used in elementary mathematics and for short intervals, the approximation is generally quite satisfactory for practical purposes. This is the basis on which we use (linear) interpolation in various tables (such as a table of logarithms).

Illustration 1 Find the real zeros of the function defined by

$$y = x^3 - 2x^2 + x - 3$$

Solution We find the table of values

x	-1	1	0	2	3
$f(x)$	-7	-3	-3	-1	9

and plot the graph as in Fig. 10.8. We see that there is a zero between 2 and 3. We cannot prove it with our present knowledge, but this is the

Graph of $y = x^3 - 2x^2 + x - 3$

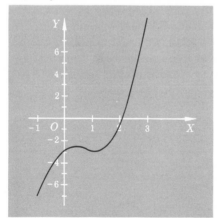

Figure 10.8 The graph suggests a root between 2 and 3.

Linear approximation in interval $2 \leq x \leq 3$

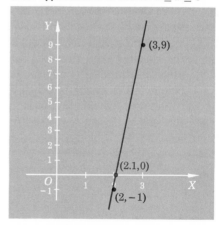

Figure 10.9 A linear approximation suggests a root near 2.1.

only real zero of this function. We plot Fig. 10.9. The line crosses the axis at exactly 2.1; therefore we calculate the following table.

x	2.0	2.1	2.2
$f(x)$	-1	-0.459	$+0.168$

Thus the zero is between 2.1 and 2.2.

Linear approximation in interval $2.1 \leq x \leq 2.2$

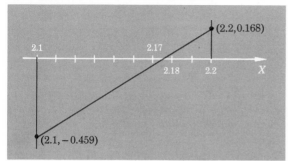

Figure 10.10
A further linear approximation suggests a root between 2.17 and 2.18.

To obtain the next decimal place, we plot Fig. 10.10.   The line crosses the axis between 2.17 and 2.18; therefore we calculate the following table.

x	2.17	2.18
$f(x)$	-0.03	0.04

Hence the zero is $2.17+$.

Repeated, this process will determine the decimal expansion of the root in question.   But note that to obtain the best approximation to, say, two decimal places we should compute the expansion to three places and then round off to two places.

Problems 10.10

Find the first two decimal places (and round to one decimal place) of the real zero of least absolute value of the function defined as follows:

1 $x^3 + 3x^2 - 2x + 5$ **2** $2x^3 - x^2 + x - 3$
3 $x^3 + x^2 - 7x + 2$ **4** $2x^3 - 4x^2 + 3x - 4$
5 $x^3 + x^2 + x + 2$ **6** $3x^3 - 4x^2 + 8x - 6$
7 $7x^3 + 19x^2 - 62x + 16$ **8** $x^3 + 4x^2 - 3x - 1$
9 $2x^4 + 7x^3 + 2x^2 - 4x - 16$ **10** $2x^4 + 7x^3 - 8x^2 + 2x - 21$

11 $1 - \dfrac{2}{x^2}$ **12** $1 - \dfrac{3}{x^2}$

13 $(x - 4)(x - 3.18)(x + 0.12)$ **14** $(x + 6.01)(x - 3.14)(2x - 7)$
15 $(x^2 - 1)(x - 5)^3$ **16** $(x^2 + x + 1)(x + 2)(x - 1)$

10.11   Complex Roots

Although the problem of finding the complex roots of complex polynomial equations is far beyond the scope of this book, there is an important special case worthy of our attention here. This is the matter of finding the roots of

$$x^n - a = 0$$

where a is a complex number; i.e., we wish to discuss the nth roots of a. In order to do this, we must expand upon our treatment of complex numbers in Chap. 2 and assume that you have an elementary knowledge of trigonometry.

We are already acquainted with the expression $a + bi$ for a complex number and with the fact that complex numbers can be represented by points in the plane (Sec. 2.15), where the real part a is measured along the horizontal axis and the imaginary part b is measured along the vertical axis. Now we wish to introduce new coordinates, r and θ, for a point P in the plane, called *polar coordinates*. As in Fig. 10.11, r measures the distance OP and θ is the angle from the positive horizontal axis to the line OP. In contrast to rectangular coordinates, the polar coordinates of a point are not unique; for if P has polar coordinates (r,θ), it also has polar coordinates $(r, \theta + 2n\pi)$, where n is any integer.

Since from trigonometry we know that (Fig. 10.11)

$$a = r \cos \theta \quad \text{and} \quad b = r \sin \theta$$

we find that

(1) $$a + ib = r(\cos \theta + i \sin \theta)$$

where $$r = |a + ib| = \sqrt{a^2 + b^2}$$

and $$\tan \theta = \frac{b}{a}$$

The real, nonnegative number $r\,(= \sqrt{a^2 + b^2})$ is called the *absolute value* (or *modulus*) of the complex number and is written $|a + ib|$. The angle θ associated with the number $a + ib$ is called the *argument* (or *amplitude*)

Polar form of $a + ib$

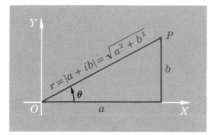

Figure 10.11

The graphical representation of complex numbers was discussed in Sec. 2.15. This figure illustrates how complex numbers can be represented in polar coordinates.

of $a + ib$. The left-hand side of (1) is the *rectangular form* and the right-hand side is the *polar form* of a complex number. A complex number is therefore a vector having both magnitude (absolute value) and direction (argument).

Addition (and subtraction) of complex numbers is best accomplished in rectangular $(a + ib)$ form. Thus

$$(a + ib) \pm (c + id) = (a \pm c) + i(b \pm d)$$

But multiplication and division are conveniently treated in polar $[r(\cos \theta + i \sin \theta)]$ form. Often, to simplify the notation, we write $r(\cos \theta + i \sin \theta)$ in the form r cis θ.

Multiplication Consider r_1 cis θ_1 and r_2 cis θ_2, two complex numbers. Their product (Fig. 10.12) is given by

r_1 cis $\theta_1 \cdot r_2$ cis θ_2
$= r_1 r_2$ cis θ_1 cis θ_2
$= r_1 r_2 (\cos \theta_1 + i \sin \theta_1)(\cos \theta_2 + i \sin \theta_2)$
$= r_1 r_2 [(\cos \theta_1 \cos \theta_2 - \sin \theta_1 \sin \theta_2) + i(\sin \theta_1 \cos \theta_2 + \cos \theta_1 \sin \theta_2)]$
$= r_1 r_2 [\cos (\theta_1 + \theta_2) + i \sin (\theta_1 + \theta_2)]$
$= r_1 r_2$ cis $(\theta_1 + \theta_2)$

Therefore the absolute value of the product is the product of the absolute values and the argument of the product is the sum of the arguments (plus or minus a multiple of 2π).

By similar reasoning

$$r_1 \text{ cis } \theta_1 \cdot r_2 \text{ cis } \theta_2 \cdot r_3 \text{ cis } \theta_3 = r_1 r_2 r_3 \text{ cis } (\theta_1 + \theta_2 + \theta_3)$$

If the three numbers $\theta_1, \theta_2, \theta_3$ are all equal to θ, and if r_1, r_2, r_3 are all equal to r, we have

$$[r \text{ cis } \theta]^3 = r^3 \text{ cis } 3\theta$$

Product of complex numbers

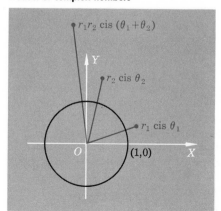

Figure 10.12

To multiply two complex numbers in polar form, add their arguments and multiply their absolute values.

And similarly

(**2**) $\qquad [r \text{ cis } \theta]^n = r^n \text{ cis } n\theta \qquad n$ a positive integer

With proper interpretations, (2) is true for any real number n, but we shall not give the proof. This is known as de Moivre's Theorem.

Theorem 7   de Moivre's Theorem $[r \text{ cis } \theta]^n = r^n \text{ cis } n\theta, \ n$ real

Exercise

A Prove de Moivre's Theorem for the case where n is a positive integer. HINT: Use induction.

Division To find the quotient of two complex numbers, write

$$\frac{r_1 \text{ cis } \theta_1}{r_2 \text{ cis } \theta_2} = \frac{r_1 \text{ cis } \theta_1}{r_2 \text{ cis } \theta_2} \times \frac{r_2 \text{ cis } (-\theta_2)}{r_2 \text{ cis } (-\theta_2)}$$

$$= \frac{r_1 r_2 \text{ cis } (\theta_1 - \theta_2)}{r_2{}^2 \text{ cis } 0}$$

$$= \frac{r_1}{r_2} \text{ cis } (\theta_1 - \theta_2)$$

Thus we see that the absolute value of the quotient of two complex numbers is the quotient of their absolute values and the argument of the quotient is the argument of the numerator minus the argument of the denominator (Fig. 10.13).

roots of complex numbers   **Roots of Complex Numbers** First, we note that the argument of a complex number is not uniquely defined. If

$$a + ib = r \text{ cis } \theta$$

it is also equal to $r[\text{cis } (\theta + 2\pi n)]$ for any integer n. Up to now this was not important, but we must use it here.

Given the complex number $r \text{ cis } \theta$, we seek to find all complex numbers whose pth powers are equal to $r \text{ cis } \theta$. These are called its pth roots. From de Moivre's Theorem we see at once that for every n

$$\left[r^{1/p} \text{ cis} \left(\frac{\theta + 2\pi n}{p} \right) \right]^p = r \text{ cis } (\theta + 2\pi n) = r \text{ cis } \theta$$

Quotient of complex numbers

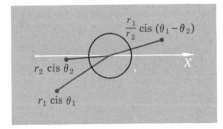

Figure 10.13

To divide two complex numbers in polar form, take the difference of their arguments and the quotient of their absolute values.

Roots of a complex number

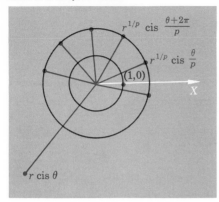

Figure 10.14
The p "pth roots" of a complex number are evenly spaced on a circle of radius $r^{1/p}$.

Therefore the numbers (Fig. 10.14)

(**3**)
$$r^{1/p} \operatorname{cis} \frac{\theta + 2\pi n}{p}$$

are pth roots of $r \operatorname{cis} \theta$. It can be shown that these comprise all the pth roots.

Exercise

B   If $[R \operatorname{cis} \phi]^p = r \operatorname{cis} \theta$, show that $R \operatorname{cis} \phi$ must have the form (3) for some value of n.

Let us examine (3) for various values of n. Letting $n = 0$, we have

$$r^{1/p} \operatorname{cis} \frac{\theta}{p}$$

In more advanced books this is called the *principal* pth root of $r \operatorname{cis} \theta$ and is denoted by the symbol $(r \operatorname{cis} \theta)^{1/p}$. Letting $n = 1$, we have

$$r^{1/p} \operatorname{cis} \frac{\theta + 2\pi}{p}$$

Each of these two (distinct) numbers is a pth root of $r \operatorname{cis} \theta$. By letting $n = 2, 3, \ldots, p - 1$, we obtain $p - 2$ other distinct pth roots. Letting $n = p$, we have

$$r^{1/p} \operatorname{cis} \frac{\theta + 2\pi p}{p} = r^{1/p} \operatorname{cis} \frac{\theta}{p}$$

which yields the same result as did $n = 0$. And $n = p + 1$ yields the same result as $n = 1$, etc. Therefore there are p (distinct), and only p, pth roots of a complex number, $a + ib = r \operatorname{cis} \theta$. These are given by

(**4**)
$$r^{1/p} \operatorname{cis} \frac{\theta + 2\pi n}{p} \qquad n = 0, 1, 2, \ldots, p - 1$$

You should memorize (4).

Cube roots of 1

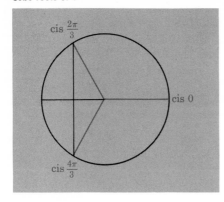

Figure 10.15

Illustration 1   Find the three cube roots of 1 (Fig. 10.15).   Since

$$1 = 1 + i0$$

$r = 1$, and $\theta = 0$.   Thus $1 = \text{cis}\,(0 + 2\pi n)$.   The cube roots are

$$1^{1/3} \text{ cis } \frac{0 + 2\pi n}{3} \qquad n = 0, 1, 2$$

or $\qquad\qquad 1 \text{ cis } 0 = 1$

$$1 \text{ cis } \frac{2\pi}{3} = \cos\frac{2\pi}{3} + i\sin\frac{2\pi}{3} = -\frac{1}{2} + i\frac{\sqrt{3}}{2}$$

$$1 \text{ cis } \frac{4\pi}{3} = \cos\frac{4\pi}{3} + i\sin\frac{4\pi}{3} = -\frac{1}{2} - i\frac{\sqrt{3}}{2}$$

To check the result, multiply out $\left(-\dfrac{1}{2} + i\dfrac{\sqrt{3}}{2}\right)^3$ and $\left(-\dfrac{1}{2} - i\dfrac{\sqrt{3}}{2}\right)^3$.
The results should be 1.

This example is equivalent to solving the equation

$$x^3 - 1 = 0$$
or $\qquad\qquad (x - 1)(x^2 + x + 1) = 0$

The roots are

$$x = 1$$

$$x = \frac{-1 \pm \sqrt{1 - 4}}{2} = -\frac{1}{2} \pm i\frac{\sqrt{3}}{2}$$

Problems 10.11

In Probs. 1 to 10 change the following to polar form:

1	$\sqrt{3} + i$	**2**	$\sqrt{3} - i$
3	$-\sqrt{3} + i$	**4**	$-\sqrt{3} - i$
5	$17i$	**6**	$-8i$
7	13	**8**	-8
9	$3 - 4i$	**10**	$-3 + 4i$

In Probs. 11 to 19 change the following to rectangular form:

11 cis 0°   **12** 4 cis 135°
13 4 cis 210°   **14** 5 cis 36°52.2′
15 10 cis 300°   **16** 8 (cos 60° + i sin 60°)
17 3 cis (−π/4)   **18** 8 cis 270°
19 4 cis 90° + 12 cis 120° + 16 cis 135°
20 Change to rectangular form and to polar form: 6 cis (−60°) + 14 cis 120°.

In Probs. 21 to 24 find the product of:

21 3 cis 30° and 2 cis 90°   **22** 2.5 cis 60° and 4 cis 135°

23 3 cis $\dfrac{\pi}{4}$ and 7 cis $\left(-\dfrac{\pi}{3}\right)$ **24** 6 cis $\dfrac{\pi}{9}$ and 13 cis $\dfrac{\pi}{6}$

In Probs. 25 to 28 find the quotient of:

25 3 cis 150° by 6 cis 30°   **26** 2 cis 170° by 5 cis 20°

27 75 cis (7π/5) by 3 cis (2π/5)   **28** 5 cis $\dfrac{2\pi}{9}$ by 3 cis $\left(-\dfrac{\pi}{6}\right)$

In Probs. 29 to 32 find the fifth power of:

29 2 cis 135°   **30** 3 cis 25°
31 2 cis (π/3)   **32** 3 cis (−4π/15)
33 Find the three cube roots of 8 cis (3π/2). Plot them and the original number.
34 Find the four fourth roots of −1. Plot them and the original number.
35 Find the four fourth roots of 1. Plot them and the original number.
36 Find the four fourth roots of i. Plot them and the original number.
37 Find the three cube roots of 27 cis π. Plot them and the original number.
38 Find all roots of the equation $x^5 - 1 = 0$.
39 Find all roots of the equation $x^5 - 32 = 0$.
40 Find all roots of the equation $4x^4 - 25 = 0$.
41 Find all roots of the equation $3x^3 - 4 = 0$.
*42 Show that the two square roots of $a + bi$ where $b \neq 0$ are

$$\pm \frac{1}{\sqrt{2}} \left[\sqrt{a + \sqrt{a^2 + b^2}} + i\frac{b}{|b|} \sqrt{-a + \sqrt{a^2 + b^2}} \right]$$

EXPONENTIAL AND LOGARITHMIC FUNCTIONS

Chapter 11

11.1 Exponential Functions

In your earlier studies you have become acquainted with powers such as 2^3, $(-3)^4$, π^5, and the like. You have also met

$$7^{-2} = \frac{1}{7^2} \qquad 4^0 = 1 \qquad \pi^{-3} = \frac{1}{\pi^3}$$

The general expression for symbols like these is a^n, where a is any real number and n is an integer. You will also recall the use of fractional exponents to represent roots, such as

$$3^{1/2} = \sqrt{3} \qquad 5^{1/3} = \sqrt[3]{5} \qquad 2^{-1/7} = \frac{1}{\sqrt[7]{2}}$$

For our present purposes we are interested in the roots of positive real numbers only, and we know that every positive real number a has a unique real nth root which we write $a^{1/n}$, where n is a positive integer.

Also, we recall that $a^{p/q}$ is defined to be $a^{(1/q)p} = (a^p)^{1/q}$, where p and q are integers and a is positive. Hence we know the meaning of the function defined by

$$y = a^x \qquad a \text{ positive, } x \text{ rational}$$

We should like to extend the domain of definition of this function to the entire set of real numbers and thus give sense to numbers such as 2^π, $\pi^{-\sqrt{3}}$, and $3^{\sqrt{2}}$. The most natural way to obtain $3^{\sqrt{2}}$, for example, is to consider the successive decimal approximations to $\sqrt{2}$, such as 1.4, 1.41, 1.414, 1.4142, etc. Then $3^{1.4}$, $3^{1.41}$, $3^{1.414}$, $3^{1.4142}$, etc., are successive approximations to $3^{\sqrt{2}}$. A complete discussion of this matter is not feasible here, for it would require a study of the real numbers in more detail than we have treated them in Chap. 2. We shall content ourselves with the remarks that this extension is possible and that the value of a number like those above can be obtained to any desired approximation by choosing an expansion of each irrational to a sufficient number of decimal places. Thus

$$2^{\pi} \approx 2^{3.1416} \approx 9.437 \qquad 3^{\sqrt{2}} \approx 3^{1.414} \approx 4.728$$

where the symbol $\approx$ means "approximately equal to".

In summary we define a new function as follows.

Definition The function f defined by $y = a^x (a > 0)$ is called the exponential function with base a. Its domain of definition is the set of real numbers. We observe that for $a \neq 1$ its range of values is $0 < y < \infty$.

$y = a^x$
$(a > 0)$

We now wish to develop some of its properties.

Theorem 1 For $a > 0$ and $b > 0$ and x and y real:

a $a^x \times a^y = a^{x+y}$
b $(a^x)^y = a^{xy}$
c $(ab)^x = a^x \times b^x$

These theorems are proved in Chap. 5 for rational values of x. We do not give the proof for irrational values of x.

Theorem 2

a $a^x > 1$ for $a > 1$, x real and > 0
b $a^x = 1$ for $a = 1$, x real and > 0
c $a^x < 1$ for $0 < a < 1$, x real and > 0

Proof Part a is immediate when x is a positive integer, for the product of two numbers each of which is greater than 1 must itself exceed 1. When $x = 1/n$ (n a positive integer), part a also is true. For if $a^{1/n}$ were to be less than 1 in this case, its nth power $(a^{1/n})^n = a$ would be less than 1. This follows from the fact that the product of two numbers each between zero and 1 must itself be less than 1. Finally, part a is true for rational x by combining the above cases. We omit the proof for irrational values of x. Part b is immediate since all powers and roots of 1 are themselves 1. Part c is proved similarly to part a.

Exercise

A Write out the details of the proof of part c for rational x.

Theorem 3 Let x and y be real numbers such that $x < y$. Then

a $a^x < a^y$ for $a > 1$
b $a^x = a^y$ for $a = 1$
c $a^x > a^y$ for $0 < a < 1$

The proof depends on Theorem 2. In all cases we know, from Theorem 1, that $a^y = a^{y-x} \cdot a^x$. By hypothesis, $y - x$ is positive. Thus, if $a > 1$, $a^{y-x} > 1$ and $a^y > a^x$, and similarly for the other cases.

Exercise

B Complete the proof of Theorem 3.

a^x is monotone   We state without proof that the exponential functions are continuous for all values of x. Theorem 3 shows that when $a > 1$, the function a^x is strictly monotone increasing (Sec. 9.6) and that, when $a < 1$, it is strictly monotone decreasing. Typical graphs are given in Figs. 11.1 and 11.2.

Graph of $y = a^x$, $a > 1$

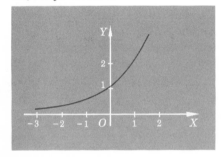

Graph of $y = a^x$, $a < 1$

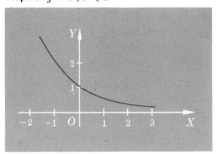

Figure 11.1 a^x is strictly monotone increasing when $a > 1$.

Figure 11.2 a^x is strictly monotone decreasing when $a < 1$.

There is an interesting symmetry between the graphs of $y = a^x$ and $y = (1/a)^x$, for $(1/a)^x = a^{-x}$, and the graph of $y = a^{-x}$ is just like the graph of $y = a^x$, with the X-axis reversed in direction.

Exercise

C Draw the graph of $y = 1^x$.

Problems 11.1

1 Show that $a^x \times a^y$ defines an exponential function.
2 Show that $a^x \div a^y$ defines an exponential function.

In Probs. 3 to 13, simplify.

3 $10^{-1}3^3$

4 $10^{-3}3^4$

5 $1/10^2 2^{-4}$

6 $2^{-3}/3^{-4}$

7 $2^6 3^{-3} (\sqrt{2})^4$

8 $3^2 4 (25)^{-1/2}$

9 $9^{5/2} 9^{-3/2}$

10 $\sqrt{4\sqrt{36}}$

11 $3(49)^{1/2}$

12 $\sqrt[3]{27} / \sqrt{25}$

13 $\sqrt{(1.44)10^4}$

In Probs. 14 to 20 simplify, but leave the answer in exponential form.

14 $a^{3x} \cdot a^{-x} / a^x$

15 $a^{6x} \cdot a^{2x} / a^{-8x}$

16 $a^n \cdot a^{-6n} \cdot b^{2n}$

17 $a^{3x/2} \cdot b^{2x} \cdot a^{x/3}$

18 $10\sqrt{10} \sqrt[4]{10} \sqrt[5]{10}$

19 $10^3 \cdot 10^{1/2} \cdot 10^{-5} \cdot 10^{x/2}$

20 $2^x + 2^{2x}$

In Probs. 21 to 24 plot the graphs $(x \geq 0)$ on the same axes.

21 $y = x^2, \; y = 2^x$

22 $y = x^3, \; y = 3^x$

23 $y = x^{1/2}, \; y = (\frac{1}{2})^x$

24 $y = x^{1/3}, \; y = (\frac{1}{3})^x$

25   Plot the graph of $y = 2^x$. Now change the scale on the Y-axis so that the graph you have drawn is that of $y = 3 \cdot 2^x$.

26   Plot the graph of $y = 3 \cdot 1^x$.

11.2   The Number e

We have defined the function $f:(x,y)$ whose values are given by $y = a^x$. Now there is a particular number $a > 1$ of great importance in mathematics; it is an irrational number and approximately equal to 2.71828. It bears the name e; thus

$$e \approx 2.71828$$

It is impossible for us to explain at this time why the number e and the associated exponential function defined by

$$y = e^x$$

are of such importance.

This exponential function is in fact so important that we speak of it as *the* exponential function and neglect to mention its base. Sometimes this function is written

$$y = \exp x$$

where no base appears at all; the base is assumed to be e unless otherwise specified. Its values are tabulated in many convenient tables. Its graph is plotted in Fig. 11.3. In this figure we have also plotted $y = 2^x$. Note that $y = e^x$, $y = 2^x$, and $y = a^x$ all pass through $(0,1)$. Of these $y = e^x$ is the one whose tangent at $(0,1)$ makes an angle of $45°$ with the Y-axis.

A convenient approximation to e^x is given by the polynomial

$$1 + x + \frac{x^2}{2!} + \frac{x^3}{3!} + \cdots + \frac{x^n}{n!}$$

Two special cases of $y = a^x$

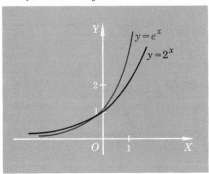

Figure 11.3
The graph of $y = e^x$ cuts the Y-axis at $(0,1)$ at an angle of $45°$.

(By definition $n! = 1 \times 2 \times \cdots \times n$ for n a positive integer, and $0! = 1$.) As n increases, the approximation becomes closer and closer; as a matter of fact, e^x is exactly equal to the infinite sum

$$e^x = 1 + x + \frac{x^2}{2!} + \cdots + \frac{x^n}{n!} + \cdots$$

Of course, we have not defined such an infinite sum. This can be done, however, and then one can give a proper definition of e and of the function e^x. Table I in the Appendix gives values of e^x and e^{-x}.

hyperbolic functions

The exponential function is also used in the definitions of some other functions which occur in elementary mathematics. These are called the *hyperbolic functions*. Their definitions are

$$\sinh x = \frac{e^x - e^{-x}}{2}$$

$$\cosh x = \frac{e^x + e^{-x}}{2}$$

$$\tanh x = \frac{e^x - e^{-x}}{e^x + e^{-x}}$$

The function $\sinh x$ is read "the hyperbolic sine of x"; $\cosh x$, "the hyperbolic cosine of x"; and $\tanh x$, "the hyperbolic tangent of x". These functions are somewhat related to the trigonometric functions with similar names.

Problems 11.2

In Probs. 1 to 10 obtain the values from Table I in the Appendix, interpolating where necessary.

1 $e^{0.05}$

2 $e^{1/2}$

3 $e^{1.8}$

4 $e^{3.1}$

5 $2e^2$

6 $3e^{4.2}$

7 e^{-2}

8 $e^{-1/2}$

9 $e^{-4.35}$

10 $e^{-0.22}$

In Probs. 11 to 20 plot the graph, using the same scale on the two axes.

11 $y = 3e^{0.1x}$

12 $y = 4e^{-0.1x}$

13 $y = -2e^{0.5x}$

14 $y = -3e^{-0.5x}$

15 $y = 6e^{-0.05x}$

16 $y = 5e^{0.05x}$

17 $y = -3e^{-1.2x}$

18 $y = e^{1.2x}/10$

19 $y = (e^x + e^{-x})/2$

20 $y = (e^x - e^{-x})/2$

11.3   Logarithmic Functions

Since $y = a^x$ defines a strictly monotone function, for $a \neq 1$, each value of y is obtained from a single x. Therefore the inverse function exists, and this is called the logarithmic function.

Definition The function inverse to that given by $y = a^x$, $a > 0$, $a \neq 1$, is written $y = \log_a x$ and is called the *logarithm* of x to the base a.

For computational purposes, the base is usually taken to be 10, so that properties of our decimal system may be used to simplify the needed tables. For theoretical purposes, the base is always taken to be e. In advanced books, this base is omitted, and log x is to be understood to mean $\log_e x$. Frequently, the notation ln x is used for $\log_e x$. Logarithms to the base 10 are called "common" logarithms; those to the base e are called "natural" logarithms. Tables of both kinds are available in most collections of elementary tables. (See Appendix, Tables II and III.)

natural
logarithms

Exercise

A Prove that $\log_a a^x = x$ and that $a^{\log_a x} = x$.

The graph of $y = \log_a x$ is obtained from that of $y = a^x$ by reflecting it in the line $y = x$. It is given in Figs. 11.4 and 11.5.

From the graphs we see that the domain and range are as follows.

Domain and Range The domain of definition of $\log_a x$ is the set of positive real numbers. Its range of values is the set of all real numbers.

Note that the logarithms of negative numbers are not defined here. In advanced books, you will learn how to extend the definition of $\log_a x$ so that x can be negative. Its value turns out to be nonreal. We do not consider this case.

Properties The logarithmic function to the base a defined by $y = \log_a x$ is strictly monotone increasing for $a > 1$, strictly monotone decreasing for $0 < a < 1$, and not defined for $a = 1$. The following theorems have useful applications.

Graph of $y = \log_3 x$

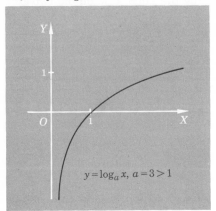

Graph of $y = \log_{1/2} x$

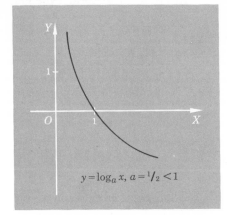

Figure 11.4   The graph of $y = \log_a x$ is strictly monotone increasing when $a > 1$.

Figure 11.5   The graph of $y = \log_a x$ is strictly monotone decreasing when $a < 1$.

Theorem 4 $\log_a xy = \log_a x + \log_a y$

Proof   Let

$$z = \log_a xy; \text{ then } a^z = xy$$
$$u = \log_a x; \text{ then } a^u = x$$
$$v = \log_a y; \text{ then } a^v = y$$

properties of logarithms

Therefore $a^z = xy = a^u \cdot a^v = a^{u+v}$

or $z = u + v$

from which the theorem follows.

Theorem 5 $\log_a \dfrac{1}{x} = -\log_a x$

Proof   Let $z = \log_a \dfrac{1}{x}$; then $a^z = \dfrac{1}{x}$, and $a^{-z} = x$.   Therefore $-z = \log_a x$.   Hence the theorem follows.

Theorem 6 $\log_a \dfrac{y}{x} = \log_a y - \log_a x$

Proof   Combine Theorems 4 and 5.

Theorem 7 $\log_a (x^y) = y \log_a x$

Proof   Let

$$z = \log_a (x^y); \text{ then } a^z = x^y$$
$$u = \log_a x; \text{ then } a^u = x$$

Therefore $(a^u)^y = a^z$ or $uy = z$

from which the theorem follows.

Exercises

B Prove $\log_b a = \dfrac{1}{\log_a b}$.

C Prove $\log_a x = \dfrac{\log_b x}{\log_b a}$. A special case of this is $\log_a x = \dfrac{\log_e x}{\log_e a}$.

Theorems 4 to 6 are useful for numerical computations involving only products and quotients. Logarithms to the base 10 are generally employed.

Illustration 1 Find $\dfrac{(33.0)(27.2)}{15.8}$.

Solution We compute

$$\log_{10} \frac{(33.0)(27.2)}{15.8} = \log_{10} 33.0 + \log_{10} 27.2 - \log_{10} 15.8$$

To find these logarithms, we use the table of common logarithms in the Appendix. We find

$$\begin{aligned}
\log_{10} 33.0 &= &1.5185\\
\log_{10} 27.2 &= &\underline{1.4346}\\
&&2.9531\\
-\log_{10} 15.8 &= &\underline{-1.1987}\\
\log_{10} \frac{(33.0)(27.2)}{15.8} &= &1.7544
\end{aligned}$$

Working backward from the table, we obtain

$$\frac{(33.0)(27.2)}{15.8} = 56.81$$

The importance of logarithms in problems of this sort is not as great as it was in former years. Calculations such as that above can be performed more rapidly on a slide rule, provided that the numbers involved do not contain more than three essential digits. When the numbers are more complicated, or when greater accuracy is desired, rapid results can be obtained from a desk computing machine. For this reason most students do not need to develop great skill in this use of logarithms.

On the other hand, logarithms must be used to compute exponentials such as $2^{1.42}$ by the use of Theorem 7.

Illustration 2 Compute $2^{1.42}$.

Solution From Theorem 7,

$$\begin{aligned}
\log_{10} 2^{1.42} &= 1.42 \log_{10} 2\\
&= (1.42)(0.3010)\\
&= 0.4274
\end{aligned}$$

Therefore $\qquad\qquad\qquad\qquad 2^{1.42} = 2.675$

Problems 11.3

In Probs. 1 to 6 compute, using Table II, Appendix:

1	$6^{2.1}$	**2**	$4^{1.5}$
3	$e^{1.2}$	**4**	$e^{2.4}$
5	$10^{2.1}$	**6**	$8^{2.4}$

In Probs. 7 to 12 compute, using Table III, Appendix:

7	$6^{2.1}$	**8**	$4^{1.5}$
9	$e^{1.2}$	**10**	$e^{2.4}$
11	$10^{2.1}$	**12**	$8^{2.4}$

In Probs. 13 to 18 evaluate or simplify:

13	$3^{\log_3 3}$	**14**	$3^{\log_2 8}$
15	$25^{\log_{25} 5}$	**16**	$5^{\log_{216} 6}$
17	$36^{\log_{64} 8}$	**18**	$10^{\log_{10} x}$

19 Show that $x^x = a^{x \log_a x}$, $x > 0$.

20 Show that $(f(x))^{g(x)} = a^{g(x) \log_a f(x)}$, $f(x) > 0$.

21 Write the function f defined by $y = a^x b^x$ as an exponential function with base a.

22 Pick out the pairs of inverse functions, and state domain and range: 5^{6x}, 6^{5x}, 2^{-x}, $5 \log_6 x$, $6 + \log_5 x$, $\log_6 (5x)$, $\frac{1}{6} \log_5 x$, $\log_2 (-x)$, $-\log_2 x$.

23 By using seven-place common logarithms, we find that $(1 + \frac{1}{10})^{10} \approx 2.594$, $(1 + \frac{1}{100})^{100} \approx 2.705$, $(1 + \frac{1}{1,000})^{1,000} \approx 2.717$. Compare with the value of e.

11.4 Graphs

With a set of the standard mathematical tables at our disposal, we can now make light work of graphing various exponential, logarithmic, and related functions.

Illustration 1 Plot the graph of the function given by $y = xe^{-x}$.

Solution Again we prepare a table of x's and corresponding values of y and sketch the graph in Fig. 11.6.

x	-3	-2	-1	0	1	2	3
e^{-x}	20.09	7.39	2.72	1	0.37	0.14	0.05
$y = xe^{-x}$	-60.27	-14.78	-2.72	0	0.37	0.28	0.15

Graph of $y = xe^{-x}$

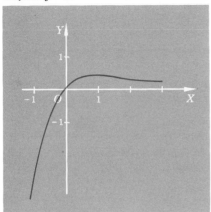

Figure 11.6
To find points on the graph of $y = xe^{-x}$, multiply the ordinates of $y = x$ and $y = e^{-x}$.

[By the methods of the calculus we can find that a maximum value of the function occurs at $x = 1$.]

Illustration 2 On the same axes and to the same scale plot the graphs of $y = e^x$, $y = \log_e x$.

Solution The functions defined by these equations are inverses of each other. We use a table of natural logarithms to prepare the following entries.

x	-3	-2	-1	0	0.2	0.5	1	2	3
$y = \log_e x$	......	......	......	$-\infty$	-1.61	-0.69	0	0.69	1.10
$y = e^x$	0.05	0.14	0.37	1	1.22	1.65	2.72	7.39	20.09

The graphs are plotted in Fig. 11.7.

Comparison of $y = e^x$ **and** $y = \log_e x$

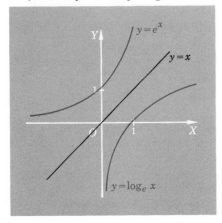

Figure 11.7
The graphs of $y = e^x$ and $y = \log_e x$ are symmetric about the line $y = x$.

Problems 11.4

In Probs. 1 to 13, sketch the graphs.

1 $y = \log_e |x|$ **2** $y = \log_e |2x|$
3 $y = |\log_e x|$ **4** $y = |\log_e 2x|$
5 $y = \log_e \sqrt{x}$ **6** $y = \log_e \sqrt{4x}$
7 $y = 10^{\sqrt{x}}$ **8** $y = \log_e x$
9 $y = \log_{10} x$ **10** $y = x \log_e x$
11 $y = e^{-x^2}$. (This graph is called the "probability curve".)
12 $y = \log_e e^x$
13 $y = e^{-2}2^x/x!$, for $x = 0, 1, 2, 3, 4$, etc. (This defines an important function in statistics; note that the domain is the positive integers and zero. It is called the "Poisson distribution function".)

In Probs. 14 to 19 solve the equation for x. Use tables in the Appendix. Do not interpolate.

14 $4^x = 10$ **15** $4.5^x = 10$
16 $4^{-x} = 16$ **17** $5^{-x} = 25$
18 $e^x - e^{-x} = 1$ HINT: Plot $y = e^x$ and $y = 1 + e^{-x}$. See where they intersect. Confirm by tables.
19 $e^x + 2e^{-x} = 3$ See hint for Prob. 18.

11.5 Applications

There are many problems in biology, chemistry, economics, etc., involving growth and decay for which the natural mathematical model is the exponential function. Our basic illustration is from the field of economics.

Illustration 1

compound interest

a An amount P dollars (principal) is invested at 100 per cent interest (rate), compounded annually. (The accrued interest is to be added to the principal.) Find the total amount A after 1 year.
b Same problem compounded monthly
c Same problem compounded daily (360 days/year)
d Same problem compounded continuously

Solution

a $A = P(1 + 1)$
b $A = P(1 + \frac{1}{12})^{12}$
c $A = P(1 + \frac{1}{360})^{360}$
d In order to arrive at something meaningful in this case we should begin with a description of what is meant by compounding "continuously".

At this time we can give only an intuitive explanation since a precise explanation involves the theory of limits. We would have an approximate answer if we compounded each second. A year (360 days) has 31,104,000 sec. The amount, at the end of 1 year, would be

$$A_{31,104,000} = P\left(1 + \frac{1}{31,104,000}\right)^{31,104,000}$$

We should like to know what, if anything, $A = P(1 + 1/n)^n$ would approach with ever-increasing n. The answer (beyond the scope of this text to develop) is Pe. That is, in technical language: "The limit of $(1 + 1/n)^n$, as n grows without bound, is e." Or, in symbols,

$$\lim_{n \to \infty} \left(1 + \frac{1}{n}\right)^n = e$$

If continuous compounding took place over a period of kt years, the amount would be given by

$$A = P \lim_{n \to \infty} \left(1 + \frac{1}{n}\right)^{nkt} = P \lim_{n \to \infty} \left[\left(1 + \frac{1}{n}\right)^n\right]^{kt} = Pe^{kt}$$

The same kind of problem arises in biology, where each of P cells in a given culture splits into two cells in a certain time t.

Illustration 2 The number of bacteria in a culture at time t was given by

$$y = N_0 e^{5t}$$

What was the number present at time $t = 0$? When was the colony double this initial size?

Solution When $t = 0$, $y = N_0 e^0 = N_0$. The colony will be $2N_0$ in size when t satisfies the equation $2N_0 = N_0 e^{5t}$, that is, when $5t = \log_e 2$ or when $t = \frac{1}{5} \log_e 2 = 0.6932/5 \approx 0.1386$ unit of time.

In chemistry, certain disintegration problems are similarly explained.

Illustration 3 Radium decomposes according to the formula $y = k_0 e^{-0.038t}$, where k_0 is the initial amount (corresponding to $t = 0$), and y is the amount undecomposed at time t (in centuries). Find the time when only one-half the original amount will remain. This is known as the "half-life" of radium.

radioactive decay

Solution We must solve $\frac{1}{2} k_0 = k_0 e^{-0.038t}$ for t.

$$\log_e \tfrac{1}{2} = -0.038t$$
$$-0.6932 = -0.038t$$
$$t = \frac{693.2}{38} = 18.24 \text{ centuries}$$

Illustration 4   Given that the half-life of a radioactive substance is 10 min, how much out of a given sample of 5 g will remain undecomposed after 20 min?

Solution   The substance decays according to the formula

$$y = 5e^{-kt}$$

First we must find k. From the given data

$$\tfrac{5}{2} = 5e^{-10k}$$

or

$$\tfrac{1}{2} = e^{-10k}$$

Taking natural logarithms of both sides, we have

$$-\log_e 2 = -10k$$

$$k = \frac{\log_e 2}{10}$$

Substituting back, we find

$$y = 5e^{-(\log_e 2)(t/10)}$$

$$= 5e^{-\frac{\log_e 2}{10} t}$$

When $t = 20$ min,

$$y = 5e^{-\frac{\log_e 2}{10} 20} = 5e^{\log_e 0.25}$$

$$= 5(0.25) = 1.25 \text{ g}$$

We could have seen this at once, for half remains after 10 min and so half of a half, or a quarter, remains after 20 min. The above method, however, will give us the answer for any time t.

Problems 11.5

In Probs. 1 to 6 solve for the unknown.

1   $3 = 5e^{0.1t}$

2   $5 = 2e^{0.5t}$

3   $2.5 = 3.4e^{-0.2t}$

4   $1.2 = 3.1e^{-0.3t}$

5   $2 = 3k4^{0.2}$

6   $3 = 2k3^{0.6}$

7   An approximation for the pressure p in millimeters of mercury at a height h km above sea level is given by the equation $p = 760e^{-0.144h}$. Find the height for which the pressure is one-half the pressure of sea level.

8   One "healing law" for a skin wound is $A = Be^{-n/10}$, where A (in square centimeters) is the unhealed area after n days and B (in square centimeters) is the original wound area. Find the number of days required to reduce the wound down to one-half the area.

9   A special case of Newton's Law for the rate r at which a hot body cools is $100 = 50e^{-0.25r}$. Find r.

10   How long will it take a sum to treble at 100 per cent interest compounded continuously?

11   Given that the half-life of a radioactive substance is 2,000 years, how much out of a given sample of 1 g will remain after 10,000 years?

12   A radioactive substance decays from 5 to 3 g in 4 days.   Find the half-life.

13   Let N be the number of π^0 mesons generated at time $t = 0$, and let

$$y = Ne^{-\left(\frac{10^{16}\log_e 2}{3}\right)t}$$

be the number at any subsequent time t (in seconds).   If only $N/2$ are present when $t = t_1$ sec, find t_1.

11.6   The Logarithmic Scale

slide rule

Ordinary addition can be performed mechanically quite simply by sliding one ruler along another as in Fig. 11.8.   We assume that the rulers are

A slide rule for addition

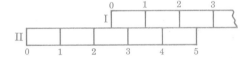

Figure 11.8
The two rules have identical linear scales.

graduated in the usual way with linear scales (see Sec. 2.11).   A linear scale is one in which the marks 1, 2, 3, . . . are placed 1, 2, 3, . . . units from one end (say the left).   Thus with ruler I in its present position we could add $2.5 + 1.5 = 4$, $2.5 + 3 = 5.5$.   A logarithmic scale is obtained by changing the coordinates x to new coordinates x' by means of the transformation $x' = \log_{10} x$, $x \geq 1$ (see Sec. 6.13).   Examine the scale in Fig. 11.9.   The

Logarithmic scale

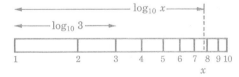

Figure 11.9
In a logarithmic scale the distance of the point labeled x from the left end is actually $\log_{10} x$.

distance from the left end to the mark 3 is not 3 units but is the logarithm (to base 10) of 3 units; that is, the distance is $\log_{10} 3 = 0.4771$ of the whole length.   Similarly the mark x is placed at a distance of $\log_{10} x$ units from the left end.   Note that the left end itself is marked 1 as it should be since $\log_{10} 1 = 0$.   If we placed two such scales side by side as in Fig. 11.10, we could add the logarithms of numbers and hence multiply the

Slide rule for multiplication

Figure 11.10
The two rules have identical logarithmic scales. Since $\log x + \log y = \log xy$, these rules can be used for multiplication.

numbers themselves. Thus $\log_{10} 2 + \log_{10} 3 = \log_{10} 6$. But since the scales are *marked* with units 2, 3, 6, etc., we *read* 2 (on the *D* scale) $\times$ 3 (on the *C* scale) $= 6$ (back on the *D* scale). In the same way we compute $2 \times 3\frac{1}{2} = 7$, $2 \times x = 2x$, etc.

Reading "backward", we perform division; thus $\dfrac{6(\text{on } D)}{3(\text{on } C)} = 2$ (on *D*, opposite 1 on *C*), etc. The usual slide rule also has scales that permit raising to powers and extraction of square and cube root. A slide rule is a useful aid in calculating where only two- or three-place accuracy is required. Instructions come with a rule.

In all of our graph work up to this point we have described and used but one type of graph paper, called rectangular coordinate paper, in which the rulings are laid out on linear scales. Many other types are available for special purposes. There is a type called polar coordinate paper which is useful for plotting curves in polar coordinates. (Sec. 10.11.)

It is appropriate at this time to mention briefly two other types that are in common use and are available at a bookstore. These are:

semilog paper
   1   Semilogarithmic (semilog) paper in which one axis has a linear scale, the other a logarithmic scale (see Fig. 11.11).

Semilog paper

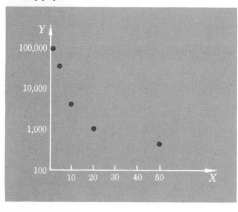

Figure 11.11
Semilog paper has a vertical logarithmic scale and a horizontal linear scale. See Illustration 1 for data.

log-log paper
   2   Double-logarithmic (log-log) paper in which each axis is marked with a logarithmic scale (see Fig. 11.12).

Exercise

A   Why is there no zero on a logarithmic scale?

Log-log paper

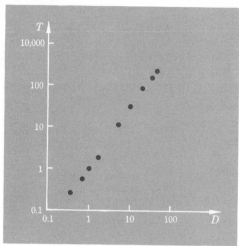

Figure 11.12
Both scales of log-log paper are log-arithmic. See Illustration 2 for data.

One use for semilog paper is for graphing functions that have both small and relatively large values in their range, such data as, for example, in the following illustration.

Illustration 1   Graph the function whose total set of elements is given by the following table.

x	50	20	8	3	1
y	500	1,000	5,000	25,000	100,000

Solution   First note that the domain is the set X of the five integers 50, 20, 8, 3, and 1. The range is the set Y of the five corresponding integers 500, 1,000, 5,000, 25,000, and 100,000. You can see that ordinary rectangular paper is inadequate because of the tremendous differences in magnitude of the values of y. These data might refer to the number of bank depositors x each writing checks (for an average month) with a total value y.

In order to get some geometric picture of this function, we resort to semilog paper with the linear scale on the X-axis and the logarithmic scale on the Y-axis. The graph is shown in Fig. 11.11. We do not connect the points, since there are no other elements of this function.

Note that a logarithmic scale goes by repetitive blocks, the set of marks representing 100, 200, 300, . . . , 900 being repeated in the same pattern for 1,000, 2,000, 3,000, . . . , 9,000. This is because of the decimal characteristics of the base 10 and is made clear by the following partial table. By blocks the mantissas repeat; the characteristics increase by 1. Semilog (and log-log) paper comes in several block styles. The one of Fig. 11.11 is three-block paper. In the printed forms the scales in each block run from 1 to 10; you will have to relabel them to fit the given problem.

y	$\log_{10} y$
.	.
.	.
.	.
100	2.0000
200	2.3010
300	2.4771
.	.
.	.
.	.
900	2.9542
1,000	3.0000
2,000	3.3010
3,000	3.4771
.	.
.	.
.	.
9,000	3.9542
.	.
.	.
.	.

curve
fitting

Another use for logarithmic paper is in the search for an equation which might be satisfied by a set of data obtained, perhaps, as the result of some kind of an experiment. In running a given experiment, a research worker may feel that there is possibly a simple "law" that the data should or might follow. If he were able to discover this law in the form of an equation, he could then use the equation for purposes of prediction.

Before we illustrate with an example we must make a few remarks about equations of the form

(1) $y = ae^{bx}$ [This defines an *exponential function*]
(2) $y = ax^b$ [This defines a *power function*]
(3) $y = ax + b$ [This defines a *linear function*]

In Sec. 6.5 we remarked that the graph of (3) is a straight line when plotted on rectangular coordinate paper. We have considered special cases of (1) and (2) before.

Let us take the logarithms of each side of (1) and (2); we do this in the following double column:

(4) $y = ae^{bx}$ $y = ax^b$
$\log_{10} y = \log_{10} a + bx \log_{10} e$ $\log_{10} y = \log_{10} a + b \log_{10} x$
        $= \log_{10} a + 0.43429\, bx$

Rewrite these in the form

(5) $Y = A + Bx$ $Y = A + bX$

where $Y = \log_{10} y$, $A = \log_{10} a$, $B = 0.43429b$, $X = \log_{10} x$.

With the introduction of the new variables X and Y, we see that each equation (4) can be written as a linear equation (5). Therefore each equation (5) will plot a straight line. Thus, if we had data such as

x	x_1	x_2	$\cdots$	x_n
y	y_1	y_2	$\cdots$	y_n

and suspected that they followed (approximately) an exponential law (power law), we could look up the logarithms and write them down in a table as follows:

	x	x_1	x_2	$\cdots$	x_n	I
	y	y_1	y_2	$\cdots$	y_n	
$X = \log_{10} x$		X_1	X_2	$\cdots$	X_n	II
$Y = \log_{10} y$		Y_1	Y_2	$\cdots$	Y_n	III

Plotting I against III on rectangular paper would yield a straight line if the original data followed an exponential law. Plotting II against III on rectangular paper would yield a straight line if the original data followed a power law.

But we can do better. If we plot x against y (original data) on semilog paper, we will get a straight line if the law is exponential since, in effect, the paper looks up the logarithms for us. Similarly x plotted against y on log-log paper will yield a straight line if the data follow a power law.

By plotting data on semilog or log-log paper it is therefore a simple matter to tell whether the law is exponential or power (or approximately so) by determining whether the points lie along a line (or nearly so). If the points lie along some other curve, we must resort to other methods to find a suitable equation. The general process is called *curve fitting*.

Illustration 2 The mean distance D of the planets from the sun and their periods of revolution (T years) are given by the table. (The distance of the earth from the sun is taken as one unit.) Discover the (approximate) law.

	Mercury	Venus	Earth	Mars	Jupiter	Saturn	Uranus	Neptune	Pluto
T	0.241	0.615	1.00	1.88	11.9	29.5	84.0	165	265
D	0.387	0.723	1.00	1.52	5.20	9.54	19.2	30.1	41.3

Solution We plot the data on rectangular, semilog, and log-log paper. The results are shown in Fig. 11.12 for log-log paper (three-block by four-block). We take D as independent variable.

The points lie (almost) on a straight line, and thus the law is (approximately) given by $T = aD^b$. It turns out (we will not compute it here) that $a = 1$ and $b = \frac{3}{2}$ (approximately) so that the final answer is $T = D^{3/2}$ or $T^2 = D^3$. Thus the square of the time is the cube of the distance. This is known as *Kepler's Law*.

Exercises

B For Illustration 2, show the graphs of the data on rectangular and semilog paper.

C Plot the graph of the equation $T = D^{3/2}$ on rectangular paper.

Problems 11.6

1 Given that the number N of bacteria in a culture at time t is given by $N = ae^{bt}$ and that $N(0) = 100$, $N(1) = 1,000$; find $N(2)$.

2 If $y = ae^{bx}$, $y(1) = 1$, and $y(2) = 4$, find $y(0)$.

3 Discover an approximate law from the following data:

x	0	1	3	5	8
y	-1	2	8	14	24

4 The speed s of a certain chemical reaction doubles every time the temperature $T°$ is raised $5°$. Make out a table of some of the elements of the function $f:(T,s)$ thus defined, and discover the type of law. Let one element of f be $(0°,1)$.

5 The total adsorption (x cubic units) of a certain gas by another chemical varied with time (t units) as follows:

x	0	1	2	3
t	0	2	16	53

Discover an appropriate law.

In Probs. 6 to 15 name the kind of paper on which the graph is a straight line.

6 $y = 3x^2$

7 $y = 6e^{-2x}$

8 $y = ee^x$

9 $y = ex^e$

10 $y = ee^e$

11 $xy = 2$

12 $x^2y = 2$

13 $xy^2 = 2$

14 $y = a2^{bx}$

15 $y = 3x - 3$

11.7   Elasticity of Demand

In this section we return to the discussion of the demand curve which was introduced in Sec. 9.8. The problem before us here is the following: Suppose that we are at a particular point of a demand curve (i.e., our commodity is selling at a certain price and the corresponding demand is determined from the demand curve); what happens to the demand if the price is *reduced* by a small amount? If the demand curve is very steep, the demand will hardly increase at all; but if the demand curve is flat, the demand will increase substantially. We might say that for a steep demand curve the demand is *inelastic* and that for a flat demand curve the demand is *elastic*.

This intuitive definition of elasticity of demand has a serious technical flaw. The flatness or steepness of the demand curve depends heavily upon the units in which q and p are measured and upon the scales used along the Q- and P-axes. For example, the demand curve for cigarettes in Figs. 9.18 and 9.21 is fairly flat at $p = 4$, but the same curve is quite steep at $p = 4$ in Fig. 9.23. This difference is explained by our use of two different scales in plotting the values of q. A change of units from dollars/ 1,000 cigarettes to dollars/10,000 cigarettes or to cents/100 cigarettes, while using the original vertical scale, would have had the same effect on the demand curve. Since our economic conclusions should not be dependent on the units used in our measurements, the elasticity of demand cannot be determined from the demand curve except in very special cases. Nevertheless, the idea of elasticity is an important one which deserves a satisfactory treatment. Our purpose in this section is to describe one way of accomplishing this; an equivalent, but more precise, theory can be obtained through the use of calculus.

log demand
curve

The idea is to replace the ordinary demand curve with the *logarithmic demand curve*. In this curve we plot log q horizontally and log p vertically. The base of logarithms is optional. Instead of the demand function $q = D(p)$, we have the *logarithmic demand function* log $q = D_L(\log p)$. Rather than look up the logarithms of p and q in a table, our procedure will be to plot the demand curve on log-log paper. This will automatically give us the logarithmic demand curve.

Illustration 1   The logarithmic demand curve for bread (see Fig. 9.17) is given in Fig. 11.13.

Illustration 2   The logarithmic demand curve for cigarettes (see Fig. 9.18) is given in Fig. 11.14.

Now suppose that the unit of price is changed so that the price p' in the new units is given by $p' = cp$, where p is in the old units. In the case of cigarettes,

Logarithmic demand curve for bread

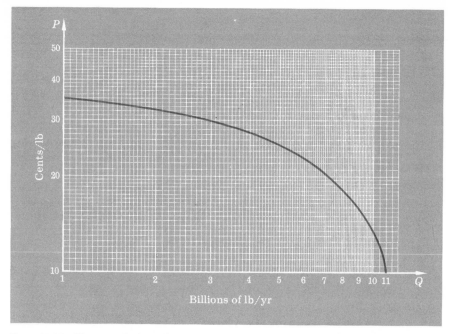

Figure 11.13 The points (q,p) satisfying $q = (75 - 2p)/5$ are plotted on log-log paper.

$$p' = \frac{\text{dollars}}{10,000 \text{ cigarettes}} \quad \text{and} \quad p = \frac{\text{dollars}}{1,000 \text{ cigarettes}}$$

so that $p' = p/10$ and $c = 1/10$. Then $\log p' = \log c + \log p$. Thus the graph of $\log p'$ vs. $\log q$ is obtained from the graph of $\log p$ vs. $\log q$ by raising the latter by $\log c$ (or by lowering it if $\log c$ is negative). Similarly, if the unit of quantity is changed, the graph will move to the right or left. Since we shall focus our attention on the *shape* of the demand curve and not on its *position* on the axes, we conclude that these changes of units do not affect the logarithmic demand curve in any essential way.

Returning to the idea of elasticity, we call the demand elastic at a point if the logarithmic demand curve is flat there and inelastic at a point if the curve is steep there. Thus from Fig. 11.13 we see that the demand for bread (at least according to our demand curve) is elastic for high prices and inelastic for low prices. This intuitive approach still requires precision. Where is the point at which elastic turns into inelastic? How flat must the curve be to be called elastic? We need some standard of elasticity.

This standard is chosen in the following way. Draw a line on the log-log paper which cuts the two axes at equal distances from the lower, left-hand corner of the paper; i.e., the line makes an angle of 45° with each axis. If possible, draw this line so that it is tangent to the logarithmic demand curve at some point U. At U the demand is said to have *unit elasticity*. The demand is *inelastic* where the curve is *steeper* than this line and is *elastic* where it is *flatter*.

unit
elasticity

Logarithmic demand curve for cigarettes

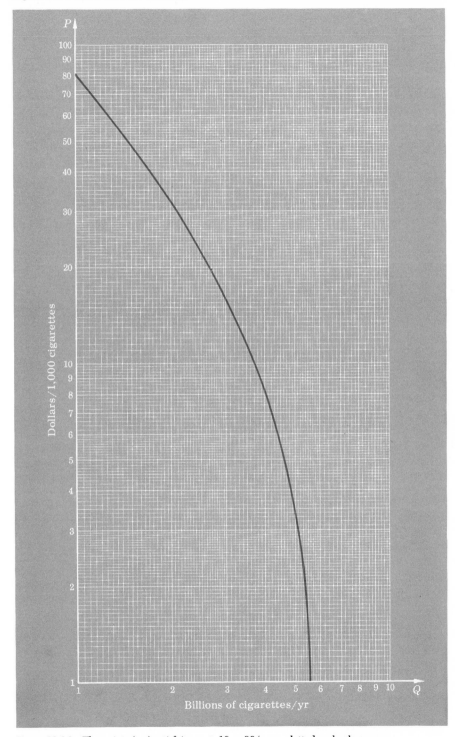

Figure 11.14 The points (q,p) satisfying $p + 16 = 96/q$ are plotted on log-log paper.

Illustration 3   Find the point of unit elasticity U for bread using the demand curve whose equation is $q = (75 - 2p)/5$. For what values of p is this demand elastic? Inelastic?

Solution   We start from Fig. 11.13 and construct the required tangent by trial and error as in Fig. 11.15. The point U is approximately $q = 7\frac{1}{2}$, $p = 18\frac{3}{4}$. The demand is elastic for $p > 18\frac{3}{4}$ and is inelastic for $p < 18\frac{3}{4}$.

Elasticity for bread

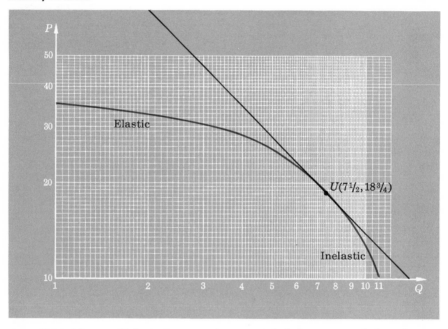

Figure 11.15   The point U of unit elasticity is the point at which the tangent to the logarithmic demand curve makes angles of $45°$ with each axis.

With the techniques available to us so far, we can determine U only by this graphical method. It can be found precisely by the methods of calculus. In the answers to the problems below, we have calculated U by calculus. Your graphical solution should be a close approximation to given answers.

Problems 11.7

For Probs. 1 to 10, Sec. 9.8, draw the logarithmic demand curve on log-log paper. By a graphical technique approximate U and indicate where the demand is elastic and where inelastic.

References

In addition to the many standard textbooks on algebra, the reader should consult the following articles in the *American Mathematical Monthly*.

Cairns, W. D.: Napier's Logarithms as He Developed Them, vol. 35, p. 64 (1928).

Cajori, Florian: History of the Exponential and Logarithmic Concepts, vol. 20, pp. 5, 20, 35, 75, 107, 148, 173, 205 (1913).

Huntington, E. V.: An Elementary Theory of the Exponential and Logarithmic Functions, vol. 23, p. 241 (1916).

Lenser, W. T.: Note on Semi-logarithmic Graphs, vol. 49, p. 611 (1942).

Sandham, H. F.: An Approximate Construction for *e*, vol. 54, p. 215 (1947).

Thomas, J. M.: Pointing Off in Slide Rule Work, vol. 55, p. 567 (1948).

PROBABILITY

Chapter 12

12.1   Random Experiment

The mathematical problems we have discussed thus far in this book have led to definite answers such as

$(-3)(-2) = 6.$
$\{1, 2, 3\} \cap \{2, 3, 4, 5, 6\} = \{2, 3\}.$
The complex numbers form a field.
No real number x satisfies the equation $x^2 + 1 = 0$.

We now wish to consider some problems for which there are no definite answers but which are, nevertheless, subject to logical analysis. A large body of such material is treated in the branch of mathematics called the *theory of probability,* the basic idea being that of a *random experiment.*

The term random experiment, or random phenomenon, is not usually defined in mathematics, but is used in reference to any and all physical experiments or other situations where nonidentical data result from what appear to be essentially identical processes. Repeated measurements on some physical object, such as measurements of length, weight, and velocity, generally lead to a set of readings not all of which are alike regardless of the care we exercise in making them. Random variations, sometimes called random errors, creep in quite beyond our precautions. A stamping machine may on occasion, and for some unknown reason, produce a faulty part. As yet we cannot control the sex of an embryo, which may therefore be considered a random phenomenon. A name taken from a list of registered voters may turn up a Democrat or Republican. A fair toss of a coin or die results in one of the outcomes head (H), tail (T), or 1, 2, 3, 4, 5, 6, entirely at random.

These notions are applicable to a wide variety of problems in games of chance and in the physical, biological, and social sciences and to many other areas of human endeavor.

12.2   The Sample Space

A random experiment may lead not only to many outcomes but also to different categories, or sets, of outcomes, and our first task is to decide what we mean by a permissible set of outcomes. Let us suppose that we draw a card from a standard deck and find that it is the five of clubs. We may focus our attention upon one or more of the following *outcomes* of this draw: (1) the card is a club; (2) the card is black; (3) the card is a five; or possibly (4) the card has a cigarette burn on it. All these are perfectly acceptable *outcomes,* but we must decide which of them is of interest to us at the moment. In making the draw we are seeking the answer to one or more of the questions: (1) What suit is the card? (2) What color is the card? (3) What is the denomination of the card? (4) What is the physical condition of the card? The possible answers to these questions form sets of outcomes which we shall call *permissible.*

Definition   A set of outcomes of a random experiment is called *permissible*
permissible
set of
outcomes

if and only if:

1   It is exhaustive; i.e., we can assign at least one element of this set to every performance of the experiment.
2   The elements of the set are mutually exclusive; i.e., we can assign no more than one element of this set to a single performance of the experiment.

Illustration 1
a   The following are permissible sets of outcomes of the experiment of drawing one card from a deck (let c, d, h, s refer to club, diamond, heart, spade, respectively): {red, black}; {c, d, h, s}; {burned, unburned}; {2, 3, 4, 5, 6, 7, 8, 9, 10, J, Q, K, A}; {$2c$, $3c$, . . . , Ac, $2d$, $3d$, . . . , Ad, $2h$, $3h$, . . . , Ah, $2s$, $3s$, . . . , As}.
b   The set {2, 3, 5} is not permissible since it is not exhaustive.
c   The set {red, black, club} is not permissible since its elements are not mutually exclusive. For instance, the three of clubs would be assigned to two elements of this set.

The technical words used to describe this situation are given below:

Definition   A *sample space* is a permissible set of outcomes of a random
sample space
simple event

experiment. A *simple event* is an element of a sample space.

Notation   We shall write e_1, e_2, e_3, etc., for simple events, and $U = \{e_1, e_2, e_3, . . . , e_n\}$ for a sample space. Thus, for a deck of cards, we might write $U = \{e_1, e_2, . . . , e_{52}\}$, where each e_i refers to a specific card.

Illustration 2 In a single toss of a coin, where we are interested only in knowing whether head or tail shows, we use as sample space $U = \{H, T\}$.

Illustration 3 In a single toss of an ordinary die, we take as sample space $U = \{1, 2, 3, 4, 5, 6\}$.

Illustration 4 An experiment consists of drawing a card from a deck and tossing a coin. $U = \{e_1H, e_2H, \ldots, e_{52}H, e_1T, e_2T, \ldots, e_{52}T\}$ is a permissible sample space that contains all the possible outcomes which are usually considered pertinent.

The situation in Illustration 4, where two distinct experiments have been combined into one, might be described as a joint experiment having a joint sample space. The importance of this point of view is explained in Sec. 12.6.

Illustration 5 An experiment consists of making random drawings from a hat containing a nickel (N), a dime (D), and a quarter (Q).

a One coin is drawn. $U_1 = \{N, D, Q\}$.
b Two coins are drawn simultaneously. $U_2 = \{ND, NQ, DQ\}$.
c One coin is drawn and examined. Then another coin is drawn. $U_3 = \{ND, DN, NQ, QN, DQ, QD\}$. Here we are interested in the *order* in which the coins are drawn.
d One coin is drawn and examined, and then it is returned to the hat. A second drawing is made. An appropriate sample space is $U_4 = \{ND, DN, NQ, QN, DQ, QD, NN, DD, QQ\}$.

Suppose that, in our card-drawing experiment, we have chosen the sample space $U = \{2, 3, 4, 5, 6, 7, 8, 9, 10, J, Q, K, A\}$ and that our interest now centers on whether the draw is below an 8. The simple events satisfying this condition form a subset of U, namely, $\{2, 3, 4, 5, 6, 7\}$. We shall call such a subset an *event*. In general, we have the definitions:

Definition An *event* is a subset of a sample space U. Two events A and

event
B are called *complementary events* if A and B are complementary subsets of U; i.e. if $A' = B$ and $B' = A$.

The subsets of $U = \{e_1, e_2, \ldots, e_n\}$ are:

$\emptyset$, the null set: this is an impossible event.
$\{e_1\}, \{e_2\}, \ldots, \{e_n\}$; each of these is a simple event.[1]
$\{e_1, e_2\} = \{e_1\} \cup \{e_2\}, \{e_1, e_3\} = \{e_1\} \cup \{e_3\}, \ldots, \{e_{n-1}, e_n\} = \{e_{n-1}\} \cup \{e_n\}$; each is the union of two simple events.

[1] Usually, we shall omit the braces in the case of a simple event, writing e_1 instead of $\{e_1\}$, etc.

$\{e_1, e_2, e_3\} = \{e_1, e_2\} \cup \{e_3\} = \{e_1\} \cup \{e_2\} \cup \{e_3\}, \ldots, \{e_{n-2},$
$e_{n-1}, e_n\} = \{e_{n-2}\} \cup \{e_{n-1}\} \cup \{e_n\}$; union of three simple events.

. . . .

$U = \{e_1, e_2, \ldots, e_n\} = \{e_1\} \cup \{e_2\} \cup \ldots \cup \{e_n\}$; the whole sample space. Our interest will be in finite sample spaces, namely those which contain only a finite number of simple events.

All the five regular solids make beautiful *dice*. Inasmuch as many of the problems in this chapter are connected with them, they are shown in Figs. 12.1 to 12.5, along with ways of constructing them.

The five regular solids

Figure 12.1   Tetrahedron

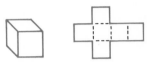

Figure 12.2   Cube

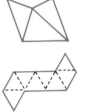

Figure 12.3   Octahedron

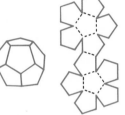

Figure 12.4   Dodecahedron

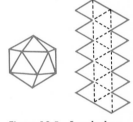

Figure 12.5   Icosahedron

Problems 12.2

In Probs. 1 to 16 set up a sample space.

1   Two coins are tossed once.
2   Three coins are tossed once.
3   A coin and a die are tossed once.
4   Two ordinary dice are tossed once.
5   A tetrahedral (regular) die is tossed once. Count the *down* face.
6   A tetrahedral die is tossed once. Count the sum of the visible faces.
7   A regular homogeneous octahedron (8 faces) is tossed once. Count the *up* face.
8   A regular dodecahedron (12 faces) is tossed once. Count the *up* face.
9   A regular icosahedron (20 faces) is tossed once. Count the *up* face.
10   One toss of a coin with two heads.
11   One toss of a die marked 1, 1, 2, 3, 4, 5.
12   One toss of a die marked 1, 1, 1, 2, 2, 3.
13   A certain radio tube consists of just three parts, a grid, a filament, and a plate, each one of which (when first made) might test defective or not.

14 A random sample of two people is drawn from a population of 1,000 Democrats and 500 Republicans. (The numbers are extraneous.)

15 Assume that, in a certain town, the population of all three-children families in which only one is a boy shows the following distribution according to the order of the male birth: male first child, 25 cases; second child, 35 cases; third child, 40 cases. (The numbers are extraneous.)

16 Our present calendar repeats in cycles of 400 years. In each cycle the 13th of the month falls on the days of the week with the frequencies $S(687)$, $M(685)$, $T(685)$, $W(687)$, $T(684)$, $F(688)$, $S(684)$. (The numbers are extraneous.)

In Probs. 17 to 20 which are permissible sets of outcomes, i.e., sample spaces? Explain.

17 Of the 10 digits one is selected at random.

   a $U = \{0, 1, 2, 3, 4, 5, 6, 7, 8, 9\}$

   b $U = \{0, 2, 4, 6, 8\}$

   c $U = \{\text{even, odd}\}$

   d $U = \{\text{less than 2, 2, greater than 2}\}$

   e $U = \{0, 1, 2, 3, \text{less than 9, 9}\}$

18 Two coins and a die are tossed.

   a $U = \{(\text{coins alike, die shows 1}), (\text{alike, 2}), \ldots, (\text{alike, 6})\}$

   b $U = \{(\text{coins alike, die is even number}), (\text{alike, odd number})\}$

   c $U = \{(\text{coins alike, die is 1}), \ldots, (\text{alike, 6}), (\text{unlike, 1}), \ldots, (\text{unlike, 6})\}$

   d $U = \{(H, H, \text{die shows number} <6), (H, T, <6), (T, H, <6), (T, T, <6)\}$

   e $U = \{HH, HT, TH, TT\}$

19 Three dice are tossed.

   a $U = \{(a, b, c)\,|\,1 \leq a \leq 6,\ 1 \leq b \leq 6,\ 1 \leq c \leq 6\}$

   b $U = \{(3 \leq a + b + c \leq 18)\,|\,1 \leq a \leq 6,\ 1 \leq b \leq 6,\ 1 \leq c \leq 6\}$

   c $U = \{\text{sum even, sum odd}\}$

   d $U = \{\text{number on first die even, number on first die odd}\}$

20 Two icosahedra (one is red, one is green) are tossed.

   a $U_1 = \{(a, b)\,|\,1 \leq a \leq 20,\ 1 \leq b \leq 20\}$

   b $U_2 = \{\text{sum even, sum odd}\}$

   c $U_3 = \{\text{neither number is prime, exactly one is prime, two primes}\}$

   d $U_4 = \{(\text{red is even, green is odd})\}$

   e $U_5 = \{\text{red is even, green is odd}\}$

   f $U_6 = \{(\text{red is even, green is odd}), (\text{red is odd, green is even})\}$

   g $U_7 = \{(\text{red is even, green is even}), (\text{red is odd, green is odd})\}$

   h $U_8 = U_6 \cup U_7$ (Union of U_6 and U_7; either U_6 or U_7, or both.)

   i $U_9 = U_6 \cap U_7$ (Intersection; both U_6 and U_7.)

   j $U_{10} = U_1 \cup U_2$

   k $U_{11} = U_1 \cup U_2'$ (U' is complement of U.)

12.3   Frequency Definition of Probability

frequency

Consider that an experiment is designed and that the category of outcomes is stated; i.e., the sample space $U = \{e_1, e_2, \ldots, e_n\}$ is specified. The experiment is now repeated N times (N large), and the number of times an event A occurs is found to be N_A. N_A is the *absolute frequency* of A, and $f_1(A) = N_A/N$ is the *relative frequency* of A in N trials. The experiment is again repeated, say, M times (M large). The relative frequency of A in M trials is $f_2(A) = M_A/M$. A sequence of such repetitions produces a sequence of relative frequencies $f_1(A), f_2(A), \ldots, f_i(A), \ldots$, and it is man's experience that these relative frequencies do not differ greatly from one set of repetitions to another, but seem to cluster around a fixed but unknown number $P(A)$. Because of this experience, we lay down the following hypothesis, the validity of which is essential if the theory of probability is to be applicable to the physical world.

Hypothesis   There exists a constant $P(A)$ around which the experimental frequencies $f_i(A)$ cluster.

Definition   The number $P(A)$ is called the *probability* of the event A.

probability

Thus the relative frequencies $f_i(A)$ are experimental approximations to $P(A)$. If we have only one such, say, $f(A)$, then the best we can do on an experimental basis is to use it for $P(A)$.

Illustration 1   The faces of a hexagonal pencil are marked 1 to 6, and some of the wooden edges are shaved off a bit here and there. This makes a biased object, a roll of which on a smooth floor constitutes a random experiment. The sample space is $U = \{1, 2, 3, 4, 5, 6\}$. If we wish to estimate $P(4)$, the probability that face 4 will turn up, we must make a long sequence of rolls and compute $f(4)$, which can then be taken as an approximation to $P(4)$. Essentially, there is no other way of getting an estimate of $P(4)$.

Illustration 2   What is the probability that the next child born in the United States will be a girl? This type of problem cannot be treated without some sensible information as to how we should assign probabilities to the two sexes. Biologically, we do not know too much about the way in which sex is determined. It is only natural that we turn to the records of some past time interval. United States vital statistics indicate that in 1957 the total number of live births was 4,254,784, of which 2,074,824 were girls. Counting these cases as a long run of a random experiment, we find that the answer to the question is about $2{,}074{,}824/4{,}254{,}784 = 0.4876$. For a number of years there have been fewer female than male births.

Exercise
A.   What is the sample space in Illustration 2?

Illustration 3 If today is your twentieth birthday, what is the probability that you will be alive on your twenty-fifth birthday? To get some kind of meaningful answer, we think of dying as a random process and search the records for a long run of such events. Mortality-experience tables used by insurance companies indicate that, of 95,991 persons (United States) alive on their twentieth birthday, 95,400 can be expected to be alive on their twenty-fifth. Your probability is therefore about 95,400/95,991 = 0.9938. But drive safely.

Exercise

B What is the sample space in Illustration 3?

Illustration 4 A taw, or die, is made from some homogeneous material in the form of a right prism with equilateral triangles for base and top and with square sides (Fig. 12.6). The base and top are marked 1, 2, and the sides 3, 4, 5. A sample space is $U = \{1, 2, 3, 4, 5\}$, and a toss is a random experiment. Undoubtedly, the die could come to rest with any number 1 to 5 *down*, and, because of the symmetries involved, we are willing to believe that $f(1) \approx f(2)$ and $f(3) \approx f(4) \approx f(5)$ in a long sequence of tosses. Even before the experiment, we should consider it appropriate to assign $P(1) = P(2)$ and $P(3) = P(4) = P(5)$, but a sensible assignment could be made only after experimentation. Another interesting sample space for this die is $U = \{\text{triangle down, square down}\}$.

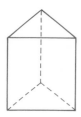

Figure 12.6

Most of our work will be with conceptual rather than actual experiments, but it will be highly instructive for you to carry out one or more of the random experiments in the problems below.

Problems 12.3

1 Tossing a thumbtack, which may fall point up or point down, is a random experiment. Estimate the probability of point up. What is the sample space? To simplify the work, put 10 like thumbtacks in a box, shake, and examine for the number showing point up. Do this 10 times. This will give a reasonable approximation to the problem of making 100 tosses with a single tack.

2 Repeat as in Prob. 1 with 10 new bottle caps to estimate the probability of top up. What is the sample space?

3 Stack eight pennies head to tail and smoothly scotch-tape. This is now a three-sided die. Set up the sample space, toss 50 times, and estimate the probability of "cylindrical side".

4 Take four paper book matches (unused) and toss them about a foot above your head and let fall on a hard, smooth floor, such as asphalt tile. You may very well find some instances where a book comes to rest on an edge. Take $U = \{$scratch side up, scratch side down, edge$\}$ as sample space and estimate the probability of edge from a sequence of 25 tosses.

5 Glue together a dime and a quarter, tail to tail. Toss 25 times and estimate the probability of dime up.

6 Cover a sheet of paper with parallel lines drawn 1 in. apart (or use graph paper so ruled). Cut a square ½ in. on a side from some reasonably stiff paper or cardboard. Drop the square from about 1 ft onto the rulings $n = 30$ times, counting the number of times it intersects a ruling. Estimate the probability P of an intersection. (Here there is a theoretical answer: $P \to 2/\pi$ as $n \to \infty$.)

12.4   Idealized Probability

Casual examination of a coin reveals that it is far from being a symmetric cylinder. Cylinders, cubes, and the like do not exist in the real world, but only in our minds. Like points and lines in geometry, they are abstractions of things we see in nature and, as such, have application to physical objects that are *almost* cylinders or *almost* cubes.

In the theory of probability it is highly desirable to idealize a random experiment—*by making suitable hypotheses*—so that we may shift attention from relative frequency, $f(A)$, directly to probability, $P(A)$. This affords economy of time and speeds applicability, whenever it can be done, and is especially helpful in problems involving games of chance.

Illustration 1   What is the probability of tossing a head with a penny? We take as sample space $U = \{H, T\}$, make a long sequence of fair tosses, and compute $f(H)$. Or we may repeat this many times, average the f's, and use the average $f(H)$ as an approximation for $P(H)$. If a sequence consisted in as many as 9^9 tosses, we have no doubt that $f(H)$ would be near ½, but that actually, because a coin *is* biased, $f(H)$ would differ from ½ in such a way as to make the assumption that $P(H) = \frac{1}{2}$ quite untenable. On the other hand, in a short game of tossing, the *assumption* that $P(H) = \frac{1}{2}$ is reasonable. We therefore idealize the coin, thinking of it as a perfect cylinder, reason that in a long run we should expect to find $f(H)$ just about ½, dispense with the testing, and *arbitrarily* assign $P(H) = \frac{1}{2}$. We also assign $P(T) = \frac{1}{2}$ for similar reasons. Note that $P(H) + P(T) = 1$. This is the common basis for action in games of chance using honest coins.

Illustration 2   What is the probability of tossing a 6 with an ordinary die which is made of homogeneous material and marked in the usual way with 1, 2, 3, 4, 5, 6 dots on the six faces, respectively? To make a dot, a small amount of material, roughly hemispherical, is removed and the hole painted. This method of manufacture obviously produces bias, and the face with only one dot, being the heaviest, should turn down most often, placing uppermost the opposite face, with six dots. Thus, calling $f(i)$ the relative frequency of the face with exactly i dots, we should expect to find, in a long sequence of tosses, $f(6) > f(1)$. With some 315,672 tosses, the English biologist Weldon verified this conjecture; his experiments showed that we are unjustified in making the hypothesis that $f(6)$ is about ⅙. [Actually, Weldon tossed 12 dice 26,306 times, counting a five or a six as a success. Ideally, $P(5 \text{ or } 6)$ would be assumed to be ⅓ for a symmetric die. He found a relative frequency of 0.3377, which is significantly too high.] However, in games of chance involving dice, this bias is usually disregarded, complete symmetry is assumed, and, arbitrarily, to each face is assigned the same probability ⅙ so that $\sum_{i=1}^{6} P(i) = P(1) + P(2) + P(3) + P(4) + P(5) + P(6) = 1$.

Illustration 3   A gadget resembling a clock face is marked uniformly around the perimeter with the integers 1 to 12. A central hand is rotated at high uniform angular velocity, and when a button is pressed, an instrument will record the next number which the sweeping hand passes. What is the probability that the number will be 5? Here again is an example of a random process. It seems reasonable to assign equal probabilities to each number so that the answer to the question should be ¹⁄₁₂, because, no doubt, the experimental method would yield a relative frequency $f(5)$ very close to ¹⁄₁₂ for large N. This means, of course, that we are assuming that $\sum_{i=1}^{12} P(i) = 1$.

Problems 12.4

Set up a sample space, idealize the random experiment, and make suitable probability assignments assuming regular solids, regular markings (or as indicated), and one toss. For a sample space $U = \{e_1, e_2, \ldots, e_n\}$ be sure that $\sum_{i=1}^{n} P(e_i) = 1$.

1 Regulation coin
2 Coin with two heads
3 Tetrahedron
4 Tetrahedron marked 1, 1, 2, 3
5 Cube marked 1, 2, 2, 3, 3, 3
6 Cube marked 1, 2, 3, 4, 5, blank
7 Octahedron
8 Octahedron marked 1, 1, 1, 1, 2, 2, 2, 3
9 Dodecahedron

10 Dodecahedron marked with eight 1's, three 2's, and one 3
11 Icosahedron
12 Icosahedron marked with seventeen H's, three T's

12.5 The Probability Distribution

These somewhat vague and intuitive notions must be formalized. Associated with every random experiment there is a sample space $U = \{e_1, e_2, \ldots, e_n\}$. To each simple event e_i there must be assigned a probability $P(e_i)$, and this must be done solely on the basis of experimentation or hypothesis. In a long sequence of N repetitions of the experiment there are certain relations that hold for relative frequencies, and these shall serve as guides in assigning probabilities. These relations are:

I Let each simple event e_i occur with absolute frequency N_i. Then, since $f(e_i) = N_i/N$, the relative frequency $f(e_i)$ satisfies the double inequalities $0 \leq f(e_i) \leq 1$. This suggests that the assignment $P(e_i)$ be made in such a way that $0 \leq P(e_i) \leq 1$.

II The fact that

$$f(e_1) + f(e_2) + \cdots + f(e_n) = \frac{N_1}{N} + \frac{N_2}{N} + \cdots + \frac{N_n}{N}$$

$$= \frac{N_1 + N_2 + \cdots + N_n}{N} = \frac{N}{N}$$

$$= 1$$

suggests that the assignments be made in such a way that

$$P(e_1) + P(e_2) + \cdots + P(e_n) = 1$$

III Let s simple events, say, $e_1, e_2, \ldots, e_s$, occur with absolute frequencies $N_1, N_2, \ldots, N_s$, respectively, and let the event A be the union of these s simple events. (There is no loss in generality in using the first s simple events, since the order of writing down the simple events is unimportant.) Thus $A = e_1 \cup e_2 \cup \ldots \cup e_s$. Then

$$f(A) = \frac{N_1 + N_2 + \cdots + N_s}{N} = \frac{N_1}{N} + \frac{N_2}{N} + \cdots + \frac{N_s}{N}$$

$$= f(e_1) + f(e_2) + \cdots + f(e_s)$$

This suggests that we take $P(A) = P(e_1) + P(e_2) + \cdots + P(e_s)$.

IV Finally, the relative frequency of an impossible event $\emptyset$ is $f(\emptyset) = 0/N = 0$. This suggests the assignment $P(\emptyset) = 0$.

These intuitive considerations supply the motivation for a formal definition.

Definition   A *probability distribution* is an assignment of a real number to each event of a sample space, $U = \{e_1, e_2, \ldots, e_n\}$, such that the following four conditions are satisfied:

probability
distribution

   I   For each simple event e_i, $0 \leq P(e_i) \leq 1$.

  II $\sum_{i=1}^{n} P(e_i) = 1$.

 III   If $A = e_1 \cup e_2 \cup \cdots \cup e_s$, then $P(A) = \sum_{i=1}^{s} P(e_i)$.

  IV $P(\emptyset) = 0$.

When probabilities are assigned in this manner to, or *distributed over,* the simple events of U, we say that a probability model, or *probability distribution over U,* has been established for the experiment.

Definition   The probability distribution in which there has been assigned

uniform
distribution

the same real number to each simple event is called the *uniform probability distribution,* or simply the uniform distribution.

Conditions I and II are the important ones in the definition because they must be followed when the probabilities are initially assigned. Fortunately, they are easily checked in most instances. Condition III simply states what shall be regarded as the probability of event A (any subset of U). Condition IV only rarely enters into numerical work, but is useful in proving theorems.

Theorem 1 $P(U) = 1$.

This is an immediate consequence of Conditions II and III, since

$$U = e_1 \cup e_2 \cup \cdots \cup e_n, \; P(U) = \sum_{i=1}^{n} P(e_i) = 1$$

Theorem 2   For any event A, $P(A)$ satisfies the double inequality $0 \leq P(A) \leq 1$.

This follows since A is a subset of U.

Exercise

A   Write out the details of the proof of Theorem 2.

Theorem 3   If A and B are any two events such that

$$A \cap B = e_1 \cup e_2 \cup \cdots \cup e_t$$

then $P(A \cap B) = \sum_{i=1}^{t} P(e_i)$.

This follows from Condition III.

Exercise

B By considering relative frequencies, prove that

$$f(A \cup B) = f(A) + f(B) - f(A \cap B)$$

The result in Exercise B suggests the next theorem.

Theorem 4 $P(A \cup B) = P(A) + P(B) - P(A \cap B)$.

Proof A is a subset of U and is the union of some simple events. If we add the probabilities of these simple events, the result is $P(A)$ by Condition III. Likewise, B is the union of simple events some of which (say, $e_j, \ldots, e_k$) overlap with those in A. If we add the probabilities of the simple events in B, the result is $P(B)$. But in $P(A) + P(B)$ we have included $P(e_j) + \cdots + P(e_k)$ twice, once in $P(A)$ and once in $P(B)$. We should therefore subtract $P(e_j \cup \cdots \cup e_k)$ from $P(A) + P(B)$ in order to get $P(A \cup B)$. But, since $e_j \cup \cdots \cup e_k = A \cap B$, it follows that $P(e_j \cup \cdots \cup e_k) = P(A \cap B)$. Hence the proof is complete.

Stated in words, Theorem 4 says that the probability that at least one of two events occurs is the sum of the probabilities of the separate events minus the probability that both events occur. This is often referred to as the *addition rule of probability*.

Definition   Two events A and B are *mutually exclusive* if and only if
mutually
exclusive $A \cap B = \emptyset$.
events

Theorem 5   If A and B are mutually exclusive events, then

$$P(A \cup B) = P(A) + P(B)$$

Proof $A \cap B = \emptyset$; therefore, $P(A \cap B) = P(\emptyset) = 0$, by Condition IV.

Exercises

C Complete the proof of Theorem 5.
D State and prove the converse of Theorem 5.

Theorem 6   If every pair A_i, A_j of r events $A_1, A_2, \ldots, A_r$ are mutually exclusive, then $P(A_1 \cup A_2 \cup \cdots \cup A_r) = \sum_{i=1}^{r} P(A_i)$.

Exercise

E Prove Theorem 6 by induction.   HINT: $P(A_1 \cup A_2 \cup \cdots \cup A_k) \cup A_{k+1}) = P(A_1 \cup A_2 \cup \cdots \cup A_k) + P(A_{k+1})$.

Theorem 7   If A and A' are complementary events, $P(A) + P(A') = 1$.

This follows from $A \cup A' = U$, $A \cap A' = \emptyset$, Condition IV, Theorem 1, and Theorem 4.

F Write out the details of the proof of Theorem 7.

In later sections we shall make use of these theorems.

Problems 12.5

In Probs. 1 to 12 set up the probability distribution for the indicated sample space derived from the random experiment of tossing the given regular solid.

1 Cube; $U = \{$odd, even$\}$
2 Cube; $U = \{$number < 4, 4, $>4\}$
3 Tetrahedron; $U = \{$odd, even$\}$
4 Tetrahedron; $U = \{$number < 3, 3, $>3\}$
5 Octahedron; $U = \{$odd, even$\}$
6 Octahedron; $U = \{$multiple of 4, otherwise$\}$
7 Dodecahedron; $U = \{1$, prime, composite$\}$
8 Dodecahedron; $U = \{$number < 3, $3 \le$ number < 7, 7, $>7\}$
9 Icosahedron; $U = \{$multiple of 3, otherwise$\}$
10 Icosahedron; $U = \{$multiple of 2, multiple of 11, otherwise$\}$
11 Icosahedron; $U = \{$multiple of 6, multiple of 7, otherwise$\}$
12 Icosahedron; $U = \{$number < 2, $2 <$ number < 18, otherwise$\}$

In Probs. 13 to 16 set up the probability distribution for the indicated sample space derived from the random experiment of drawing one card from a full deck.

13 $U = \{c, d, h, s\}$
14 $U = \{$red, black$\}$
15 $U = \{2, 3, 4, 5, 6, 7, 8, 9, 10, J, Q, K, A\}$
16 $U = \{e_1, e_2, \ldots, e_{52}\}$

Problems 17 to 20 refer to drawing one ball from a bag containing five red balls marked 1, 2, 3, 4, 5, four green balls marked 1, 2, 3, 4, and three black balls marked 1, 2, 3. Set up the probability distribution for the indicated sample space derived from this random experiment.

17 $U = \{$red, green, black$\}$
18 $U = \{$red $\cup$ green, black$\}$
19 $U = \{$odd, even$\}$
20 $U = \{$red and odd, green and even, otherwise$\}$

12.6 Joint Distributions

We shall have immediate need of a certain argument known as the *multiplication principle*.

Multiplication Principle   If there are n ways of doing one operation and, after that, m ways of doing a second operation, then there are $n \times m$ ways of doing the two operations together.

Illustration 1   If there are five airlines flying from San Francisco to New York and eight airlines flying from New York to London, then there are $5 \times 8 = 40$ air routes from San Francisco to London.

Exercise

A   Generalize this principle for n operations.

Think of the toss of a symmetric die as one experiment and the toss of a symmetric coin as another.   The two uniform distributions are $U_1 = \{1, 2, 3, 4, 5, 6\}, P(1) = P(2) = \cdots = P(6) = \frac{1}{6}$ and $U_2 = \{H, T\}$, $P(H) = P(T) = \frac{1}{2}$.   If the two experiments are performed jointly, there is a joint sample space $U = U_1 \times U_2$ which is the cartesian product $\{1, 2, 3, 4, 5, 6\} \times \{H, T\} = \{1H, 2H, 3H, 4H, 5H, 6H, 1T, 2T, 3T, 4T, 5T, 6T\}$.   It is helpful to display this joint sample space in an array of rows and columns (something like a coordinate system) as in Fig. 12.7. Because of this representation, simple events in a sample space are sometimes called *sample points.*

A joint sample space is thus determined, but as yet there is no probability distribution because no probabilities have been assigned to the sample points of the joint sample space.   While any assignment consistent with the requirements for a probability distribution is possible, there is one which, in this case, seems more reasonable than any of the others.   To any simple event such as $\{1H\}$, say, the assignment $P(1H) = P(1) \cdot P(H)$ is supported by the following argument.   Let n_1, n_2, n_3, n_4, n_5, n_6 be the number of times 1, 2, 3, 4, 5, 6 turn up, respectively, in a large number N of tosses of the die.   By reason of symmetry we *assume*

Joint sample space

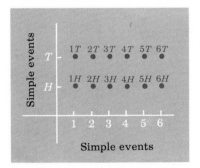

Joint sample space

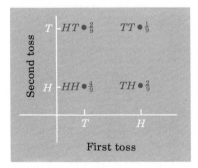

Figure 12.7   The simple events correspond to the toss of a symmetric die and the toss of a symmetric coin.

Figure 12.8   Each simple event corresponds to a toss of a biased coin, where $P(H) = \frac{2}{3}$ and $P(T) = \frac{1}{3}$.

that $n_1 \approx n_2 \approx n_3 \approx n_4 \approx n_5 \approx n_6 \approx N/6$, where $\sum_{i=1}^{6} n_i = N$. Similarly, let m_1, m_2 be the number of times H and T show on the die in a large number M of tosses. Again we assume that $m_1 \approx m_2 \approx M/2$, where $m_1 + m_2 = M$. Then, by the Multiplication Principle, $n_1 \times m_1$ gives the number of times $1H$ occur together in a joint toss of coin and die. The relative frequency of $1H$ is $n_1 m_1/NM$; that is,

$$f(1H) = \frac{n_1 m_1}{NM} = \frac{n_1}{N} \cdot \frac{m_1}{M} \approx \frac{N/6}{N} \cdot \frac{M/2}{M} = \frac{1}{6} \cdot \frac{1}{2} = \frac{1}{12}$$

We therefore arbitrarily assign $\frac{1}{12}$ to each simple event in the joint sample space following the rule that $P(1H) = P(1) \cdot P(H)$, etc.

Exercise

B Carry out the above argument, and assign probabilities in the joint sample space of two tosses of a symmetric coin.

Symmetry is not an essential feature of the above considerations. Two tosses are made with a coin so biased that $P(H) = \frac{2}{3}$, $P(T) = \frac{1}{3}$. A joint probability distribution over the joint sample space

$$U = \{HH,\ HT,\ TH,\ TT\}$$

is to be made. $U = U_1 \times U_1 = \{H, T\} \times \{H, T\}$ (Fig. 12.8). What is a sensible way of assigning the probabilities? First, we must understand what the statement "$P(H) = \frac{2}{3}$" means. This implies that, if n_1 and n_2 are the number of times H and T turn up, respectively, in a large number N of tosses, then $n_1 \approx \frac{2}{3} N$, $n_2 \approx \frac{1}{3} N$, $n_1 + n_2 = N$. Similarly, if m_1 and m_2 are the number of times H and T show in a large number M of tosses, then $m_1 \approx \frac{2}{3} M$, $m_2 \approx \frac{1}{3} M$, $m_1 + m_2 = M$. By the Multiplication Principle, $n_1 \times m_1$ gives the number of times HH occurs in one toss followed by another. (Or we might have two such coins and toss them once simultaneously.) The relative frequency of HH is $n_1 m_1/NM$; that is,

$$f(HH) = \frac{n_1 m_1}{NM} = \frac{n_1}{N} \cdot \frac{m_1}{M} \approx \frac{\frac{2}{3} N}{N} \cdot \frac{\frac{2}{3} M}{M} = \frac{2}{3} \cdot \frac{2}{3} = \frac{4}{9}$$

We therefore arbitrarily assign $\frac{4}{9}$ to the simple event HH in the joint sample space following the rule $P(HH) = P(H) \cdot P(H)$. In like manner $P(HT) = P(H) \cdot P(T) = \frac{2}{3} \cdot \frac{1}{3} = \frac{2}{9}$, etc. Note that the distributions are not uniform.

Exercise

C Show that Condition II is satisfied by this assignment.

In general, if two probability distributions are given,[1]

---

[1] Here the primes are used merely to indicate that a second set of simple events e_j are being considered and are not associated with complements.

$$U_1 = \{e_1, e_2, \ldots, e_n\} \qquad P(e_1), P(e_2), \ldots, P(e_n)$$

and $\quad U_2 = \{e_1', e_2', \ldots, e_m'\} \qquad P(e_1'), P(e_2'), \ldots, P(e_m')$

then the joint distribution is

$$U = U_1 \times U_2 \qquad P(e_i e_j') \qquad i = 1, 2, \ldots, n; j = 1, 2, \ldots, m$$

Our only concern will be with distributions where we assign to the simple events $e_i e_j'$ in the joint distribution the probabilities $P(e_i e_j')$ in accordance with the rule $P(e_i e_j') = P(e_i) \cdot P(e_j')$. This is called the case of *independent experiments*.

independent experiments

That this assignment satisfies Condition II is clear from the following computations. The cartesian product $U_1 \times U_2$ may be written in rows and columns, thus exhibiting the simple events of the joint sample space.

$$U = U_1 \times U_2 = \{e_1, e_2, \ldots, e_n\} \times \{e_1', e_2', \ldots, e_m'\}$$

$$\begin{array}{cccc}
e_1 e_1' & e_1 e_2' & \cdots & e_1 e_m' \\
e_2 e_1' & e_2 e_2' & \cdots & e_2 e_m' \\
\cdots\cdots\cdots\cdots\cdots\cdots\cdots \\
e_n e_1' & e_n e_2' & \cdots & e_n e_m'
\end{array}$$

Now consider a *fixed* simple event in U_1, say, e_i, and add the probabilities assigned to the elements in the ith row of the joint distribution.

$$\sum_{j=1}^{m} P(e_i e_j') = \sum_{j=1}^{m} P(e_i) \cdot P(e_j') = P(e_i)$$

since, for U_2, $\sum_{j=1}^{m} P(e_j') = 1$. Now if we take the sum of the probabilities of all the rows, we have $\sum_{i=1}^{n} P(e_i)$, which shows that Condition II is satisfied.

Exercises

D Write out the above proof using $\sum_{i=1}^{n} \sum_{j=1}^{m} P(e_i e_j')$ notation.

E Argue that the proof follows from $\sum_{i=1}^{n} \sum_{j=1}^{m} P(e_i) \cdot P(e_j') = [P(e_1) + P(e_2) + \cdots + \cdots + P(e_n)] \cdot [P(e_1') + P(e_2') + \cdots + P(e_m')]$.

Generalization to more than two basic probability distributions is immediate. Thus, for U_1, U_2, U_3, we have $U_1 \times U_2 \times U_3$ and assume that

$$P(e_i e_j' e_k'') = P(e_i) \cdot P(e_j') \cdot P(e_k'')$$
$$i = 1, \ldots, n; j = 1, \ldots, m; k = 1, \ldots, p$$

Exercise

F Show that Condition II is satisfied by this assignment

Illustration 2 Three symmetric coins are tossed. We take $P(HHT) = P(H) \cdot P(H) \cdot P(T) = \frac{1}{2} \cdot \frac{1}{2} \cdot \frac{1}{2} = \frac{1}{8}$, etc. The joint distribution is uniform.

Illustration 3   A biased coin with $P(H) = \frac{2}{3}$ is tossed, a symmetric die marked 1, 1, 1, 2, 2, 3 is tossed, and a card is drawn from a full deck. To the simple event (H, 1, K spades) in the joint sample space we assign $P(H, 1, K \text{ spades}) = \frac{2}{3} \cdot \frac{1}{2} \cdot \frac{1}{52} = \frac{1}{156}$, etc.   The joint distribution is not uniform.

This disposes of the assignment of probabilities to the simple events in a joint probability distribution.   Theorems 3 and 4 must be used in computing $P(A \cap B)$ and $P(A \cup B)$, where A and B are any events in either a simple or joint distribution.   We turn to these tasks in the next two sections.

Problems 12.6

In Probs. 1 to 20 set up a joint probability distribution assuming symmetry, ordinary markings (or as indicated), and one toss.

1   Two tetrahedra
2   Two octahedra
3   Two dodecahedra
4   Two icosahedra
5   One tetrahedron, one octahedron
6   One tetrahedron, one dodecahedron
7   One tetrahedron, one icosahedron
8   One octahedron, one dodecahedron
9   One octahedron, one icosahedron
10   One dodecahedron, one icosahedron
11   Two tetrahedra marked 1, 1, 2, 3 and 1, 2, 2, 3.
12   Two octahedra each marked 1, 2, 2, 3, 3, 3, 3, 4.
13   Two dodecahedra each marked with eleven 1's and one 2.
14   Two icosahedra each marked with eighteen H's and two T's.
15   Two "regular" n-gons.   Devise a gadget simulating one.
16   One "regular" n-gon and one "regular" m-gon.   Devise gadgets simulating them.
17   Three coins
18   Three tetrahedra
19   Three tetrahedra marked 1, 1, 1, 2.
20   Four coins

In Probs. 21 to 26 set up a probability distribution for the experiment of drawing:

21   A card from a deck, examining it, replacing it, and drawing a second card.
22   A card from a deck, not examining it at the moment, not replacing it, drawing a second card, and examining both, without knowing which was drawn first.
23   A ball from a bag containing four red balls, five black balls, examining it, replacing it, and drawing a second ball.

24 A ball from a bag containing four red balls, five black balls, not examining it at the moment, not replacing it, drawing a second ball, and examining both, without knowing which was drawn first.

25 A ball from a bag containing four red balls, five black balls, not ever examining it, not replacing it, and drawing a second ball.

26 Four balls from a bag containing four red balls, five black balls, throwing them away *without examining them,* and drawing one more ball and *examining it.*

27 A College Entrance Examination Board test has 50 questions with five choices for each answer, one and only one being correct. If an answer is chosen at random for each question, what is the probability that all questions will be answered correctly? How many simple events are there in the sample space? Is the distribution uniform?

28 Of 100 people, 60 are males, of whom 30 smoke; 15 of the females smoke. One person is selected at random. Set up a probability distribution. Is it uniform?

29 If 3 per 1,000 males are stillborn and 1 per 1,000 females is still-born, what is the probability that the next child born is female and alive? Assume the ratio male: female $= 51:49$.

30 The following fictitious data refer to a population of 100,000 people and polio shots.

	Number of polio shots				
	0	*1*	*2*	*3*	*Total*
Contracted polio	5	1	1	0	7
Did not contract polio	293	700	1,000	98,000	99,993
Total	298	701	1,001	98,000	100,000

From the 100,000, one is selected at random. For this experiment, set up a probability distribution.

31 Each of three identical boxes has two identical compartments each of which has one coin as follows: box I, silver, silver; box II, silver, gold; box III, gold, gold. A box is selected, then one of its two compartments. What is the probability distribution?

12.7 Probability of Both A and B. $P(A \cap B)$

In order to find the probability that both A and B will occur, we must turn to the sample space, determine the sample points in $A \cap B$, and

$P(A \cap B)$ compute $P(A \cap B)$ directly. This is the essence of Theorem 3, Sec. 12.5.

Illustration 1 A symmetric coin and a symmetric die are tossed. Let A be the event "H on coin", and let B be the event "3 on die". Find $P(A \cap B)$. The uniform distribution is assumed. $U = \{H1, H2, H3, H4, H5, H6, T1, T2, T3, T4, T5, T6\}$, and each of the simple events has probability $\frac{1}{12}$.

$$A = \{H1, H2, H3, H4, H5, H6\}$$
$$B = \{H3, T3\}$$
$$A \cap B = \{H3\}$$
$$P(A \cap B) = \frac{1}{12}$$

Exercise
A Show that $P(A \cap B) = P(A) \cdot P(B)$ in this case.

Illustration 2 A symmetric penny and a symmetric nickel are tossed. Let A be the event "coins match", and let B be "penny falls head". Find $P(A \cap B)$. We assume the uniform distribution over $U = \{HH, HT, TH, TT\}$, where HT means H on penny, T on nickel, etc.

$$A = \{HH, TT\}$$
$$B = \{HH, HT\}$$
$$A \cap B = \{HH\}$$
$$P(A \cap B) = \frac{1}{4}$$

Exercise
B Show that $P(A \cap B) = P(A) \cdot P(B)$ in this case.

Illustration 3 A penny and a nickel are tossed. Let A be the event "coins match", and let B be "at least one head shows". Find $P(A \cap B)$. Again we use the uniform distribution over $U = \{HH, HT, TH, TT\}$, with the following results:

$$A = \{HH, TT\}$$
$$B = \{HH, HT, TH\}$$
$$A \cap B = \{HH\}$$
$$P(A \cap B) = \frac{1}{4}$$

Exercise
C Show that $P(A \cap B) \neq P(A) \cdot P(B)$ in this case.

Illustration 4 Data and events are the same as in Illustration 2, except that penny and nickel are similarly biased with $P(H) = \frac{2}{3}$. Find $P(A \cap B)$. Here we have a nonuniform joint distribution. $P(HH) = \frac{4}{9}$, $P(HT) = \frac{2}{9}$, $P(TH) = \frac{2}{9}$, $P(TT) = \frac{1}{9}$.

$$A \cap B = \{HH\}$$
$$P(A \cap B) = \frac{4}{9}$$

Exercise
D Show that $P(A \cap B) \neq P(A) \cdot P(B)$ in this case.

Illustration 5   A die marked 1, 2, 3, 4 in red, 5, 6 in green, is tossed. Let A be "number is even" and B be "number is red". Find $P(A \cap B)$. The uniform distribution is over $U = \{1R, 2R, 3R, 4R, 5G, 6G\}$.

$$A = \{2R, 4R, 6G\}$$
$$B = \{1R, 2R, 3R, 4R\}$$
$$A \cap B = \{2R, 4R\}$$
$$P(A \cap B) = \tfrac{1}{3}$$

Exercise

E   Show that $P(A \cap B) = P(A) \cdot P(B)$ in this case.

Illustration 6   A die marked 1, 2, 3 in red, 4, 5, 6 in green, is tossed. Let A be "number is even" and B be "number is red". Find $P(A \cap B)$. The uniform distribution is over $U = \{1R, 2R, 3R, 4G, 5G, 6G\}$.

$$A = \{2R, 4G, 6G\}$$
$$B = \{1R, 2R, 3R\}$$
$$A \cap B = \{2R\}$$
$$P(A \cap B) = \tfrac{1}{6}$$

Exercise

F   Show that $P(A \cap B) \neq P(A) \cdot P(B)$ in this case.

Exercises A to F indicate that in some cases $P(A \cap B) = P(A) \cdot P(B)$, and in some cases, not.

Definition   Two events A and B are said to be *independent* if and only if
independent $P(A \cap B) = P(A) \cdot P(B)$.
events

The notion of independent events is important in advanced probability theory.

Problems 12.7

In Probs. 1 to 14 find $P(A \cap B)$, where A, B refer to either solid if alike; otherwise A refers to first.   One toss is made.

1   Two tetrahedra. A, odd number; B, even.

2   Two octahedra. A, odd number; B, perfect square.

3   Two dodecahedra. A, 1 or prime; B, perfect cube.

4   Two icosahedra. A, number exceeds 17; B, positive power of 3.

5   One tetrahedron, one octahedron. A, odd; B, even.

6   One tetrahedron, one dodecahedron. A, 3; B, prime or even.

7   One tetrahedron, one icosahedron. A, number ≥ 1; B, perfect cube.

8   One octahedron, one dodecahedron. A, number ≥ 7; B, number ≤ 3.

9   One octahedron, one icosahedron. $A = \{x \mid (x - 1)(x - 2) = 0\}$; B, 1 or prime.

10 One dodecahedron, one icosahedron. $A = \{x \mid (x - 4)(x - 7) = 0\}$; B, positive power of 4.

11 Two tetrahedra, marked 1, 1, 2, 3 and 1, 2, 2, 3. A, 1; B, 2.

12 Two octahedra, each marked 1, 2, 2, 3, 3, 3, 3, 4. A, 2; B, 3.

13 Two dodecahedra, each marked with eleven 1's, one 2. A, 2; B, 1.

14 Two icosahedra, each marked with eighteen H's, two T's. A, H; B, T.

In Probs. 15 to 20 three coins are tossed once. Find $P(A \cap B)$.

15 A, coins match; B, at least one H.

16 A, coins match; B, at least two H.

17 A, coins match; B, at least three H.

18 A, coins match; B, not more than one H.

19 A, coins match; B, not more than two H.

20 A, coins match; B, not more than three H.

21 An octahedron, marked 1, 2, 3, 4, 5, 6 in red, 7, 8 in green, is tossed. Find $P(A \cap B)$, where A is "number is even", B is "number is green".

22 A dodecahedron, marked 1, 2, 3, 4, 5, 6, 7, 8 in red, 9, 10, 11, 12 in green, is tossed. Find $P(A \cap B)$, where A is "number is even", B is "number is green".

23 In bag I there are three red and four black balls; in bag II there are four red and five black balls. One ball is drawn from each bag. Find $P(R, R)$.

24 In bag I there are three red and four black balls; in bag II there are four red and five black balls. One ball is drawn from bag I, and if (and only if) it is red, it is put in bag II. Then a ball is drawn from bag II. Find $P(R,R)$.

25 Four random strikes on the letter keys of a typewriter are made. What is the probability of typing the word "good"?

26 The letter keys of a typewriter are repeatedly struck at random 26 times. What is the probability of typing "idontunderstandthis-problem"?

27 Find the probability that three people chosen at random have the same birthday (February 29 is to be omitted.)

28 Prove that if A and B are independent events, then A and B' are independent. (So are A' and B, A' and B'.)

In Probs. 29 to 32 find $P(1,1,1)$ in one toss of the three given dice.

29 Three tetrahedra

30 Three octahedra

31 Three dodecahedra

32 Three icosahedra

33 A bag contains five red balls and two black balls; one is drawn. A tetrahedron marked 1, 1, 2, 3 is tossed. And a card is drawn from a full deck. Find $P(\text{black}, 1, \text{king})$.

34   A coin, a die, and a tetrahedron are tossed.   Find $P(H,3,2)$.

35   What is the probability of H on the first of:

   a   Two tosses of a coin?   **b**   Three tosses?   **c** n tosses?

36   Three bags contain red, blue, and white balls as follows: $I(7,4,2)$, $II(3,1,5)$, and $III(6,1,4)$.   If a bag is chosen at random and a ball drawn:

   a   What is the probability of getting a white ball?

   b   If the ball is replaced after each draw, what is the probability of drawing a white ball two times in succession? n times?

37   A bag contains 10 counters marked with the integers 1 to 10.   A counter is drawn and replaced.   A counter is again drawn.   Find the probability that:

   a   The same counter was drawn the second time.

   b   Each counter was an odd number.

38   Four persons in turn each draw a card from one and the same full deck without replacements.   Find the probability that:

   a   Each suit is represented.   **b**   All are of the same suit.

   c   No two are of the same value.

12.8   Probability of Either A or B or Both. $P(A \cup B)$

$P(A \cup B)$ Theorem 4, Sec. 12.5, furnishes the method of computing the probability that of two events at least one occurs:

$$P(A \cup B) = P(A) + P(B) - P(A \cap B)$$

The first step is to find $P(A \cap B)$.

Illustration 1   Find the probability of at most two tails or at least two heads on a toss of three coins.   The uniform distribution is over $U = \{HHH, HHT, HTH, THH, HTT, THT, TTH, TTT\}$.   Let A be the event "at most two tails", B be "at least two heads".

$$A = \{HHH, HHT, HTH, THH, HTT, THT, TTH\}$$
$$P(A) = \frac{7}{8}$$
$$B = \{HHH, HHT, HTH, THH\}$$
$$P(B) = \frac{4}{8}$$
$$A \cap B = B$$
$$P(A \cap B) = \frac{4}{8}$$
$$A \cup B = A$$
$$P(A \cup B) = \frac{7}{8} + \frac{4}{8} - \frac{4}{8}$$
$$= \frac{7}{8}$$

Illustration 2   A coin is tossed three times.   Find the probability of $A = \{HHH\}$ or $B = \{HHT\}$ or $C = \{HHT\}$.   In this case the events are mutually exclusive and Theorem 4, Sec. 12.5, applies.

$$P(A \cup B \cup C) = \frac{3}{8}$$

Illustration 3 If A is "H on coin" and if B is "3 on die", find $P(A \cup B)$.

$$A = \{H1, H2, H3, H4, H5, H6\}$$
$$P(A) = \tfrac{6}{12}$$
$$B = \{H3, T3\}$$
$$P(B) = \tfrac{2}{12}$$
$$A \cap B = \{H3\}$$
$$P(A \cap B) = \tfrac{1}{12}$$
$$P(A \cup B) = \tfrac{6}{12} + \tfrac{2}{12} - \tfrac{1}{12}$$
$$= \tfrac{7}{12}$$

Illustration 4 A penny and a nickel are tossed. If A is "coins match" and if B is "at least one head shows", find $P(A \cup B)$.

$$A = \{HH, TT\}$$
$$B = \{HH, HT, TH\}$$
$$A \cap B = \{HH\}$$
$$P(A \cap B) = \tfrac{1}{4}$$
$$P(A \cup B) = \tfrac{2}{4} + \tfrac{3}{4} - \tfrac{1}{4}$$
$$= 1$$

Illustration 5 A coin is tossed five times. Find the probability of getting at least one head. The answer is $P(\text{one } H) + P(\text{two } H\text{'s}) + P(\text{three } H\text{'s}) + P(\text{four } H\text{'s}) + P(\text{five } H\text{'s})$, but this involves many calculations. The better approach is to use Theorem 7, Sec. 12.5. The event complementary to "at least one head" is "all tails", the probability of which is clearly $(\tfrac{1}{2})^5$. The answer to the problem is simply $1 - (\tfrac{1}{2})^5 = \tfrac{31}{32}$.

Problems 12.8

In Probs. 1 to 8 one toss is made. Compute the probability indicated.

1 Two coins. A, (HH); B, (TT). $P(A \cup B)$.
2 Two coins. A, (HT); B, (TH). $P(A \cup B)$.
3 Two coins. A, (at least one H); B(no T). $P(A \cup B)$.
4 Two coins. A, (at most one H); B, (no T). $P(A \cup B)$.
5 Three coins. A, (at least two H); B, (no T). $P(A \cup B)$.
6 Three coins. A, (at least two H); B, (exactly one T). $P(A \cup B)$.
7 Three coins. A, (HHH); B, (TTT); C, (exactly one H). Compute $P(A \cup B \cup C)$.
8 Three coins. A, (HHH); B, (TTT); C, (exactly two H). Compute $P(A \cup B \cup C)$.
*9 Prove: $P(A \cup B \cup C) = P(A) + P(B) + P(C) - P(A \cap B) - P(A \cap C) - P(B \cap C) + P(A \cap B \cap C)$. HINT: Consider $P(A \cup (B \cup C))$; apply Theorem 4, Sec. 12.5, and the distributive law for $A \cap (B \cup C)$ (See Prob. 17, Sec. 1.5).

In Probs. 10 to 25 one toss is made. Compute the probability indicated.

10 Three coins. A, (exactly one H); B, (at least one T); C, (HTH). $P(A \cup B \cup C)$. Use result of Prob. 9.

11 Three coins. A, (HHH); B, (HHT, THH); C, (at least two T). $P(A \cup B \cup C)$. Use result of Prob. 9.

12 Two dice. A, (1,2); B, (6,3). $P(A \cup B)$.

13 Two dice. A, $\{(1,2), (1,3)\}$; B, (3,1). $P(A \cup B)$.

14 Two dice. A, (sum $= 6$). $P(A)$.

15 Two dice. A, (sum ≥ 4). $P(A)$.

16 Two dice. A, (sum < 4); B, (sum > 5). $P(A \cup B)$.

17 Two dice. A, (sum ≤ 4); B, $\{(2,1), (3,1), (4,1)\}$. [Here (3,1) is to mean 3 on die 1 and on die 2.] $P(A \cup B)$.

18 Two dice. A, $\{(1,2), (2,1), (3,4)\}$; B, $\{(1,6), (2,1)\}$; C, $\{(1,2)\}$. Use Prob. 9.

19 Two dice. A, (sum ≤ 2); B, (sum ≤ 3); C, (sum ≤ 4). $P(A \cup B \cup C)$. Use Prob. 9.

20 One coin, one die. A, (H); B, $\{3,4\}$. $P(A \cap B)$, $P(A \cup B)$.

21 One coin, one die. A, (H); B, (even). $P(A \cap B)$, $P(A \cup B)$.

22 Two dice. A, (sum ≤ 10). $P(A)$. HINT: Use A'.

23 Two dice. A, (sum ≤ 11). $P(A)$. HINT: Use A'.

24 Two dice. A, (sum ≤ 10); B, (sum even). $P(A \cup B)$. HINT: Use $(A \cup B)' = A' \cap B'$.

25 Two dice. A, (sum < 10); B, (sum odd). $P(A \cup B)$. HINT: Use $(A \cup B)' = A' \cap B'$.

26 A special deck of cards consists of one A, two K's, three Q's, and four J's. A die is marked 1, 1, 1, 2, 3, 3. A card is drawn, and the die tossed. A, $\{K3, Q2\}$; B, $\{2\}$. $P(A \cup B)$.

27 A bag contains four dimes and four nickels. If three coins are drawn at random, what is the probability that they will make change for a quarter?

28 In a single throw of two dice, what is the probability that there will be:

   a No doublet? **b** Neither doublet nor a six?

   c No even number?

29 In three throws of two dice, what is the probability of throwing doublets not more than two times?

30 Bag A contains five red and three black balls. Bag B contains four red and six black balls. If a bag is chosen at random and a ball drawn, what is the probability that it will be black?

31 Bag A has four dimes and three nickels; bag B has three dimes and two nickels. A bag has been selected, and a dime has been drawn. What is the probability that it comes from bag B?

32 On the average, a trackman can pole-vault 12 ft two times in five trials and broad-jump 24 ft one time in three.

a On a given day what is the probability that he will succeed in both if given but one trial in each event?

b If he is to enter but one event, in which he will have two trials, and chooses this event by tossing a coin, what is his probability of success?

33 A narrow stream has one high bank and one low bank. A boy can (safely) jump from high to low four times out of five, from low to high two times in five. If he chooses to start from a bank by tossing a coin and makes two jumps:

a What is the probability that he ends up on the opposite bank? (If he misses a jump, he crawls out on the bank from which this jump was made.)

b What is the probability of his landing in the drink at least once?

34 The following data come from the *Monthly Weather Review*, vol. 81, no. 3, p. 54 (1953).

Probabilities of storm occurrences per month

Month	July	Aug.	Sept.	Oct.
At least one storm	0.39	0.75	0.92	0.83

What is the probability of the following:

a At least one storm in September and at least one in October?

b At least one in September or at least one in October, but not both?

c At least one in September or at least one in October or both?

35 A small college has 100 students each of whom plays at least one sport as follows:

49 play football
38 play basketball
35 play track
 7 play football and basketball
 9 play football and track
 8 play basketball and track
 2 play all three

One is picked at random. What is the probability that he plays:

a At least one sport?   **b** Only one sport?

36 Of 100 readers:

60 read the *Times*
40 read the *Guardian*
37 read the *Sun*
20 read the *Times* and the *Guardian*
15 read the *Times* and the *Sun*
 7 read the *Guardian* and the *Sun*
 5 read all three

One is picked at random. What is the probability that he reads:

a At least two?   **b** Not more than two?

12.9   Conditional Probability of A, Given B. $P(A \mid B)$

Thus far the probability questions have been directly related to events (subsets) of a sample space. For example, an honest die is tossed, and an unqualified question is asked: What is the probability of 6, namely, $P(6)$? The only information we have is that the die is tossed; the question is unconditional. But there is a whole class of problems where the question is a conditional one based on additional information. For example, the die is tossed, and we are told that an even number shows. What is the probability of 6 now that we have the additional information that the number showing is even? Surely this changes things since we know that 1, 3, 5 are ruled out and that only 2, 4, 6 are possibilities.

In searching for an answer to the question we return to an examination of relative frequencies in a long sequence N of tosses. If n_i is the number of times the face i turns up in N tosses, then $n_i \approx N/6$ and the absolute frequency of the event *both 6 and even* is about $N/6$. The absolute frequency of the event *even* is about $3N/6$, and, consequently, the relative frequency of *both 6 and even* when *even* is about $(N/6)/(3N/6) = \frac{1}{3}$. Therefore a plausible answer is $\frac{1}{3}$. This also seems reasonable upon examining, first, the uniform distribution over $U = \{1, 2, 3, 4, 5, 6\}$, $P(i) = \frac{1}{6}$, $i = 1, 2, \ldots, 6$, and then, when we are told that the die shows even, upon examining the uniform distribution over the (*reduced*) sample space $U(\text{reduced}) = \{2, 4, 6\}$, $P(i) = \frac{1}{3}$, $i = 1, 2, 3$.

In general, for a long sequence of N repetitions of an experiment, let $N_{A \cap B}$ and N_B be the absolute frequencies of the events $A \cap B$ and B, respectively. The relative frequencies are $f(A \cap B) = N_{A \cap B}/N$ and $f(B) = N_B/N$. Now the relative frequency with which $A \cap B$ occurs when B occurs is

$$\frac{N_{A \cap B}}{N_B} = \frac{N_{A \cap B}/N}{N_B/N}$$

$$= \frac{f(A \cap B)}{f(B)} \qquad f(B) \neq 0$$

This makes the following definition a reasonable one.

Definition   The probability of A, *given* B, is $P(A \cap B)/P(B)$, $P(B) \neq 0$.

$P(A \mid B)$ **Notation**   Another symbol for the probability of A, given B, is $P(A \mid B)$, so that

$$P(A \mid B) = \frac{P(A \cap B)}{P(B)} \qquad P(B) \neq 0$$

conditional
probability   which is also called the *conditional* probability of event A subject to the condition that event B has happened.

Illustration 1 A tetrahedron is tossed twice. Find $P(A|B)$ if A is "down faces match", B is "sum of the down faces exceeds 5". That is, given that the sum of the down faces (on the two throws) exceeds 5, find the probability that the down faces match. The sample space is

Reduced sample space

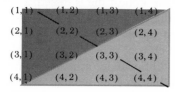

(1,1) (1,2) (1,3) (1,4)
(2,1) (2,2) (2,3) (2,4)
(3,1) (3,2) (3,3) (3,4)
(4,1) (4,2) (4,3) (4,4)

Figure 12.9 It is given that the sum of the numbers in each pair exceeds 5. These pairs lie in the *reduced sample space*, which is colored in the figure.

$U = \{1, 2, 3, 4\} \times \{1, 2, 3, 4\}$, which is listed in rows and columns in Fig. 12.9. Simple events making up A are down the main diagonal; those in B are enclosed in the triangle.

$$A = \{(1,1), (2,2), (3,3), (4,4)\}$$
$$B = \{(2,4), (3,3), (3,4), (4,2), (4,3), (4,4)\}$$
$$P(B) = \tfrac{6}{16}$$
$$A \cap B = \{(3,3), (4,4)\}$$
$$P(A \cap B) = \tfrac{2}{16}$$
$$P(A|B) = \frac{P(A \cap B)}{P(B)} = \frac{\tfrac{2}{16}}{\tfrac{6}{16}}$$
$$= \tfrac{2}{6}$$

(On occasion we may not reduce certain fractions so that we may the better see the principle involved.)

This problem may be worked from another point of view. It is given that the event B has occurred; consider B as a *reduced* sample space. Since the original distribution was uniform, it seems reasonable to assign a uniform distribution to B, which contains six simple events, each of which will now have probability $\tfrac{1}{6}$. Since there are just two simple events in B answering the description "faces match", it follows that $P(A|B) = \tfrac{2}{6}$. This checks with the previous result.

It is easy to prove that the reduced-sample-space method always works for uniform distributions and, when modified as in Exercise A below, for nonuniform distributions as well. Let the original distribution be uniform over U, containing n points, each with probability $1/n$. Let B contain n_1 of these, and let $A \cap B$ contain n_2. Then

$$P(A|B) = P(A \cap B)/P(B) = (n_2/n)(n_1/n) = n_2/n_1$$

Now set up a uniform distribution over B, considered as a reduced sample space. Each point in B now has probability $1/n_1$; also $P(A \cap B) = n_2/n_1$, and $P(B) = 1$. Hence $P(A|B) = (n_2/n_1)/1$, and this agrees with our previous result.

Exercise

A Let the distribution over U be nonuniform. Prove that, if in the reduced sample space B the assigned probabilities are proportional to the original ones in B (satisfying Condition II), then the reduced-sample-space method works.

B   In Illustration 1 above, assume bias; let $P(1) = P(2) = \frac{1}{8}$, $P(3) = \frac{3}{8}$, $P(4) = \frac{1}{8}$.   Find $P(A \mid B) = \dfrac{20\!/\!64}{44\!/\!64}$ directly from the original nonuniform sample space.   Now find $P(A \mid B) = \dfrac{20\!/\!44}{1}$ from the results of Exercise A.

Illustration 2   A tetrahedron is tossed twice.   Find $P(A \mid B)$ if A is "the number on second toss is 2", B is "the number on the first toss is 4". By the reduced-sample-space method, we find $P(A \mid B) = \frac{1}{4}$.   From the original sample space,

$$A = \{(1,2), (2,2), (3,2), (4,2)\}$$
$$B = \{(4,1), (4,2), (4,3), (4,4)\}$$
$$P(B) = \frac{4}{16}$$
$$A \cap B = \{(4,2)\}$$
$$P(A \cap B) = \frac{1}{16}$$
$$P(A \mid B) = \frac{P(A \cap B)}{P(B)} = \frac{\frac{1}{16}}{\frac{4}{16}}$$

Although $P(A)$ is not involved, note that $P(A \mid B) = P(A)$.   This means that the information furnished by knowing that the number 4 has occurred on the first toss did not affect the probability of 2 on the second toss.   Whenever $P(A \mid B) = P(A)$, it follows that $P(A \cap B) = P(A) \cdot P(B)$ and events A and B are independent by definition (Sec. 12.8).

Exercise
C   Prove that, if A and B are independent and if $P(A) \neq 0$, $P(B) \neq 0$, then $P(A \mid B) = P(A)$, and $P(B \mid A) = P(B)$.

Problems 12.9

In Probs. 1 to 20 the event B is given (has happened).   Find $P(A \mid B)$.

1   A coin is tossed three times. B, HH on first two tosses; A, H on third toss.
2   A die is tossed three times. B, $(6,5)$ on first two tosses; A, 4 on third toss.
3   A coin is tossed three times. B, at most two H; A, at least two H.
4   A coin is tossed three times. B, at least one H; A, at most two H.
5   A tetrahedron is tossed once. B, sum of faces *showing* exceeds 5; A, sum of faces *showing* exceeds 6.
6   A tetrahedron is tossed once. B, sum of the down face and the smallest face *showing* exceeds 3; A, sum of the down face and the smallest face *showing* exceeds 4.
7   A tetrahedron is tossed twice. B, first down face exceeds 2; A, second down face is 2.

8 A tetrahedron is tossed twice. B, sum of the down faces exceeds 5; A, first down face is even.

9 Two coins are tossed once. B, one coin is H; A, one coin is T.

10 Two coins are tossed once. B, no H; A, no T.

11 Two coins and two tetrahedra are tossed once. B, at least one T and sum 4; A, at least one H and sum ≤ 4.

12 Two coins and two tetrahedra are tossed once. B, at least one T and sum ≤ 3; A, two T and sum ≥ 3.

13 Two bags contain red and black balls as follows: $I(4,4)$, $II(2,6)$. B, a ball has been drawn from one of the bags at random and found to be black; A, black ball came from bag I.

14 Three bags contain red and black balls as follows: $I(4,4)$, $II(2,6)$, $III(7,1)$. B, a ball has been drawn from one of the bags at random and found to be black; A, black ball came from bag I.

15 Each of three identical boxes has two identical compartments each of which has one coin as follows: box I, silver, silver; box II, silver, gold; box III, gold, gold. A box is selected, then one of its two compartments. B, coin is gold; A, box II.

16 A person is picked at random. B, his birthday is the 13th of the month; A, Friday. (See Sec. 12.2, Prob. 16.)

17 Mother, father, and son line up at random for a family picture. B, son on one end; A, father in middle.

18 Two boys and two girls line up at random. B, girl on an end; A, girls are separated.

19 A bridge club is made up of three married couples, two single men, four single women. One member is chosen at random. B, man; A, married.

20 A committee is composed of eight Democrats, of whom two are women, and six Republicans, of whom five are men. A member is chosen at random. B, man; A, Republican.

21 Prove $P(A'|B) = 1 - P(A|B)$.

22 Prove $P(A|B') \cdot (1 - P(B)) = P(A) - P(A \cap B)$.

12.10 Counting Procedures

In the work thus far in this chapter the emphasis has been on theory, the illustrative material requiring but little computation. This was by design. The numerical details were held to a minimum in order to eliminate any possible confusion arising from arithmetic difficulties. Thus the sample spaces have had only a few simple events and have otherwise been elementary. However, in more realistic situations, for example, in the applications of probability to problems arising in the sciences, it may well be that sample spaces involve many elements, and it then becomes a real problem in itself to make an accurate count of these. We therefore turn to a study of some of the standard counting procedures so that more complicated examples may be examined.

12.11 Permutations

A permutation is defined as follows.

Definition A *permutation* of n objects is an arrangement of these objects
permutations   into a particular order.

The question before us here is, How many permutations can be
formed with a set of n objects? First, we consider an illustration.

Illustration 1 A mail carrier has six letters in his hand and has six mail-
boxes in front of him. In how many ways can he distribute the six
boxes so that just one letter goes into each box?

Solution He can put any one of the six letters in the first box and any
one of the remaining five letters in the second box. By the Multiplication Principle he has 6×5 ways of placing letters in the first two boxes.
By a repetition of the argument it follows that he can distribute the six
letters in $6 \times 5 \times 4 \times 3 \times 2 \times 1 = 720$ ways.

In order to state Theorem 8 below, we introduce two new symbols.

Definition The symbol $P_{n,n}$ stands for the number of permutations of n
$P_{n,n}$ distinct elements taken all at a time.

Definition The symbol n! (for n an integer ≥ 1), read "n factorial", stands
factorial   for the product

$$n! = 1 \times 2 \times 3 \times \cdots \times n$$

Further, 0! is defined to be 1. Factorials are not defined for negative
integers or for other real numbers.

Exercises

A Compute 5!, 6!, 7!.
B Show that $n!/n = (n-1)!$.
C Compute $8!/5!$; $n!/(n-1)!$; $n!/r!$, where $r \leq n$.

Theorem 8 $P_{n,n} = n!$

Proof The first choice can be made in n ways, the second in $(n-1)$
ways, etc. Applying the Multiplication Principle, we get at once

$$P_{n,n} = n(n-1)(n-2)\cdots 3 \times 2 \times 1 = n!$$

We may be interested in arranging only $r < n$ of our n elements and
forgetting about those left over. In Illustration 1, the mail carrier might
have had only four boxes for his six letters. Then he would have had
six choices for the first box, five for the second, four for the third, and
three for the fourth. In this case we speak of the "permutations of n
elements taken r at a time".

Definition The symbol $P_{n,r}$ $(r \leq n)$ stands for the number of permutations
$P_{n,r}$ of n distinct elements taken r at a time.

Theorem 9 $P_{n,r} = \dfrac{n!}{(n-r)!}$ $r \leq n$

Proof   The proof is similar to that of Theorem 8. From the Multiplication Principle,

$$P_{n,r} = \underbrace{n(n-1)\cdots(n-r+1)}_{r \text{ factors}}$$

This can be written

$$P_{n,r} = \frac{n(n-1)(n-r+1)(n-r)\cdots 2 \cdot 1}{(n-r)\cdots 2 \cdot 1}$$

or $P_{n,r} = \dfrac{n!}{(n-r)!}$

In the foregoing we have assumed that all n elements were distinct. It may happen that some of them are identical, and in this case the above reasoning will not hold.

Illustration 2   In how many ways can three red flags, one white flag, and one blue flag be arranged on a staff?

Solution   There are five flags present. If all five were distinct, the answer would be $P_{5,5} = 5!$. Let us label the red flags: R_1, R_2, R_3. In our 5! arrangements we have counted the permutations $R_1R_2R_3WB$, $R_2R_1R_3WB$, $R_3R_1R_2WB$, etc., as distinct arrangements. According to the statement of the problem, this is an error. There are, in fact, 3! ways of arranging $R_1R_2R_3$, all of which are to be counted as identical. Therefore the correct answer is $5!/3! = 20$.

Definition   The symbol $P_{n,n}^{r_1,r_2,\cdots}$ (for $r_1 + r_2 + \cdots = n$) stands for the number of permutations of n elements taken all at a time where there are r_1 identical elements of one kind, r_2 identical elements of another kind, etc.

Theorem 10 $P_{n,n}^{r_1,r_2,\cdots} = \dfrac{n!}{r_1!r_2!\cdots}$

$P_{n,n}^{r_1,r_2,\cdots}$

Proof   Suppose that we list all the permutations thus described. Then let us think of the r_1 identical elements as actually different and carry out the $r_1!$ arrangements in each permutation in our list, leaving all the other elements fixed. Then do the same for the r_2 identical elements, etc. When we have finished, we shall have obtained all the permutations of n distinct elements. By the Multiplication Principle,

$$P_{n,n}^{r_1,r_2,\cdots} \times r_1! \times r_2! \times \cdots = P_{n,n} = n!$$

From this the theorem follows.

Exercise

D Write out the permutations of the elements *AABC*, and verify Theorem 10.

Problems 12.11

1   If there are 20 steamers plying between New York and London, in how many ways could the round trip from New York be made if the return was made on:
a The same ship?   **b** A different ship?

2   A company has 15 planes to fly its executives between office and construction site. In how many ways could the round trip from the office be made if the return was made on:
a The same plane?   **b** A different plane?

3   Seven travelers stop overnight in a town where there are seven hotels. If no two persons are to stay at the same hotel, in how many ways may they choose hotels?

4   Seven travelers stop overnight in a town where there are six hotels. In how many ways can the travelers choose hotels?

5   How many six-letter nonsense words can be made with the letters *b*, *c*, *d*, *t*, *v*, and *z*:
a With no repetitions?   **b** With repetitions?

6   A cake mix is to be made with eight ingredients. In how many orders can these be put together?

7   A semaphore has three arms, and each arm has five distinct positions. What is the total number of signals that can be made?

8   Of a boat crew of eight men, three can row only on bow side and two others can row only on stroke side. In how many ways can the crew be assigned to rowing positions (four on a side)?

9   Each of 10 socks is a different color. How many color combinations can a person wear if left and right foot arrangements are:
a Considered distinct?   **b** Unimportant?

10   In how many of the distinct permutations of the letters in "syzygy" do the three *y*'s come together?

11   How many permutations of 9 letters can be made out of the letters of the word "Tennessee"?

12   How many permutations of 11 letters can be made out of the letters of the word "Mississippi"?

13   In how many of the distinct permutations of the letters in "Mississippi" do the four *i*'s not come together?

14   At a political rally there are five speakers. In how many orders may they speak if:
a *A* is to precede *B*?   **b** *A* is to precede *B* immediately?

15   In how many ways can the letters of "abstemiously" be rearranged without changing the order (with respect to each other) of the vowels a, e, i, o, u?

16   How many football teams could be formed using just 11 players switching positions)?

17 How many football teams could be formed using 15 players (switching positions)?

18 How many six-place automobile tags could be made up using one and only one letter of the alphabet (omitting *I* and *O*) and the 10 digits, allowing a letter and five digits? [Assume that 0 (zero) may not be used in the first (left-hand) position.]

*19 How many identification tags, each in the form of a matrix with two rows and three columns, can be made using two letters (omitting *I* and *O*) and four digits (allowing repetitions)?

*20 How many integers *x*, such that $1{,}111 \leq x \leq 4{,}321$, can be formed with the digits 1, 2, 3, 4 (allowing repetitions)?

*21 How many natural numbers not exceeding 4,321 can be formed with the digits 1, 2, 3, 4 (allowing repetitions)?

22 A circular conference table seats 10 persons. In how many ways may 10 persons seat themselves? (Consider only positions relative to each other, not with respect to the table.)

23 In how many ways can *n* keys be put on a key ring?

24 If $21P_{n,4} = 7P_{n,5}$, find *n*.

25 In how many ways can 10 different things be divided equally among 5 persons?

26 Each face of a regular tetrahedron is to be painted a solid and different color. How many tetrahedra can thus be painted with:
a Four colors? **b** Five colors?

27 Each face of a cube is to be painted a solid and different color. How many cubes can thus be painted with six colors?

*28 A daring young lady has three shades of nail polish with which to paint her fingernails. How many ways can she do this (each nail is to be one solid color) if:
a There are no other restrictions?
b There are not more than two shades on one hand?

29 In how many ways can *n* identifiable letters be stuffed into *m* linearly arranged pigeonholes?

30 In how many ways can *n* unidentifiable letters be stuffed into *m* linearly arranged pigeonholes?

12.12 Combinations

combinations

In permutations the *order* of the elements is important. If we ignore the order of the elements, we speak of "combinations". A combination of a set of elements is a subset of these elements taken without regard to the order of the elements. We wish to know how many subsets consisting of *r* elements each can be formed from the elements of a set containing *n* elements. This is called the number of combinations of *n* elements taken *r* at a time.

Illustration 1   How many committees of five can be chosen from a group of eight persons?

Solution This is a *combination* problem since *order* of the members of the committee is unimportant. Suppose that we call this number $C_{8,5}$. For each combination we can form 5! permutations by arranging the members of the committee in different orders. This gives us $P_{8,5}$ permutations. From the Multiplication Principle,

$$C_{8,5} \times 5! = P_{8,5} = 8!/3!$$

Therefore
$$C_{8,5} = 8!/5!3! = 56$$

Definition The symbol $C_{n,r}$ ($0 \le r \le n$) stands for the number of combinations of n distinct elements taken r at a time, and $C_{n,0} = 1$. The

$C_{n,r}$ simpler symbol $\binom{n}{r}$ is widely used for $C_{n,r}$.

Theorem 11 $\quad \binom{n}{r} = \dfrac{n!}{r!(n-r)!}$

Proof   By reasoning like that of Illustration 1,

$$\binom{n}{r} \times r! = P_{n,r} = \frac{n!}{(n-r)!}$$

Problems 12.12

1 Show that $\binom{n}{r} = \binom{n}{n-r}$.

2 Show that $\binom{n}{r} = P_{n,n}^{r,n-r}$. Can you prove this directly from the definitions?

3 If $\binom{n}{9} = \binom{n}{8}$, find $\binom{n}{17}$.

4 If $P_{n,r} = 12$ and $\binom{n}{r} = 6$, find n and r.

5 An airline had only 7 seats left on a given flight when a party of 10 applied for tickets.
   a In how many ways could the 7 seats be filled?
   b Suppose a second flight had just 3 vacant seats. In how many ways could the 10 passengers then be accommodated?

6 A man buys five new tires and tubes. In how many ways could the tires and tubes be paired?

7 A gardener has 20 different plants, but places for only 17. In how many ways may the planting be made?

8 A woman has seven dresses, three of which she proposes to wear daily (one for breakfast, one for lunch, one for dinner). For how many days can she do this before repeating some daily selection?

9 A bookstore stocks works A, B, C, D bound singly, in pairs, in triplets, and all together. In how many editions can A be bought?

10 If four straight lines are drawn in a plane, no two parallel, no three concurrent, how many triangles will be formed?

11 How many triangles can be formed if each side is to be 3 or 4 or 5 or 6 in. long?

12 Each night the Sultan selects some 4 of his 14 storytellers to meet with him as a group, each group of 4 differing in at least one storyteller. On how many nights can they meet without repeating a group?

13 Two identical pieces of chalk and one eraser are to be issued to four pupils. In how many ways can this be done if:
a No pupil is to receive more than one object?
b No pupil is to receive both pieces of chalk?

14 At a luncheon for six there is a choice of three beverages, coffee, tea, or milk. How many different selections are possible (such as two coffees, one tea, three milks)? Assume that each person takes one and only one beverage.

15 An automobile dealer has just four automobiles. How many different selections (of at least one) can be made?

16 A post office keeps four kinds of stamps. How many different selections (of at least one but not more than three) can be made?

17 Two opposite faces of a cube are to be left unpainted. The other four faces are to be painted each a solid color. How many ways can the cube be painted if there are four colors to choose from but no color is to be used for more than two faces?

18 Two opposite faces of a regular dodecahedron are to be left unpainted. The other 10 faces are to be painted each a solid and a different color. How many ways can the dodecahedron be painted if there are 15 colors to choose from?

19 A travel agency advertises round trips to 30 countries. In how many ways could a tourist make two such trips (nonrepeating) per month per year?

20 How many different sums of money can be formed from a cent, a nickel, a dime, a quarter, a half dollar, and a dollar?

21 In how many ways can a party of 10 persons be divided into:
a Two groups of 7 and 3 each?
b Two named groups (A and B) of 5 each?
c Two unnamed groups of 5 each?

22 In how many ways can at least four, but not more than six, persons be chosen from nine?

23 We wish to select six persons from eight, but if A is chosen, then B must also be chosen. In how many ways can the selection be made?

24 From a group consisting of 15 Republicans and 10 Democrats, how many committees can be formed consisting of either 4 Republicans and 6 Democrats or 6 Republicans and 4 Democrats?

25 A committee of five is to be selected from a club consisting of 60 men and 40 women. In how many ways can this be done so that the committee women outnumber the committee men?

26 A plane is divided into how many regions by n lines, no two parallel and no three concurrent? HINT: Use induction.

27 A super checkerboard has $2n$ rows and $2n$ columns. In how many ways can two squares of different color be selected?

28 How many natural numbers less than 1,000 do not contain the digit 9?

*29 In how many ways can three distinct natural numbers, each less than 100, be chosen so that their sum is divisible by 3? HINT: Consider the subsets: $\{1, 4, 7, \ldots, 97\}$; $\{2, 5, 8, \ldots, 98\}$; $\{3, 6, 9, \ldots, 99\}$.

*30 A room has seven lights, each controlled by a separate switch. In how many ways can three be turned on so that no two are turned on together more than once?

12.13 Binomial Theorem

We previously encountered the Binomial Theorem in Sec. 3.4, where we gave a proof of it based upon mathematical induction. Now it is possible for us to approach this theorem from another point of view. For reference, here is the statement of the theorem.

Theorem 12 Binomial Theorem Let n be a positive integer. Then

$$(a + b)^n = a^n + \binom{n}{1} a^{n-1}b + \binom{n}{2} a^{n-2}b^2 + \cdots$$

$$+ \binom{n}{r} a^{n-r}b^r + \cdots + \binom{n}{n-2} a^2 b^{n-2} + \binom{n}{n-1} ab^{n-1} + b^n$$

Or, in the expansion of $(a + b)^n$, the coefficient of $a^{n-r}b^r$ is $\binom{n}{r}$.

In order to give another proof of this theorem, we consider

$$(a + b)^n = \underbrace{(a + b)(a + b)\cdots(a + b)}_{n \text{ factors}}$$

$$= a^n + \cdots + Ca^{n-r}b^r + \cdots + b^n$$

where C is the coefficient of a typical term $a^{n-r}b^r$ $(0 \leq r \leq n)$. Now, in multiplying out the product $(a + b)(a + b)\cdots(a + b)$, $a^{n-r}b^r$ will be obtained by selecting b from each of some r factors (and a from the remaining $n - r$ factors). Since a selection of r things from n can be made in $\binom{n}{r}$ ways, it appears that $C = C_{n,r} = \binom{n}{r}$. This proves the result.

Exercise

A Show that $\binom{n}{r} = P^{r,n-r}_{n,n}$. Can you prove this directly from the definitions?

B Show that the coefficient of $x^r y^s z^t$ (where $r + s + t = n$) in the expansion of $(x + y + z)^n$ is $P^{r,s,t}_{n,n}$.

12.14 Binomial Probability Distribution

The binomial theorem is most useful in probability theory. Consider an experiment in which our interest centers on an event A and its complement A'. Since one of these will be thought of as the successful event and the other as the event of failure, we might as well call them S and F. An appropriate probability distribution is $U_1 = \{F, S\}$; $P(F)$, $P(S)$. For simplicity, let $P(S) = p$ and $P(F) = q(= 1 - p)$. Further, let $P(0)$ be the probability of exactly 0 successes and $P(1)$ the probability of exactly 1 success in one performance (or trial) of the experiment. Thus

binomial distribution

$$P(0) = q \qquad P(1) = p$$

Note that

$$(q + p)^1 = q + p$$

If the experiment is performed a second time (two trials), the joint distribution is over

$$U_2 = \{F, S\} \times \{F, S\} = \{FF, FS, SF, SS\}$$

with respective probabilities

$$P(FF) = qq \qquad P(FS) = qp \qquad P(SF) = pq \qquad P(SS) = pp$$

From these it follows that the probabilities $P(0)$, $P(1)$, $P(2)$ of exactly 0, 1, 2 successes in two trials are, respectively,

$$P(0) = q^2 \qquad P(1) = qp + pq \qquad P(2) = p^2$$
$$= 2qp$$

Note that

$$(q + p)^2 = q^2 + 2qp + p^2$$

The reason for adding qp and pq to obtain $2qp$ is that we have no interest in whether the first or the second trial resulted in the one success.

If the experiment is performed a third time, the distribution is over
$$U_3 = \{F, S\} \times \{F, S\} \times \{F, S\}$$
$$= \{FFF, FFS, FSF, SFF, FSS, SFS, SSF, SSS\}$$

with probabilities

$$qqq, \; qqp, \; qpq, \; pqq, \; qpp, \; pqp, \; ppq, \; ppp$$

Therefore $P(0)$, $P(1)$, $P(2)$, $P(3)$ of exactly 0, 1, 2, 3 successes in three trials are

$$P(0) = q^3$$

$$P(1) = qqp + qpq + pqq \qquad P(2) = qpp + pqp + ppq$$
$$= 3q^2p \qquad\qquad\qquad = 3qp^2$$

$$P(3) = p^3$$

Note that

$$(q + p)^3 = q^3 + 3q^2p + 3qp^2 + p^3$$

Again, like terms q^2p have been added and like terms qp^2 have been added because the order of the successes and failures is considered unimportant.

These three cases suggest that there is a general rule which gives, in n trials, $P(r)$ the probability of exactly r successes (and, of course, $n - r$ failures) and that this number $P(r)$ is exactly the $(r - 1)$st term in the binomial expansion of $(q + p)^n$. The reasoning is precisely that of the initial argument used in proving Theorem 12, Sec. 12.13. Consequently,

$$P(r) = \binom{n}{r} q^{n-r}p^r$$

$$(q + p)^n = q^n + \binom{n}{1} q^{n-1}p + \cdots + \binom{n}{r} q^{n-r}p^r + \cdots + p^n$$

$$= P(0) + P(1) + \cdots + P(r) + \cdots + P(n)$$

Exercises

A   Why is $\sum_{i=0}^{n} P(i) = 1$?

B   Is the distribution over $\{F, S\}$ necessarily uniform?

Where the order of S and F is unimportant, we write S^aF^b for the event of a successes and b failures. Thus $P(S^rF^{n-r}) = P(r)$. Also, we have agreed to write $\binom{n}{0} = 1$. This leads to the following useful definition:

Definition   The distribution over $U = \{F^n, S^1F^{n-1}, \ldots, S^rF^{n-r}, \ldots, S^n\}$ with probabilities $P(S^rF^{n-r}) = \binom{n}{r} q^{n-r}p^r$, $r = 0, 1, \ldots, n$, is called the *Binomial Probability Distribution*.

Illustration 1   A coin is tossed 100 times. What is the probability of getting 47 heads? ANSWER: $P(47\ H\text{'s}) = \binom{100}{47} (\frac{1}{2})^{100-47}(\frac{1}{2})^{47} = \binom{100}{47}/2^{100}$.

Illustration 2   A die is tossed 100 times. What is the probability of getting exactly 47 3's? Here $p = \frac{1}{6}$, $q = \frac{5}{6}$. ANSWER: $P(47\ 3\text{'s}) = \binom{100}{47} (\frac{5}{6})^{100-47}(\frac{1}{6})^{47} = 5^{53} \binom{100}{47}/6^{100}$.

Illustration 3   A biased coin with $p = P(H) = \frac{2}{3}$ is tossed 100 times. What is the probability of 47 or 48 heads?   ANSWER: $P(47 \text{ or } 48 \ H\text{'s}) = \binom{100}{47}(\frac{1}{3})^{53}(\frac{2}{3})^{47} + \binom{100}{48}(\frac{1}{3})^{52}(\frac{2}{3})^{48}$.

Illustration 4   An icosahedron is tossed 100 times. What is the probability of not more than 47 16's? Here $p = \frac{1}{20}$.   ANSWER:

$$P(\leq 47 \ 16\text{'s}) = \binom{100}{0}(\tfrac{19}{20})^{100} + \binom{100}{1}(\tfrac{19}{20})^{99}(\tfrac{1}{20}) + \cdots$$

$$+ \binom{100}{47}(\tfrac{19}{20})^{53}(\tfrac{1}{20})^{47}$$

$$= \sum_{r=0}^{47}\binom{100}{r}(\tfrac{19}{20})^{100-r}(\tfrac{1}{20})^{r}$$

Illustration 5   Two dice are tossed 100 times. What is the probability that the sum on a given toss will not exceed 7 more than 97 times?

Solution   For a single toss of two dice the probability of getting a sum of 2 or 3 or 4 or 5 or 6 or 7 is

$$p = \tfrac{1}{36} + \tfrac{2}{36} + \tfrac{3}{36} + \tfrac{4}{36} + \tfrac{5}{36} + \tfrac{6}{36} = \tfrac{21}{36} = \tfrac{7}{12} \qquad q = \tfrac{15}{36} = \tfrac{5}{12}$$

The answer to the problem is therefore

$$P(\text{sum} \leq 7 \text{ not more than } 97 \text{ times}) = \sum_{r=0}^{97}\binom{100}{r}q^{100-r}p^{r}$$

$$= \sum_{r=0}^{97}\binom{100}{r}(\tfrac{5}{12})^{100-r}(\tfrac{7}{12})^{r}$$

$$= 1 - \sum_{r=98}^{100}\binom{100}{r}(\tfrac{5}{12})^{100-r}(\tfrac{7}{12})^{r}$$

Exercise
C   Explain the last line in the answer above.

In each of the above five illustrations there would be great numerical difficulties in reducing the answer directly to simple decimal form. There are easy methods for getting good approximations, but these will not be considered in this book. (The idea is to use the normal probability distribution as an approximation to the binomial distribution.)

Illustration 6   From a bag containing m red balls and n black balls, k balls are drawn. Given that they are all of one color, what is the probability that they are black?

Solution   First an adequate sample space must be chosen. A simple event will be any combination of k balls taken from the $m + n$ balls. The total number of such is $\binom{m+n}{k}$, and, for a uniform distribution, the probability of each simple event is $1/\binom{m+n}{k}$. The events of special interest are $A(k$ black$)$, $B(k$ black or k red$)$, and $(A \cap B) = A$. The number of simple events in A is $\binom{n}{k}$ and in B is $\binom{n}{k} + \binom{m}{k}$. We are asked to find $P(A|B)$.

$$P(A|B) = \frac{P(A \cap B)}{(P(B)} = \frac{P(A)}{P(B)}$$

$$= \frac{\dfrac{1}{\binom{m+n}{k}} \cdot \binom{n}{k}}{\dfrac{1}{\binom{m+n}{k}} \cdot \left[\binom{n}{k} + \binom{m}{k}\right]}$$

$$= \frac{\binom{n}{k}}{\binom{n}{k} + \binom{m}{k}}$$

The final form of the answer gives a short cut to the solution.

Problems 12.14

1 A batter is now batting 250. What is the probability that he will make:
   a A hit next time at bat?
   b Exactly two hits out of his next three trips to bat?
   c At least two hits out of his next three trips to bat?
2 The probability that a marksman will hit a target is given as $\frac{1}{10}$. What is his probability of at least 1 hit in 10 shots?
3 The probability of winning a certain contest is $\frac{1}{5}$. What is the contestant's probability of winning at least four out of six such contests?
4 What is the probability of obtaining:
   a Exactly one head in tossing a coin six times?
   b Two heads?   c Three heads?   d At least three heads?
5 If four dice are tossed, what is the probability that:
   a Exactly two will turn up aces?   b At least two?
6 What is the probability of obtaining the sum 7 at least once in three throws of two dice?
7 A die is marked 0, 0, 0, 0, 2, 3. What is the probability of obtaining the sum 12 in five tosses?
8 A tetrahedron is tossed 10 times. Find $P(\text{sum} = 34)$.

In Probs. 9 to 12 a coin is tossed 500 times.   Find:

9 $P(\text{at least } 2T)$
10 $P(\text{at most } 2T)$
11 $P(\text{at least } 2T \text{ but not more than } 5T)$
12 $P(\text{at least } 200T \text{ but not more than } 300T)$

In Probs. 13 to 16 a die marked x, x, y, y, z, z is tossed five times.   Find:

13 $P(\text{at least } 4y\text{'s})$
14 $P(\text{at most } 4y\text{'s})$
15 $P(\text{at least } 1 \text{ but not more than } 3y\text{'s})$
16 $P(2y\text{'s or } 3z\text{'s})$

In Probs. 17 to 20 a die marked x, x, x, y, y, z is tossed five times.   Find:

17 $P(\text{at least } 4x\text{'s})$
18 $P(\text{at most } 4z\text{'s})$
19 $P(\text{at least } 1x \text{ and not more than } 1z)$
20 $P(\text{at least } 1y \text{ and not more than } 1z)$
21   In one toss of two dice, consider a success to be either a 7 or an 11 or a 12.   What is the probability of success?
22   From a deck consisting of the cards 2, 3, 4, 5, 6, 7, a single draw is made, success being either a 3 or a 5.   Find the probability of x successes in seven draws (with replacements).
23   A batter has a probability of 0.25 of making a hit in a single trip to the plate.   Find the probability of x hits in 25 times at bat.
24   Suppose that the probability of catching a trout on a single cast is 0.01.   Find the probability of catching x trout in 250 casts.
25   Two letters fall out of the word "little".   If they are replaced at random, what is the probability that it still reads *little*?   HINT: Write down the fall-out sample space $\{l_1i, l_1t_1, l_1t_2, \ldots, l_2e\}$; treat l_1l_2, t_1t_2 separately.
26   A red card is removed from a pack.   Thirteen cards are then drawn and found to be all of one color.   What is the probability that the color is black?   (See Illustration 6 above.)
27   What is the probability that two squares marked at random on a chessboard have contact only at a corner?
28   What is the probability that a hand of five cards contains:
   a   No ace?   **b**   Exactly one ace?
29   What is the probability that a hand of five cards contains:
   a   No pair?   **b**   At least two aces?
30   What is the probability of throwing a doublet with:
   a   Two true dice?   **b**   One true and one biased die?
31   Find the probability of throwing a sum of 14 with:
   a   Three dice.   **b**   Eleven dice.
32   The integers 1, 2, . . . , 20 are written down in a row so that no two of the integers 1, 2, 3, 4 come together.   What is the probability that 2 is next to 7?

33   The integers 1, 2, . . . , 20 are written down in a ring so that no two of the integers 1, 2, 3, 4 come together.   What is the probability that 1 is next to 8?

34   Ten people stand in a line.   What is the probability that A stands next to B?

35   Ten people sit at a round table.   What is the probability that A sits next to B?

36   Five men speak at a political rally.   If it is known that A speaks after B, what is the probability that it is immediately after?

12.15   Testing Hypotheses

Suppose we have tossed a coin 10 times and have obtained 10 heads. What may we conclude as to whether or not the coin is biased?   The procedure is to make a hypothesis about the coin and to establish a means of accepting or rejecting this hypothesis.   We may make any hypothesis we please and then carry out the test, but we must make a specific hypothesis.

hypothesis
To be definite, we shall make the hypothesis that the coin is true, i.e., that the probability of a head on a single toss is $p = \frac{1}{2}$.   On the basis of this hypothesis, the probability of 10 heads in 10 tosses is

$$\left(\frac{1}{2}\right)^{10} = \frac{1}{1,024} = 0.00098$$

Hence, if the coin were true, it is very unlikely that the observed result would have taken place.   We are therefore inclined to reject the hypothesis that the coin is true.

Suppose the coin is biased with $p = P(\text{head}) = \frac{2}{3}$.   On the basis of this hypothesis, the probability of 10 heads in 10 tosses is

$$\left(\frac{2}{3}\right)^{10} = \frac{1,024}{59,049} = 0.017$$

This is still not very likely, but is certainly more likely than in the hypothesis that $p = \frac{1}{2}$.   We may be in doubt as to acceptance or rejection of this hypothesis.

Finally, if we have a badly warped coin and make the hypothesis that $p = 0.9$, the probability of 10 heads in 10 tosses is $(0.9)^{10} = 0.34$.   This is so likely that we should certainly accept this hypothesis.   We should similarly accept the hypotheses that $p = 0.99$ and $p = 1.0$.

We need to understand clearly what is meant by the acceptance of a hypothesis.   We have chosen this hypothesis because of evidence not connected with our particular experiment, and we wish to know whether the experiment confirms or denies the hypothesis.   If we reject the hypothesis, we certainly must abandon it.   If we accept the hypothesis, we are essentially saying that we have no valid reason for rejecting it. We cannot be at all certain that an accepted hypothesis is true.   Thus we may accept a number of hypotheses on the basis of the same experiment. The decision as to which of these is correct (if indeed any is) cannot be made with certainty, and we proceed on the assumption that any of them may be correct.

In order to have a definite rule for deciding whether to accept or reject a hypothesis, we must decide in advance on some level of probability which will separate these two situations. In common practice this value is taken to be 0.05 or 0.01 or sometimes 0.005. For consistency in this book we arbitrarily use the value of 0.01. We now adopt the following procedure:

0.01 level

Rule for Testing Hypotheses (Connected with a Binomial Distribution)

1 Make the hypothesis that the probability of success in a single trial of the event under consideration is equal to some definite value p.
2 Using this value of p, in a binomial distribution compute the probability $\overline{p}$ of the observed number of successes in the stated number of trials n.
3 If $\overline{p} < 0.01$, reject the hypothesis (at the 0.01 level).
   If $\overline{p} \geq 0.01$, accept the hypothesis (at the 0.01 level).

In these terms we then state the conclusions to the questions raised at the beginning of this section and we use the following language:

At the 0.01 level we reject the hypothesis that $p = \frac{1}{2}$.
At the 0.01 level we accept the hypotheses that p equals $\frac{2}{3}$, 0.9, 0.99, and 1.0.

Note that we cannot conclude from our observations what the correct value of p actually is. The best that we can do is to reject certain values of p and to accept others. Even when we follow this rule, we may make errors. We may reject a true hypothesis, but this will happen in only a few cases. We may also accept a false hypothesis. We must run the risk of these errors if we are to come to any conclusions at all. Statistical inference may therefore be called "probable inference" in distinction to the exact inferences we draw in other types of reasoning.

Problems 12.15

1 A coin is tossed five times, and three heads are obtained. Let p be the probability of obtaining a head on a single toss. At the 0.01 level, test for acceptance or rejection the hypothesis:
   a $p = \frac{1}{10}$ **b** $p = \frac{1}{3}$ **c** $p = \frac{1}{2}$ **d** $p = \frac{2}{3}$
   e $p = \frac{3}{4}$
2 Toss a coin 10 times and count the number of heads. Assuming the coin to be true, compute the probability of obtaining this number of heads. Do you accept or reject (at the 0.01 level) the hypothesis that your coin is true?

3 A coin is tossed 10 times and gives 8 heads and 2 tails. Test the hypothesis that the coin is true (using the 0.01 level).

4 A die is rolled six times, in which the two-spot turns up twice. Test the hypothesis (at the 0.01 level) that the die is true.

5 A pair of dice are rolled five times, and on each roll the total number of spots is seven. In case the dice belong to your opponent (who is a stranger), decide whether or not you should shoot him, taking into account the laws of the land and of probability.

6 Six thumbtacks are tossed simultaneously onto the floor, and exactly three land with their points straight up. Test (at the 0.01 level) the hypothesis that the probability that a tack will land point up is:
a ⅓ **b** ½

7 An automatic bolt machine has been making 3 faulty bolts in 100 on the average. An inspector takes five bolts at random and finds one defective among them. Test (at the 0.01 level) the hypothesis that the machine is out of adjustment.

8 A random sample of five people reveals that one is left-handed and four are right-handed. At the 0.01 level test the hypothesis that:
a One person in eight is left-handed.
b One person in nine is left-handed.

9 On the average 5 persons in 100 still get disease X even though they have been inoculated against it. The use of a new drug with a sample of 100 persons reduces the number who contract the disease to just 3. Test (at the 0.01 level) the hypothesis that the new drug is better. (Use logarithms for the calculations.)

10 A seed house sells seed in packets of 100. Experience indicates that, without a special treatment, only 90 per cent of the seed will germinate, on the average. After treatment a random packet tested 95 per cent viable. At the 0.01 level, test the hypothesis that the treatment is effective. (Use logarithms for the calculations.)

12.16 Expectation

Let us suppose that to each simple event e_i of a sample space $U = \{e_1, e_2, \ldots, e_n\}$ we assign a real number (positive, zero, or negative) called its value $V(e_i)$. The motivation for doing this comes from various games of chance; but as suggested in the illustrations below, the process is one of considerable generality.

Illustration 1 Consider a dice game in which a single die is rolled. If it turns up 1, 2, or 3, the player wins nothing. If it turns up 4 or 5, he wins $1. If it turns up 6, he wins $2. This information can be assembled in the table:

Event	1	2	3	4	5	6
Value	$V(1) = 0$	$V(2) = 0$	$V(3) = 0$	$V(4) = 1$	$V(5) = 1$	$V(6) = 2$

Illustration 2   Let the sample space refer to a deck of cards and consist of $\{A, 2, 3, 4, 5, 6, 7, 8, 9, 10, J, Q, K\}$. Then we can assign the values

$$V(A) = 1, \; V(2) = 2, \; V(3) = 3, \ldots,$$
$$V(9) = 9, \; V(10) = 10, \; V(J) = 10, \; V(Q) = 10, \; V(K) = 10$$

Of course, there are many other ways of assigning values to the simple events of a sample space.

Illustration 3   When the sample space is a set of actual measurements $\{x_1, \ldots x_n\}$, we often take $V(x_i)$ to be x_i itself. More generally, we can choose an arbitrary function $f(x)$ and take $V(x_i)$ to be $f(x_i)$. An important special case is the choice $V(x_i) = (x_i)^2$.

Given a sample space, a set of values $V(e_i)$ for its simple events, and a probability distribution $P(e_i)$ over the space, we can define the corresponding *expectation* or *expected value* as follows:

Definition   The *expectation* E of a set of values $V(e_i)$ assigned to a sample

expectation   space $U = \{e_1, e_2, \ldots, e_n\}$ carrying a probability distribution $P(e_i)$ is

$$E = \sum_{i=1}^{n} V(e_i)P(e_i)$$

When e_i is a number x_i and when $V(x_i)$ is given by $V(x_i) = f(x_i)$, the *expectation of $f(x)$*, $E[f(x)]$, is given by

$$E[f(x)] = \sum_{i=1}^{n} f(x_i)P(x_i)$$

Illustration 4   Returning to the dice game of Illustration 1, we choose the uniform probability distribution and obtain the table:

Event	1	2	3	4	5	6
V	0	0	0	1	1	2
P	$\frac{1}{6}$	$\frac{1}{6}$	$\frac{1}{6}$	$\frac{1}{6}$	$\frac{1}{6}$	$\frac{1}{6}$

Then $E = 1(\frac{1}{6}) + 1(\frac{1}{6}) + 2(\frac{1}{6}) = \$ \frac{2}{3}$.

If a man were to play this game a large number of times, he would thus expect to win an average of $\$ \frac{2}{3}$ each time. If the game is fair, this is the amount the player should be charged to play it.

Definition   When the sample space is a set of numbers x_i with a probability distribution $P(x_i)$, the expectation of x, namely,

$$E(x) = \sum_{i=1}^{n} x_i P(x_i)$$

mean   is called the *mean m* of the probability distribution.

Further the expectation of $(x - m)^2$, namely,

$$E(x - m)^2 = \sum_{i=1}^{n} (x_i - m)^2 P(x_i)$$

variance

is called the *variance* σ^2 of the distribution. Its square root σ is called the

standard deviation

standard deviation.

Theorem 13 $E(x - m)^2 = E(x^2) - [E(x)]^2$

Proof   To prove this, we recall that:

$$E(x - m)^2 = \sum_{i=1}^{n} (x_i - m)^2 P(x_i)$$

$$= \sum_{i=1}^{n} (x_i^2 - 2mx_i + m^2) P(x_i)$$

$$= \sum_{i=1}^{n} x_i^2 P(x_i) - 2m \sum_{i=1}^{n} x_i P(x_i) + m^2 \sum_{i=1}^{n} P(x_i)$$

$$= E(x^2) - 2m^2 + m^2$$
$$= E(x^2) - m^2$$
$$= E(x^2) - [E(x)]^2$$

Illustration 5   The values of x_i and the probability distribution $P(x_i)$ are given in the following table.   Find the mean, variance, and standard deviation.

x_i	-3	2	1	4
$P(x_i)$	0.3	0.4	0.2	0.1

$$m = E(x) = [0.3 \times (-3)] + [0.4 \times 2] + [0.2 \times 1] + [0.1 \times 4]$$
$$= 0.5$$
$$E(x^2) = [0.3 \times 9] + [0.4 \times 4] + [0.2 \times 1] + [0.1 \times 16]$$
$$= 5.1$$
$$\sigma^2 = E(x^2) - [E(x)]^2 = 5.1 - (0.5)^2$$
$$= 4.85$$
$$\sigma = \sqrt{4.85}$$

Theorem 14   For the Binomial Distribution, $m = np$.

Proof   As in Sec. 12.14 the probability of r successes and $n - r$ failures in

n trials is $P(r) = \binom{n}{r} q^{n-r} p^r$. Hence

$$m = E(r) = \sum_{r=1}^{n} r \binom{n}{r} q^{n-r} p^r$$

$$= \sum_{r=1}^{n} r \frac{n!}{r!(n-r)!} q^{n-r} p^r$$

$$= \sum_{r=1}^{n} \frac{n!}{(r-1)!(n-r)!} q^{n-r} p^r$$

$$= np \sum_{r=1}^{n} \frac{(n-1)!}{(r-1)!(n-r)!} q^{n-r} p^{r-1}$$

Put $s = r - 1$ so that $n - r = n - 1 - s$. Then

$$m = np \sum_{s=0}^{n-1} \frac{(n-1)!}{s!(n-1-s)!} q^{n-1-s} p^s$$

$$= np$$

since the summation on the right is equal to $\sum_{s=0}^{n-1} P(s) = 1$.

Problems 12.16

1 Everything above a 5 is removed from a deck of cards, including the ace. The player draws one card from those cards which remain. The banker is to pay face value in dollars. What is the player's expectation? Assume the uniform probability distribution.

2 A set of 100 lottery tickets, numbered 1, 2, . . . , 100, are shuffled. If a person drawing one ticket is to pay face value in cents for it, what is his expectation (in this case the amount he is to pay)? Assume the uniform probability distribution.

3 Each time you flip a head with a coin you win a prize and flip again. The game ends when you flip tails.
  a What is your expectation if the prize is $1 for each head?
  b What is your expectation if, for the nth head, the prize is 2^n dollars?

4 If six coins are thrown, what is the expected number of heads that will turn up?

5 A man draws at random two coins from a purse which contains five silver dollars and seven half-dollars. Assuming the uniform distribution, what is his expectation?

6 A gambler will receive $36 if in a throw of two dice he obtains a total of 8 or 5. He loses $18 if he fails. Assuming the uniform distribution, find his expectation.

7 On a multiple-choice test containing 100 questions there are four possible answers to each question, only one of which is correct. A score of 1 is assigned to each correct answer and a score of $-\frac{1}{3}$ to each incorrect answer. No question may be omitted. A completely unprepared student answers the questions at random. What is his expected score?

8 A die and a coin are tossed simultaneously. If the coin turns up heads, the player receives, in dollars, the number of spots on the die. If the coin turns up tails, he receives nothing. Assuming the uniform distribution, find his expectation.

In Probs. 9 to 12 the values of x_i and the probability distribution $P(x_i)$ are given.   Find the mean, variance, and standard deviation.

9

x_i	1	2	3	4
$P(x_i)$	0.4	0.2	0.3	0.1

10

x_i	-3	-2	2	3
$P(x_i)$	0.1	0.4	0.3	0.2

11

x_i	5	7	9	11	13
$P(x_i)$	0.2	0.2	0.1	0.4	0.1

12

x_i	10	20	40
$P(x_i)$	0.3	0.4	0.3

13   Eight coins are tossed.   What is the mean of the corresponding Binomial Distribution?   The variance?

14   Sixteen coins are tossed.   What is the mean of the corresponding Binomial Distribution?   The standard deviation?

15   Let the sample space be $\{1, 2, 3, \ldots, n\}$, and assume the uniform probability distribution.   Find the mean and variance.

16   Prove that $E(x + c) = E(x) + c$.   **17**   Prove that $E(kx) = kE(x)$.

18   Let Var $[f(x)] = E[f(x)]^2 - [E(f(x))]^2$.   Show that

$$\text{Var } (kx) = k^2 \text{ Var } (x)$$

19   Prove that Var $(x + c) = $ Var x.   (See Prob. 18 for notation.)

*20   Prove that the variance of the Binomial Distribution is npq.   Model your proof on that of Theorem 14.

12.17   Two-by-Two Games

The mathematical theory of games was originated by John von Neumann in 1928 and has developed into a full-blown mathematical subject in recent years.   There is an unconfirmed rumor that Professor von Neumann invented this theory in the hope of cutting his losses at poker, but unfortunately poker and other popular games such as checkers and chess are too complicated for analysis by this method.   The theory, however, has been applied to the analysis of the rational behavior of men and nations who are engaged in business, economics, or warfare.   Game theory is particularly important as a mathematical model (Sec. 1.13) of human behavior; it is still imperfect, but it does help us to understand many of the activities of man.

Since game theory is an extensive and complicated mathematical subject, we cannot hope to do more here than to give you the flavor of a particularly simple portion of the theory. This section deals with what are technically known as 2×2, zero-sum, two-person games—which hereafter we shall simply call *games*. We begin by describing several games of this type.

Illustration 1   Matching Pennies. There are two players, Player 1 and Player 2. At a given signal each player displays one face of a penny, heads or tails. If the faces match (both heads or both tails) Player 2 pays Player 1 one cent. If the faces do not match, Player 1 pays Player 2 one cent. The play is repeated a large number of times. The problem is to determine how each player should proceed in order to maximize his return from the game.

It is customary to record the data for a game in a game matrix, as written below for this game.

		Player 2 H	T
Player 1	H	1	−1
	T	−1	1

The entries in this matrix record payments *from* Player 2 *to* Player 1 as positive numbers. Payments in the opposite direction are, therefore, negative. The game is completely determined by its game matrix.

Illustration 2   Player 1 holds two cards, a 2 and a 7, and Player 2 holds a 3 and a 5. At a given signal each exposes one of his cards. If the sum of the numbers on the cards is *even*, Player 2 pays this sum to Player 1. If the sum is *odd*, Player 1 pays this sum to Player 2. The game matrix is

		Player 2 3	5
Player 1	2	−5	−7
	7	10	12

In general, a game matrix is written

		Player 2 t_1	t_2
Player 1	s_1	a_{11}	a_{12}
	s_2	a_{21}	a_{22}

game matrix

where s_1 and s_2 are the two choices available to Player 1 and t_1 and t_2 are the two choices available to Player 2. The entries a_{ij} are the assigned payoffs. For example, if Player 1 chooses s_1 and Player 2 chooses t_2,

Player 2 pays Player 1 the amount a_{12} if a_{12} is positive. If a_{12} is negative, the payment is from Player 1 to Player 2.

Since the format of this table is standard, we shall usually abbreviate it to

a_{11}	a_{12}
a_{21}	a_{22}

or to

a	b
c	d

Any 2×2 matrix can be interpreted as the *game matrix* for some game.

Strictly Determined Games Let us analyze the game of Illustration 2 whose matrix is

	3	5
2	-5	-7
7	10	12

It is evident to Player 1 that he will be better off in choosing a 7 no matter which choice is made by Player 2. Player 2 assumes that Player 1 is rational and that he has in fact made this choice. So Player 2 seeks to minimize his losses by choosing a 3. Thus in each play of the game Player 2 pays 10 to Player 1. This payment, 10, is called the *value* of the game. Clearly, Player 2 would not willingly play this game if he had made this analysis.

Games like this are called *strictly determined* and have relatively little interest. You can verify that 10 is the minimum of its row and the maximum of its column. This leads to the general definition:

Definition A game is *strictly determined* if and only if there is an entry u which is the minimum of its row and the maximum of its column.

For a strictly determined game, the rational outcome is:
Player 1 chooses the row in which u lies.
Player 2 chooses the column in which u lies.
The value of the game is u.

Illustration 3 The following games are strictly determined:

1	3	3	-1	4	4
0	5	4	-3	1	3

The following games are not strictly determined:

5	1	-1	3	1	-1
2	4	2	0	-1	1

Nonstrictly Determined Games   When a game is not strictly determined, the problem of how to play it is more complicated.   Consider the game of Matching Pennies in Illustration 1. If Player 1 always plays H, Player 2 will soon catch on and always play T.   This is not to the advantage of Player 1.   Similarly, Player 2 can win if Player 1 always plays T.

Common sense, or experience with this game, tells us that Player 1 should "mix them up" and sometimes play H and sometimes T.   This is called a *mixed strategy.*

Definition   In a *mixed strategy,* Player 1 chooses the first row with probability p and the second with probability $1 - p$.   Similarly, Player 2 chooses the first column with probability q and the second column with probability $1 - q$.   Moreover, we require that p, q, $1 - p$, and $1 - q$ be positive.

mixed
strategy

In practice, the players will make their choices by using some random device (coins, dice, etc.) which operates with the chosen probability p or q.

The problem now is to determine what p and q should be.   Let us suppose that the game matrix is

$$\begin{array}{cc} a & b \\ c & d \end{array}$$

We shall compute Player 1's expectation when (1) Player 2 chooses the first column and (2) when Player 2 chooses the second column.   These are

(1) $$pa + (1 - p)c = p(a - c) + c$$

(2) $$pb + (1 - p)d = p(b - d) + d$$

Of course, Player 1 does not know how Player 2 will play, for Player 2 may choose either column or use a mixed strategy with probability q.   The "safe" position for Player 1 is to choose p so that his expectation is the same regardless of how Player 2 plays, for then he is sure of his expected return from the game.   He can do this by equating the expectations (1) and (2) above and solving for p:

$$p(a - c) + c = p(b - d) + d$$
or $$p(a - b - c + d) = d - c$$
or

(I) $$p = \frac{d - c}{a - b - c + d}$$

Then

(II) $$1 - p = \frac{a - b}{a - b - c + d}$$

This common expectation for Player 1 is

$$E_1 = \frac{d - c}{a - b - c + d}(a - c) + c$$

$$= \frac{ad - bc}{a - b - c + d}$$

E_1 remains the same even if Player 2 uses a mixed strategy with probability q, for in this case the expectation for Player 1 is

$$E = q[p(a - c) + c] + (1 - q)[p(b - d) + d]$$
$$= q(E_1) + (1 - q)(E_1)$$
$$= E_1$$

We now turn to Player 2. He has two expectations: (1) when Player 1 chooses the first row and (2) when Player 1 chooses the second row. These are

(1) $$qa + (1 - q)b = q(a - b) + b$$

(2) $$qc + (1 - q)d = q(c - d) + d$$

Equating these and solving for q as above, we find

$$q(a - b) + b = q(c - d) + d$$
$$q(a - b - c + d) = d - b$$

(III) $$q = \frac{d - b}{a - b - c + d}$$

and

(IV) $$1 - q = \frac{a - c}{a - b - c + d}$$

The common value of these two expectations for Player 2 is

$$E_2 = \frac{d - b}{a - b - c + d}(a - b) + b$$
$$= \frac{ad - bc}{a - b - c + d}$$
$$= E_1$$

The common value of E_1 and E_2 is called v, the *value* of the game. That is,

(V) $$v = \frac{ad - bc}{a - b - c + d}$$

By choosing the mixed strategy with p as in (I) above, Player 1 will expect to receive an amount of v from the game. Similarly, by choosing q as in (III), Player 2 will expect to pay an amount v. If $v = 0$, the game is called *fair*.

Before this discussion is concluded, there are some fine points that should be investigated. We must be sure that the denominators $a - b - c + d$ are not zero. Also, it is essential that the values for p, $1 - p$, q, and $1 - q$ be positive. To show that these conclusions follow from our assumption that the game is nonstrictly determined, we prove the following theorem.

Theorem 15   A 2×2 game with matrix

$$\begin{array}{cc} a & b \\ c & d \end{array}$$

is nonstrictly determined if and only if one of the following is true:

(**1**) $$a < b, a < c, d < b, \text{ and } d < c$$

(**2**) $$a > b, a > c, d > b, \text{ and } d > c$$

Proof If (1) holds, there is no element that is both a minimum for its row and a maximum for its column. Hence the game is not strictly determined. For a is the minimum of its row, but not the maximum of its column, and d is the minimum of its row and not the maximum of its column. A similar argument shows that when (2) holds, the game is not strictly determined.

To prove the converse, assume that the game is not strictly determined and consider element a. If a is the minimum of its row, then a is not the maximum of its column, so $a < c$, and c is the maximum of its column. Since c is not the minimum of its row, $d < c$, and d is the minimum of its row. Hence d is not the maximum of its column, and $d < b$. Thus b is the maximum of its column and cannot be the minimum of its row, so $a < b$. This gives us the set of inequalities (1).

If we assume that a is the maximum of its column and carry out a similar analysis, we obtain the inequalities (2).

Theorem 15 gives the answer to our worries. Assume that (1) is true. Then $a - b, d - c$, and $d - b$, and $a - c$ are all negative. From these $a - b - c + d = (a - b) + (d - c)$ is also negative. Hence $p, 1 - p, q$, and $1 - q$ are all positive. When (2) is true, we obtain the same result. Hence our queries are answered satisfactorily.

Finally, we wish to show that the choices of p and q in (I) and (III) above are better than any other choices. Suppose that instead of the given value for p we chose $p_1 > p$. Suppose also that case 1 of Theorem 15 is true, namely, $a - c$ is positive and $b - d$ is negative. Then Player 1's two expectations are

(**1**) $$p_1(a - c) + c > p(a - c) + c = v$$

(**2**) $$p_1(b - d) + d < p(b - d) + d = v$$

If Player 2 is aware of this choice, he can improve his performance by playing column 2 only. Then Player 1 will receive an amount less than the value v that he would have received by using the probability p of formula (I). Since this analysis is typical of the other possible deviations from formulas (I) and (III), we see that the values of p and q given there are *optimal*. Hence they lead to what are called *optimal strategies*.

optimal
strategies

Let us summarize with some illustrations of 2×2 games. By a *solution* of a game, we mean the specification of what each player should do and the determination of the value of the game.

Illustration 4 Solve the game whose matrix is

$$\begin{array}{cc} 4 & 2 \\ 5 & 3 \end{array}$$

Since 3 is a minimum for its row and a maximum for its column, the game is strictly determined. The solution is
Player 1 plays row 2.
Player 2 plays column 2.
The value is 3.

Illustration 5 Solve the game whose matrix is

$$\begin{array}{cc} 5 & 1 \\ -1 & 3 \end{array}$$

This game is nonstrictly determined, so we seek the optimal mixed strategies and the value. We can substitute in formulas (I) to (V), but it is usually better to use the method by which they were derived. For Player 1 we equate his two expectations and obtain

$$5p - 1(1 - p) = p + 3(1 - p)$$

Hence $p = \frac{1}{2}$; $1 - p = \frac{1}{2}$; $v = 5(\frac{1}{2}) - 1(\frac{1}{2}) = 2$. For Player 2 we obtain

$$5q + 1(1 - q) = (-1)q + 3(1 - q)$$

Hence $q = \frac{1}{4}$; $1 - q = \frac{3}{4}$; v (to check) $= 5(\frac{1}{4}) + 1(\frac{3}{4}) = 2$.

Games exist with square and nonsquare matrices of any size. The methods which we have used to solve 2×2 games above must undergo substantial changes when applied to these other games. For this more elaborate theory you must consult advanced treatments of this subject.

Problems 12.17

1 State which of the following games are strictly determined:

a $\begin{array}{cc} 1 & -2 \\ 3 & 6 \end{array}$ **b** $\begin{array}{cc} 5 & 0 \\ -1 & 3 \end{array}$ **c** $\begin{array}{cc} -3 & 4 \\ 1 & 2 \end{array}$

2 State which of the following games are strictly determined:

a $\begin{array}{cc} 3 & 2 \\ 1 & 4 \end{array}$ **b** $\begin{array}{cc} 0 & -1 \\ 1 & 0 \end{array}$ **c** $\begin{array}{cc} 2 & -1 \\ 4 & 3 \end{array}$

3 Show that the game $\begin{array}{cc} a & b \\ a & d \end{array}$ is strictly determined for every set of values for a, b, and d.

4 For what values of a and b is the game $\begin{array}{cc} a & 0 \\ 0 & b \end{array}$ strictly determined?

In Probs. 5 to 10 solve the game whose matrix is given. Which of these are fair?

5
$$\begin{array}{cc} 3 & 4 \\ 2 & 100 \end{array}$$

6
$$\begin{array}{cc} 4 & 1 \\ 2 & 6 \end{array}$$

7
$$\begin{array}{cc} 4 & -2 \\ -2 & 1 \end{array}$$

8
$$\begin{array}{cc} -1 & 3 \\ -2 & 1 \end{array}$$

9
$$\begin{array}{cc} 5 & 1 \\ 2 & 3 \end{array}$$

10
$$\begin{array}{cc} 6 & -3 \\ -2 & 1 \end{array}$$

In Probs. 11 to 15 set up the game matrix and find its solution.

11 In a football game the offense huddles and decides to run or pass not knowing what the defensive pattern will be. The defense sets its pattern without knowing what the play will be. How should each make its choice under the assumption that the position on the field, number of downs remaining, score, and time remaining are to be ignored? Assume the following data:

A running play against a running defense yields 3 yd.

A running play against a pass defense yields 8 yd.

A pass play against a running defense yields 20 yd.

A pass play against a pass defense yields 0 yd.

12 Solve Prob. 11 if the data are:

A running play against a running defense yields 3 yd.

A running play against a pass defense yields 5 yd.

A pass play against a running defense yields 15 yd.

A pass play against a pass defense loses 5 yd.

13 Each player holds up one or two fingers. Player 2 pays Player 1 a sum equal to the total number of fingers shown.

14 Each player holds up one or two fingers. If the sum is even, Player 2 pays Player 1 a sum equal to the total number of fingers shown. If the sum is odd, Player 1 pays this amount to Player 2.

15 In a war involving the use of marine transport, the enemy can try to harm us by mining our harbors in the hope of sinking our ships. We can counter this by using mine sweepers in an attempt to explode the mines harmlessly. We do not know what the enemy is doing, and he does not know what countermeasures we are taking. The following data apply:

Enemy does not lay mines and we do not sweep: no cost to us or enemy.

Enemy does not lay mines, but we do sweep: cost to us = $10 million; no cost to enemy.

Enemy does lay mines, and we do not sweep: cost to us = $105 million; cost to enemy = $5 million; net cost to us = $100 million.

Enemy does lay mines and we do sweep: Cost to us = $10 million; cost to enemy = $5 million; net cost to us = $5 million.

16 In the situation of Prob. 15 suppose that our intelligence tells us that the enemy is mining with probability ½. What should we do?

17 In the situation of Prob. 15 suppose that our intelligence tells us that the enemy is mining with probability ½₀. What should we do?

18 Prove that for the nonstrictly determined game $\begin{array}{|cc|} a & b \\ c & d \end{array}$ the vector $(p, 1 - p, v) = k(a,c,-1) \wedge (b,d,-1)$ and find k. Recall that $\wedge$ represents the outer product (Sec. 7.8). Similarly, prove that the vector $(q, 1 - q, v) = k(a,b,-1) \wedge (c,d,-1)$ with the same k.

19 For some larger games the methods of this section apply. For example, consider the familiar game of "rock, scissors, and paper". The two players simultaneously signal one of these choices. Then rock beats scissors, scissors beats paper, and paper beats rock. The winner receives one cent from the loser. Nothing is paid in case of ties. The game matrix is

	R	S	P
R	0	1	-1
S	-1	0	1
P	1	-1	0

Let Player 1 play R with probability p_1, S with probability p_2, P with probability p_3, where $p_1 + p_2 + p_3 = 1$. Find an optimal mixed strategy: p_1, p_2, p_3. Similarly, find Player 2's optimal mixed strategy. What is the value of the game?

20 Solve the generalization of Prob. 19 whose matrix is

0	a	$-b$
$-a$	0	c
b	$-c$	0

where a, b, and c are positive. First verify that the game is nonstrictly determined.

References

Feller, William: "Probability Theory and Its Applications", Wiley, New York (1950).

Mosteller, Frederick, Robert E. K. Rourke, and George B. Thomas, Jr.: "Probability with Statistical Applications", Addison-Wesley, Reading, Mass. (1961).

Appendix

Table I **Values of e^x and e^{-x}**

x	e^x	e^{-x}	x	e^x	e^{-x}
0.00	1.0000	1.00000	2.10	8.1662	0.12246
0.01	1.0101	0.99005	2.20	9.0250	0.11080
0.02	1.0202	0.98020	2.30	9.9742	0.10026
0.03	1.0305	0.97045	2.40	11.023	0.09072
0.04	1.0408	0.96079	2.50	12.182	0.08208
0.05	1.0513	0.95123	2.60	13.464	0.07427
0.06	1.0618	0.94176	2.70	14.880	0.06721
0.07	1.0725	0.93239	2.80	16.445	0.06081
0.08	1.0833	0.92312	2.90	18.174	0.05502
0.09	1.0942	0.91393	3.00	20.086	0.04979
0.10	1.1052	0.90484	3.10	22.198	0.04505
0.20	1.2214	0.81873	3.20	24.533	0.04076
0.30	1.3499	0.74082	3.30	27.113	0.03688
0.40	1.4918	0.67032	3.40	29.964	0.03337
0.50	1.6487	0.60653	3.50	33.115	0.03020
0.60	1.8221	0.54881	3.60	36.598	0.02732
0.70	2.0138	0.49659	3.70	40.447	0.02472
0.80	2.2255	0.44933	3.80	44.701	0.02237
0.90	2.4596	0.40657	3.90	49.402	0.02024
1.00	2.7183	0.36788	4.00	54.598	0.01832
1.10	3.0042	0.33287	4.10	60.340	0.01657
1.20	3.3201	0.30119	4.20	66.686	0.01500
1.30	3.6693	0.27253	4.30	73.700	0.01357
1.40	4.0552	0.24660	4.40	81.451	0.01228
1.50	4.4817	0.22313	4.50	90.017	0.01111
1.60	4.9530	0.20190	4.60	99.484	0.01005
1.70	5.4739	0.18268	4.70	109.95	0.00910
1.80	6.0496	0.16530	4.80	121.51	0.00823
1.90	6.6859	0.14957	4.90	134.29	0.00745
2.00	7.3891	0.13534	5.00	148.41	0.00674

Table II **Common Logarithms (Base 10)**

N.	0	1	2	3	4	5	6	7	8	9
10	0000	0043	0086	0128	0170	0212	0253	0294	0334	0374
11	0414	0453	0492	0531	0569	0607	0645	0682	0719	0755
12	0792	0828	0864	0899	0934	0969	1004	1038	1072	1106
13	1139	1173	1206	1239	1271	1303	1335	1367	1399	1430
14	1461	1492	1523	1553	1584	1614	1644	1673	1703	1732
15	1761	1790	1818	1847	1875	1903	1931	1959	1987	2014
16	2041	2068	2095	2122	2148	2175	2201	2227	2253	2279
17	2304	2330	2355	2380	2405	2430	2455	2480	2504	2529
18	2553	2577	2601	2625	2648	2672	2695	2718	2742	2765
19	2788	2810	2833	2856	2878	2900	2923	2945	2967	2989
20	3010	3032	3054	3075	3096	3118	3139	3160	3181	3201
21	3222	3243	3263	3284	3304	3324	3345	3365	3385	3404
22	3424	3444	3464	3483	3502	3522	3541	3560	3579	3598
23	3617	3636	3655	3674	3692	3711	3729	3747	3766	3784
24	3802	3820	3838	3856	3874	3892	3909	3927	3945	3962
25	3979	3997	4014	4031	4048	4065	4082	4099	4116	4133
26	4150	4166	4183	4200	4216	4232	4249	4265	4281	4298
27	4314	4330	4346	4362	4378	4393	4409	4425	4440	4456
28	4472	4487	4502	4518	4533	4548	4564	4579	4594	4609
29	4624	4639	4654	4669	4683	4698	4713	4728	4742	4757
30	4771	4786	4800	4814	4829	4843	4857	4871	4886	4900
31	4914	4928	4942	4955	4969	4983	4997	5011	5024	5038
32	5051	5065	5079	5092	5105	5119	5132	5145	5159	5172
33	5185	5198	5211	5224	5237	5250	5263	5276	5289	5302
34	5315	5328	5340	5353	5366	5378	5391	5403	5416	5428
35	5441	5453	5465	5478	5490	5502	5514	5527	5539	5551
36	5563	5575	5587	5599	5611	5623	5635	5647	5658	5670
37	5682	5694	5705	5717	5729	5740	5752	5763	5775	5786
38	5798	5809	5821	5832	5843	5855	5866	5877	5888	5899
39	5911	5922	5933	5944	5955	5966	5977	5988	5999	6010
40	6021	6031	6042	6053	6064	6075	6085	6096	6107	6117
41	6128	6138	6149	6160	6170	6180	6191	6201	6212	6222
42	6232	6243	6253	6263	6274	6284	6294	6304	6314	6325
43	6335	6345	6355	6365	6375	6385	6395	6405	6415	6425
44	6435	6444	6454	6464	6474	6484	6493	6503	6513	6522
45	6532	6542	6551	6561	6571	6580	6590	6599	6609	6618
46	6628	6637	6646	6656	6665	6675	6684	6693	6702	6712
47	6721	6730	6739	6749	6758	6767	6776	6785	6794	6803
48	6812	6821	6830	6839	6848	6857	6866	6875	6884	6893
49	6902	6911	6920	6928	6937	6946	6955	6964	6972	6981
50	6990	6998	7007	7016	7024	7033	7042	7050	7059	7067
51	7076	7084	7093	7101	7110	7118	7126	7135	7143	7152
52	7160	7168	7177	7185	7193	7202	7210	7218	7226	7235
53	7243	7251	7259	7267	7275	7284	7292	7300	7308	7316
54	7324	7332	7340	7348	7356	7364	7372	7380	7388	7396
N.	0	1	2	3	4	5	6	7	8	9

Table II **Common Logarithms (Continued)**

N.	0	1	2	3	4	5	6	7	8	9
55	7404	7412	7419	7427	7435	7443	7451	7459	7466	7474
56	7482	7490	7497	7505	7513	7520	7528	7536	7543	7551
57	7559	7566	7574	7582	7589	7597	7604	7612	7619	7627
58	7634	7642	7649	7657	7664	7672	7679	7686	7694	7701
59	7709	7716	7723	7731	7738	7745	7752	7760	7767	7774
60	7782	7789	7796	7803	7810	7818	7825	7832	7839	7846
61	7853	7860	7868	7875	7882	7889	7896	7903	7910	7917
62	7924	7931	7938	7945	7952	7959	7966	7973	7980	7987
63	7993	8000	8007	8014	8021	8028	8035	8041	8048	8055
64	8062	8069	8075	8082	8089	8096	8102	8109	8116	8122
65	8129	8136	8142	8149	8156	8162	8169	8176	8182	8189
66	8195	8202	8209	8215	8222	8228	8235	8241	8248	8254
67	8261	8267	8274	8280	8287	8293	8299	8306	8312	8319
68	8325	8331	8338	8344	8351	8357	8363	8370	8376	8382
69	8388	8395	8401	8407	8414	8420	8426	8432	8439	8445
70	8451	8457	8463	8470	8476	8482	8488	8494	8500	8506
71	8513	8519	8525	8531	8537	8543	8549	8555	8561	8567
72	8573	8579	8585	8591	8597	8603	8609	8615	8621	8627
73	8633	8639	8645	8651	8657	8663	8669	8675	8681	8686
74	8692	8698	8704	8710	8716	8722	8727	8733	8739	8745
75	8751	8756	8762	8768	8774	8779	8785	8791	8797	8802
76	8808	8814	8820	8825	8831	8837	8842	8848	8854	8859
77	8865	8871	8876	8882	8887	8893	8899	8904	8910	8915
78	8921	8927	8932	8938	8943	8949	8954	8960	8965	8971
79	8976	8982	8987	8993	8998	9004	9009	9015	9020	9025
80	9031	9036	9042	9047	9053	9058	9063	9069	9074	9079
81	9085	9090	9096	9101	9106	9112	9117	9122	9128	9133
82	9138	9143	9149	9154	9159	9165	9170	9175	9180	9186
83	9191	9196	9201	9206	9212	9217	9222	9227	9232	9238
84	9243	9248	9253	9258	9263	9269	9274	9279	9284	9289
85	9294	9299	9304	9309	9315	9320	9325	9330	9335	9340
86	9345	9350	9355	9360	9365	9370	9375	9380	9385	9390
87	9395	9400	9405	9410	9415	9420	9425	9430	9435	9440
88	9445	9450	9455	9460	9465	9469	9474	9479	9484	9489
89	9494	9499	9504	9509	9513	9518	9523	9528	9533	9538
90	9542	9547	9552	9557	9562	9566	9571	9576	9581	9586
91	9590	9595	9600	9605	9609	9614	9619	9624	9628	9633
92	9638	9643	9647	9652	9657	9661	9666	9671	9675	9680
93	9685	9689	9694	9699	9703	9708	9713	9717	9722	9727
94	9731	9736	9741	9745	9750	9754	9759	9763	9768	9773
95	9777	9782	9786	9791	9795	9800	9805	9809	9814	9818
96	9823	9827	9832	9836	9841	9845	9850	9854	9859	9863
97	9868	9872	9877	9881	9886	9890	9894	9899	9903	9908
98	9912	9917	9921	9926	9930	9934	9939	9943	9948	9952
99	9956	9961	9965	9969	9974	9978	9983	9987	9991	9996
N.	0	1	2	3	4	5	6	7	8	9

Table III   Natural Logarithms (Base *e*)

	.00	.01	.02	.03	.04	.05	.06	.07	.08	.09
1.0	0.0000	0.0100	0.0198	0.0296	0.0392	0.0488	0.0583	0.0677	0.0770	0.0862
1.1	0.0953	0.1044	0.1133	0.1222	0.1310	0.1398	0.1484	0.1570	0.1655	0.1740
1.2	0.1823	0.1906	0.1989	0.2070	0.2151	0.2231	0.2311	0.2390	0.2469	0.2546
1.3	0.2624	0.2700	0.2776	0.2852	0.2927	0.3001	0.3075	0.3148	0.3221	0.3293
1.4	0.3365	0.3436	0.3507	0.3577	0.3646	0.3716	0.3784	0.3853	0.3920	0.3988
1.5	0.4055	0.4121	0.4187	0.4253	0.4318	0.4383	0.4447	0.4511	0.4574	0.4637
1.6	0.4700	0.4762	0.4824	0.4886	0.4947	0.5008	0.5068	0.5128	0.5188	0.5247
1.7	0.5306	0.5365	0.5423	0.5481	0.5539	0.5596	0.5653	0.5710	0.5766	0.5822
1.8	0.5878	0.5933	0.5988	0.6043	0.6098	0.6152	0.6206	0.6259	0.6313	0.6366
1.9	0.6419	0.6471	0.6523	0.6575	0.6627	0.6678	0.6729	0.6780	0.6831	0.6881
2.0	0.6932	0.6981	0.7031	0.7080	0.7129	0.7178	0.7227	0.7275	0.7324	0.7372
2.1	0.7419	0.7467	0.7514	0.7561	0.7608	0.7655	0.7701	0.7747	0.7793	0.7839
2.2	0.7885	0.7930	0.7975	0.8020	0.8065	0.8109	0.8154	0.8198	0.8242	0.8286
2.3	0.8329	0.8373	0.8416	0.8459	0.8502	0.8544	0.8587	0.8629	0.8671	0.8713
2.4	0.8755	0.8796	0.8838	0.8879	0.8920	0.8961	0.9002	0.9042	0.9083	0.9123
2.5	0.9163	0.9203	0.9243	0.9282	0.9322	0.9361	0.9400	0.9439	0.9478	0.9517
2.6	0.9555	0.9594	0.9632	0.9670	0.9708	0.9746	0.9783	0.9821	0.9858	0.9895
2.7	0.9933	0.9969	1.0006	1.0043	1.0080	1.0116	1.0152	1.0188	1.0225	1.0260
2.8	1.0296	1.0332	1.0367	1.0403	1.0438	1.0473	1.0508	1.0543	1.0578	1.0613
2.9	1.0647	1.0682	1.0716	1.0750	1.0784	1.0818	1.0852	1.0886	1.0919	1.0953
3.0	1.0986	1.1019	1.1053	1.1086	1.1119	1.1151	1.1184	1.1217	1.1249	1.1282
3.1	1.1314	1.1346	1.1378	1.1410	1.1442	1.1474	1.1506	1.1537	1.1569	1.1600
3.2	1.1632	1.1663	1.1694	1.1725	1.1756	1.1787	1.1817	1.1848	1.1878	1.1909
3.3	1.1939	1.1969	1.2000	1.2030	1.2060	1.2090	1.2119	1.2149	1.2179	1.2208
3.4	1.2238	1.2267	1.2296	1.2326	1.2355	1.2384	1.2413	1.2442	1.2470	1.2499
3.5	1.2528	1.2556	1.2585	1.2613	1.2641	1.2669	1.2698	1.2726	1.2754	1.2782
3.6	1.2809	1.2837	1.2865	1.2892	1.2920	1.2947	1.2975	1.3002	1.3029	1.3056
3.7	1.3083	1.3110	1.3137	1.3164	1.3191	1.3218	1.3244	1.3271	1.3297	1.3324
3.8	1.3350	1.3376	1.3403	1.3429	1.3455	1.3481	1.3507	1.3533	1.3558	1.3584
3.9	1.3610	1.3635	1.3661	1.3686	1.3712	1.3737	1.3762	1.3788	1.3813	1.3838
4.0	1.3863	1.3888	1.3913	1.3938	1.3962	1.3987	1.4012	1.4036	1.4061	1.4085
4.1	1.4110	1.4134	1.4159	1.4183	1.4207	1.4231	1.4255	1.4279	1.4303	1.4327
4.2	1.4351	1.4375	1.4398	1.4422	1.4446	1.4469	1.4493	1.4516	1.4540	1.4563
4.3	1.4586	1.4609	1.4633	1.4656	1.4679	1.4702	1.4725	1.4748	1.4771	1.4793
4.4	1.4816	1.4839	1.4861	1.4884	1.4907	1.4929	1.4951	1.4974	1.4996	1.5019
4.5	1.5041	1.5063	1.5085	1.5107	1.5129	1.5151	1.5173	1.5195	1.5217	1.5239
4.6	1.5261	1.5282	1.5304	1.5326	1.5347	1.5369	1.5390	1.5412	1.5433	1.5454
4.7	1.5476	1.5497	1.5518	1.5539	1.5560	1.5581	1.5602	1.5623	1.5644	1.5665
4.8	1.5686	1.5707	1.5728	1.5748	1.5769	1.5790	1.5810	1.5831	1.5851	1.5872
4.9	1.5892	1.5913	1.5933	1.5953	1.5974	1.5994	1.6014	1.6034	1.6054	1.6074
5.0	1.6094	1.6114	1.6134	1.6154	1.6174	1.6194	1.6214	1.6233	1.6253	1.6273
5.1	1.6292	1.6312	1.6332	1.6351	1.6371	1.6390	1.6409	1.6429	1.6448	1.6467
5.2	1.6487	1.6506	1.6525	1.6544	1.6563	1.6582	1.6601	1.6620	1.6639	1.6658
5.3	1.6677	1.6696	1.6715	1.6734	1.6752	1.6771	1.6790	1.6808	1.6827	1.6845
5.4	1.6864	1.6882	1.6901	1.6919	1.6938	1.6956	1.6974	1.6993	1.7011	1.7029

Table III   **Natural Logarithms (Continued)**

	.00	.01	.02	.03	.04	.05	.06	.07	.08	.09
5.5	1.7047	1.7066	1.7084	1.7102	1.7120	1.7138	1.7156	1.7174	1.7192	1.7210
5.6	1.7228	1.7246	1.7263	1.7281	1.7299	1.7317	1.7334	1.7352	1.7370	1.7387
5.7	1.7405	1.7422	1.7440	1.7457	1.7475	1.7492	1.7509	1.7527	1.7544	1.7561
5.8	1.7579	1.7596	1.7613	1.7630	1.7647	1.7664	1.7681	1.7699	1.7716	1.7733
5.9	1.7750	1.7766	1.7783	1.7800	1.7817	1.7834	1.7851	1.7868	1.7884	1.7901
6.0	1.7918	1.7934	1.7951	1.7967	1.7984	1.8001	1.8017	1.8034	1.8050	1.8066
6.1	1.8083	1.8099	1.8116	1.8132	1.8148	1.8165	1.8181	1.8197	1.8213	1.8229
6.2	1.8245	1.8262	1.8278	1.8294	1.8310	1.8326	1.8342	1.8358	1.8374	1.8390
6.3	1.8405	1.8421	1.8437	1.8453	1.8469	1.8485	1.8500	1.8516	1.8532	1.8547
6.4	1.8563	1.8579	1.8594	1.8610	1.8625	1.8641	1.8656	1.8672	1.8687	1.8703
6.5	1.8718	1.8733	1.8749	1.8764	1.8779	1.8795	1.8810	1.8825	1.8840	1.8856
6.6	1.8871	1.8886	1.8901	1.8916	1.8931	1.8946	1.8961	1.8976	1.8991	1.9006
6.7	1.9021	1.9036	1.9051	1.9066	1.9081	1.9095	1.9110	1.9125	1.9140	1.9155
6.8	1.9169	1.9184	1.9199	1.9213	1.9228	1.9242	1.9257	1.9272	1.9286	1.9301
6.9	1.9315	1.9330	1.9344	1.9359	1.9373	1.9387	1.9402	1.9416	1.9430	1.9445
7.0	1.9459	1.9473	1.9488	1.9502	1.9516	1.9530	1.9544	1.9559	1.9573	1.9587
7.1	1.9601	1.9615	1.9629	1.9643	1.9657	1.9671	1.9685	1.9699	1.9713	1.9727
7.2	1.9741	1.9755	1.9769	1.9782	1.9796	1.9810	1.9824	1.9838	1.9851	1.9865
7.3	1.9879	1.9892	1.9906	1.9920	1.9933	1.9947	1.9961	1.9974	1.9988	2.0001
7.4	2.0015	2.0028	2.0042	2.0055	2.0069	2.0082	2.0096	2.0109	2.0122	2.0136
7.5	2.0149	2.0162	2.0176	2.0189	2.0202	2.0215	2.0229	2.0242	2.0255	2.0268
7.6	2.0281	2.0295	2.0308	2.0321	2.0334	2.0347	2.0360	2.0373	2.0386	2.0399
7.7	2.0412	2.0425	2.0438	2.0451	2.0464	2.0477	2.0490	2.0503	2.0516	2.0528
7.8	2.0541	2.0554	2.0567	2.0580	2.0592	2.0605	2.0618	2.0631	2.0643	2.0656
7.9	2.0669	2.0681	2.0694	2.0707	2.0719	2.0732	2.0744	2.0757	2.0769	2.0782
8.0	2.0794	2.0807	2.0819	2.0832	2.0844	2.0857	2.0869	2.0882	2.0894	2.0906
8.1	2.0919	2.0931	2.0943	2.0956	2.0968	2.0980	2.0992	2.1005	2.1017	2.1029
8.2	2.1041	2.1054	2.1066	2.1078	2.1090	2.1102	2.1114	2.1126	2.1138	2.1150
8.3	2.1163	2.1175	2.1187	2.1199	2.1211	2.1223	2.1235	2.1247	2.1259	2.1270
8.4	2.1282	2.1294	2.1306	2.1318	2.1330	2.1342	2.1353	2.1365	2.1377	2.1389
8.5	2.1401	2.1412	2.1424	2.1436	2.1448	2.1459	2.1471	2.1483	2.1494	2.1506
8.6	2.1518	2.1529	2.1541	2.1552	2.1564	2.1576	2.1587	2.1599	2.1610	2.1622
8.7	2.1633	2.1645	2.1656	2.1668	2.1679	2.1691	2.1702	2.1713	2.1725	2.1736
8.8	2.1748	2.1759	2.1770	2.1782	2.1793	2.1804	2.1815	2.1827	2.1838	2.1849
8.9	2.1861	2.1872	2.1883	2.1894	2.1905	2.1917	2.1928	2.1939	2.1950	2.1961
9.0	2.1972	2.1983	2.1994	2.2006	2.2017	2.2028	2.2039	2.2050	2.2061	2.2072
9.1	2.2083	2.2094	2.2105	2.2116	2.2127	2.2138	2.2148	2.2159	2.2170	2.2181
9.2	2.2192	2.2203	2.2214	2.2225	2.2235	2.2246	2.2257	2.2268	2.2279	2.2289
9.3	2.2300	2.2311	2.2322	2.2332	2.2343	2.2354	2.2364	2.2375	2.2386	2.2396
9.4	2.2407	2.2418	2.2428	2.2439	2.2450	2.2460	2.2471	2.2481	2.2492	2.2502
9.5	2.2513	2.2523	2.2534	2.2544	2.2555	2.2565	2.2576	2.2586	2.2597	2.2607
9.6	2.2618	2.2628	2.2638	2.2649	2.2659	2.2670	2.2680	2.2690	2.2701	2.2711
9.7	2.2721	2.2732	2.2742	2.2752	2.2762	2.2773	2.2783	2.2793	2.2803	2.2814
9.8	2.2824	2.2834	2.2844	2.2854	2.2865	2.2875	2.2885	2.2895	2.2905	2.2915
9.9	2.2925	2.2935	2.2946	2.2956	2.2966	2.2976	2.2986	2.2996	2.3006	2.3016

Table III Natural Logarithms (Continued)

N	Nat Log	N	Nat Log	N	Nat Log	N	Nat Log	N	Nat Log
0	$-\infty$	40	3.68 888	80	4.38 203	120	4.78 749	160	5.07 517
1	0.00 000	41	3.71 357	81	4.39 445	121	4.79 579	161	5.08 140
2	0.69 315	42	3.73 767	82	4.40 672	122	4.80 402	162	5.08 760
3	1.09 861	43	3.76 120	83	4.41 884	123	4.81 218	163	5.09 375
4	1.38 629	44	3.78 419	84	4.43 082	124	4.82 028	164	5.09 987
5	1.60 944	45	3.80 666	85	4.44 265	125	4.82 831	165	5.10 595
6	1.79 176	46	3.82 864	86	4.45 435	126	4.83 628	166	5.11 199
7	1.94 591	47	3.85 015	87	4.46 591	127	4.84 419	167	5.11 799
8	2.07 944	48	3.87 120	88	4.47 734	128	4.85 203	168	5.12 396
9	2.19 722	49	3.89 182	89	4.48 864	129	4.85 981	169	5.12 990
10	2.30 259	50	3.91 202	90	4.49 981	130	4.86 753	170	5.13 580
11	2.39 790	51	3.93 183	91	4.51 086	131	4.87 520	171	5.14 166
12	2.48 491	52	3.95 124	92	4.52 179	132	4.88 280	172	5.14 749
13	2.56 495	53	3.97 029	93	4.53 260	133	4.89 035	173	5.15 329
14	2.63 906	54	3.98 898	94	4.54 329	134	4.89 784	174	5.15 906
15	2.70 805	55	4.00 733	95	4.55 388	135	4.90 527	175	5.16 479
16	2.77 259	56	4.02 535	96	4.56 435	136	4.91 265	176	5.17 048
17	2.83 321	57	4.04 305	97	4.57 471	137	4.91 998	177	5.17 615
18	2.89 037	58	4.06 044	98	4.58 497	138	4.92 725	178	5.18 178
19	2.94 444	59	4.07 754	99	4.59 512	139	4.93 447	179	5.18 739
20	2.99 573	60	4.09 434	100	4.60 517	140	4.94 164	180	5.19 296
21	3.04 452	61	4.11 087	101	4.61 512	141	4.94 876	181	5.19 850
22	3.09 104	62	4.12 713	102	4.62 497	142	4.95 583	182	5.20 401
23	3.13 549	63	4.14 313	103	4.63 473	143	4.96 284	183	5.20 949
24	3.17 805	64	4.15 888	104	4.64 439	144	4.96 981	184	5.21 494
25	3.21 888	65	4.17 439	105	4.65 396	145	4.97 673	185	5.22 036
26	3.25 810	66	4.18 965	106	4.66 344	146	4.98 361	186	5.22 575
27	3.29 584	67	4.20 469	107	4.67 283	147	4.99 043	187	5.23 111
28	3.33 220	68	4.21 951	108	4.68 213	148	4.99 721	188	5.23 644
29	3.36 730	69	4.23 411	109	4.69 135	149	5.00 395	189	5.24 175
30	3.40 120	70	4.24 850	110	4.70 048	150	5.01 064	190	5.24 702
31	3.43 399	71	4.26 268	111	4.70 953	151	5.01 728	191	5.25 227
32	3.46 574	72	4.27 667	112	4.71 850	152	5.02 388	192	5.25 750
33	3.49 651	73	4.29 046	113	4.72 739	153	5.03 044	193	5.26 269
34	3.52 636	74	4.30 407	114	4.73 620	154	5.03 695	194	5.26 786
35	3.55 535	75	4.31 749	115	4.74 493	155	5.04 343	195	5.27 300
36	3.58 352	76	4.33 073	116	4.75 359	156	5.04 986	196	5.27 811
37	3.61 092	77	4.34 381	117	4.76 217	157	5.05 625	197	5.28 320
38	3.63 759	78	4.35 671	118	4.77 068	158	5.06 260	198	5.28 827
39	3.66 356	79	4.36 945	119	4.77 912	159	5.06 890	199	5.29 330
40	3.68 888	80	4.38 203	120	4.78 749	160	5.07 517	200	5.29 832

Table IV Squares, Cubes, Roots

n	n^2	$\sqrt{n}$	n^3	$\sqrt[3]{n}$	n	n^2	$\sqrt{n}$	n^3	$\sqrt[3]{n}$
1	1	1.000	1	1.000	51	2,601	7.141	132,651	3.708
2	4	1.414	8	1.260	52	2,704	7.211	140,608	3.733
3	9	1.732	27	1.442	53	2,809	7.280	148,877	3.756
4	16	2.000	64	1.587	54	2,916	7.348	157,464	3.780
5	25	2.236	125	1.710	55	3,025	7.416	166,375	3.803
6	36	2.449	216	1.817	56	3,136	7.483	175,616	3.826
7	49	2.646	343	1.913	57	3,249	7.550	185,193	3.849
8	64	2.828	512	2.000	58	3,364	7.616	195,112	3.871
9	81	3.000	729	2.080	59	3,481	7.681	205,379	3.893
10	100	3.162	1,000	2.154	60	3,600	7.746	216,000	3.915
11	121	3.317	1,331	2.224	61	3,721	7.810	226,981	3.936
12	144	3.464	1,728	2.289	62	3,844	7.874	238,328	3.958
13	169	3.606	2,197	2.351	63	3,969	7.937	250,047	3.979
14	196	3.742	2,744	2.410	64	4,096	8.000	262,144	4.000
15	225	3.873	3,375	2.466	65	4,225	8.062	274,625	4.021
16	256	4.000	4,096	2.520	66	4,356	8.124	287,496	4.041
17	289	4.123	4,913	2.571	67	4,489	8.185	300,763	4.062
18	324	4.243	5,832	2.621	68	4,624	8.246	314,432	4.082
19	361	4.359	6,859	2.668	69	4,761	8.307	328,509	4.102
20	400	4.472	8,000	2.714	70	4,900	8.367	343,000	4.121
21	441	4.583	9,261	2.759	71	5,041	8.426	357,911	4.141
22	484	4.690	10,648	2.802	72	5,184	8.485	373,248	4.160
23	529	4.796	12,167	2.844	73	5,329	8.544	389,017	4.179
24	576	4.899	13,824	2.884	74	5,476	8.602	405,224	4.198
25	625	5.000	15,625	2.924	75	5,625	8.660	421,875	4.217
26	676	5.099	17,576	2.962	76	5,776	8.718	438,976	4.236
27	729	5.196	19,683	3.000	77	5,929	8.775	456,533	4.254
28	784	5.292	21,952	3.037	78	6,084	8.832	474,552	4.273
29	841	5.385	24,389	3.072	79	6,241	8.888	493,039	4.291
30	900	5.477	27,000	3.107	80	6,400	8.944	512,000	4.309
31	961	5.568	29,791	3.141	81	6,561	9.000	531,441	4.327
32	1,024	5.657	32,768	3.175	82	6,724	9.055	551,368	4.344
33	1,089	5.745	35,937	3.208	83	6,889	9.110	571,787	4.362
34	1,156	5.831	39,304	3.240	84	7,056	9.165	592,704	4.380
35	1,225	5.916	42,875	3.271	85	7,225	9.220	614,125	4.397
36	1,296	6.000	46,656	3.302	86	7,396	9.274	636,056	4.414
37	1,369	6.083	50,653	3.332	87	7,569	9.327	658,503	4.431
38	1,444	6.164	54,872	3.362	88	7,744	9.381	681,472	4.448
39	1,521	6.245	59,319	3.391	89	7,921	9.434	704,969	4.465
40	1,600	6.325	64,000	3.420	90	8,100	9.487	729,000	4.481
41	1,681	6.403	68,921	3.448	91	8,281	9.539	753,571	4.498
42	1,764	6.481	74,088	3.476	92	8,464	9.592	778,688	4.514
43	1,849	6.557	79,507	3.503	93	8,649	9.644	804,357	4.531
44	1,936	6.633	85,184	3.530	94	8,836	9.695	830,584	4.547
45	2,025	6.708	91,125	3.557	95	9,025	9.747	857,375	4.563
46	2,116	6.782	97,336	3.583	96	9,216	9.798	884,736	4.579
47	2,209	6.856	103,823	3.609	97	9,409	9.849	912,673	4.595
48	2,304	6.928	110,592	3.634	98	9,604	9.899	941,192	4.610
49	2,401	7.000	117,649	3.659	99	9,801	9.950	970,299	4.626
50	2,500	7.071	125,000	3.684	100	10,000	10.000	1,000,000	4.642
n	n^2	$\sqrt{n}$	n^3	$\sqrt[3]{n}$	n	n^2	$\sqrt{n}$	n^3	$\sqrt[3]{n}$

Answers to
Odd-numbered Problems

Problems 1.3

1   (a) and (e); (b) and (c)   3 $\emptyset$, $\{4\}$, $\{7\}$, $\{4, 7\}$; all except $\{4, 7\}$

5   $\emptyset$, $\{2\}$, $\{4\}$, $\{6\}$, $\{2, 4\}$, $\{2, 6\}$, $\{4, 6\}$, $\{2, 4, 6\}$; all except $\{2, 4, 6\}$

7   $\emptyset$, $\{a\}$, $\{b\}$, $\{c\}$, $\{d\}$, $\{a, b\}$, $\{a, c\}$, $\{a, d\}$, $\{b, c\}$, $\{b, d\}$, $\{c, d\}$, $\{a, b, c\}$, $\{a, c, d\}$, $\{a, b, d\}$, $\{b, c, d\}$, $\{a, b, c, d\}$; all except $\{a, b, c, d\}$

9   A subset with n elements has 2^n subsets.

11   $A = \{(2,2), (4,4), (2,4), (4,2)\}$ 13 $C = \{(1,1), (2,2), (3,3), (4,4)\}$

15   $E = \emptyset$ 17 $G = \{(1,1), (1,3)\}$

19   $I = \{(1,1), (1,2), (1,3), (1,4), (2,1), (2,2), (2,3), (2,4), (3,1), (3,2), (3,3), (3,4), (4,1), (4,2), (4,3), (4,4)\}$

21   $I = J$; All of $A \ldots H$ are proper subsets of I and of J; E is a proper subset of A, B, C, D, F, G, H, I, J; $H \subset A$; $G \subset B$; $A \subset F$; $B \subset F$; $C \subset F$; $H \subset F$.

23   $2n \leftrightarrow 3n$

25   Not possible, for numbers are not equal.

27   Not possible

29   Assign each official to the office to which he is elected. Assume no vacancies and no disputes over outcomes of elections.

Problems 1.5

1   $\{2\}$ 3 $\{10\}$ 5 $\{5, 6, 9\}$

7   $\{(2,2), (4,4)\}$ 9 $\{(1,4), (4,1), (2,3), (3,2)\}$

11   $\{(2,2), (4,4), (2,4), (4,2)\}$ 13 $\{(1,2), (1,4), (2,1), (2,3), (3,2), (3,4), (4,1), (4,3)\}$

15   $\{(1,1), (2,2), (4,4)\}$ 17   Both sides are $\{(2,2), (4,4)\}$

19   $A \cap B = A$, $A \cup B = B$ 21 $A = B$ 23 $B \subset A$

25   $A = B$ 27 $A \subset B$ 29 $A \subset B$

31   (a); (c); (d)   33 $2x \neq 4$, $\{1, 3, 4, 5\}$

35   $(x - 3)(x - 4) \neq 0$, $\{1, 2, 5\}$

37   $x^2 \neq 9$ or $x + 2 \neq 4$, $\{1, 2, 4, 5\} \cup \{1, 3, 4, 5\} = \{1, 2, 3, 4, 5\}$

Problems 1.8

1   $P = \{-3, 3\}$, $Q = \{3\}$. False since $P \not\subseteq Q$.

3   $P = \{3\}$, $Q = \{3\}$. True since $P \subseteq Q$.

5   $P = \emptyset$ since no integer satisfies this equation; $Q = \{5\}$. True since $P \subseteq Q$.

7   Converse: For all integers x: If $2x$ is divisible by 6, then x is divisible by 3.
Inverse: For all integers x: If x is not divisible by 3, then $2x$ is not divisible by 6.
Contrapositive: For all integers x: If $2x$ is not divisible by 6, then x is not divisible by 3.

9   Converse: For all triangles T: If T is isosceles, then T is equilateral.
Inverse: For all triangles T: If T is not equilateral, then T is not isosceles.
Contrapositive: For all triangles T: If T is not isosceles, then T is not equilateral.

11   Converse: For all pairs of lines L and M: If L and M do not intersect, then L and M are parallel.

Inverse: For all pairs of lines L and M: If L and M are not parallel, then L and M intersect.

Contrapositive: For all pairs of lines L and M: If L and M intersect, then L and M are not parallel.

13   The converse   **15**   The contrapositive

17   (a) For all integers x: If $x \neq 3$, then x is not positive or $x^2 \neq 9$.
(b) For all integers x: If $x \neq 3$ and $x^2 = 9$, then x is negative.
(c) For all integers x: If x is positive and $x \neq 3$, then $x^2 \neq 9$

19   For all triples of lines L, M, N:

(a) If M and N are not identical, then   (b) If M and N are not identical, and
    L and M intersect, or L and N do not intersect, and
    L and N intersect, or L and M are coplanar, and
    L and M are not coplanar, or L and N are coplanar, and
    L and M are not coplanar, or M and N do intersect, then
    M and N do not intersect. L and M do intersect.

(c) The same with M and N interchanged.

(d) If L and M do not intersect, and   (e) The same with M and N interchanged.
    L and N do not intersect, and
    M and N are not identical, and   (f) If L and M do not intersect, and
    L and N are coplanar, and L and N do not intersect, and
    M and N do not intersect, then L and M are coplanar, and
    L and M are not coplanar. L and N are coplanar, and
                                      M and N are not identical, then
                                      M and N do not intersect.

21   (Fig. 1.8.21)

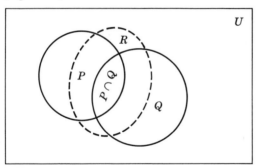

Figure 1.8.21

23   (Fig. 1.8.23)   (a) To show: If $(P \cap Q) \subset R$, then $(P \cap R') \subset Q'$. $P \cap R'$ is shaded and lies outside Q. Hence it is contained in Q'.   (b) To show: If $(P \cap R') \subset Q'$, then $(P \cap Q) \subset R$. $P \cap Q$ is shaded and lies outside R'. Hence it is contained in R.

25   A sufficient condition that a triangle be isosceles is that its base angles be equal.

27   A sufficient condition that two lines be parallel is that they be perpendicular to the same line.

29   A sufficient condition that $x = 1$ is that $3x + 2 = x + 4$.

31   A necessary condition that a triangle be inscribed in a semicircle is that it be a right triangle.

33   A necessary condition that a body be in static equilibrium is that the vector sum of all forces acting on it is zero.

35   A necessary condition that two forces be in equilibrium is that they be equal, opposite, and collinear.

37   A triangle is inscribed in a semicircle only if it is a right triangle.

39   A body is in static equilibrium only if the vector sum of all forces acting on it is zero.

41   Two forces are in equilibrium only if they are equal, opposite, and collinear.

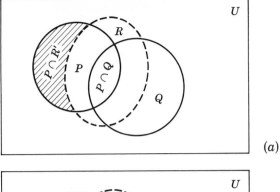

(a)

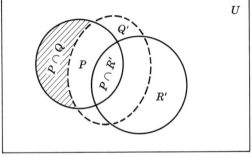

(b)

Figure 1.8.23

43 A necessary condition that a triangle be isosceles is that its base angles be equal. A sufficient condition that the base angles of a triangle be equal is that the triangle be isosceles.

45 A necessary condition that two lines be parallel is that they be perpendicular to the same line. A sufficient condition that two lines be perpendicular to the same line is that they be parallel.

47 A necessary condition that $x = 1$ is that $3x + 2 = x + 4$. A sufficient condition that $3x + 2 = x + 4$ is that $x = 1$.

49 A triangle is a right triangle only if it is inscribed in a semicircle.

51 The vector sum of all forces acting on a body is zero only if it is in static equilibrium.

53 Two forces are equal, opposite, and collinear only if they are in equilibrium.

55 A necessary and sufficient condition that two lines be parallel is that they be equidistant.

57 A necessary and sufficient condition that three concurrent forces be in equilibrium is that their vector sum be zero.

59 No. Promise is converse of that required for her to win.

Problems 1.12

1 In triangle ABC, assume $a \neq b$ *and* $\angle A = \angle B$. This contradicts the theorem that if the base angles of a triangle are equal, it is isosceles.

3 In quadrilateral $ABCD$ assume that the diagonals are unequal *and* that it is a rectangle. This contradicts a known property of a rectangle.

5 Assume there are two distinct coplanar lines perpendicular to L at P. Then the sum of the angles on one side of L at P is greater than 180—impossible.

7 Assume that two distinct circles have three distinct points in common. Then they are the same circle, for three noncollinear points determine a circle.

9 Let the tangent intersect the circle at $Q \neq P$. Then triangle OPQ is isosceles, one base angle is a right angle, and so it has two right angles—impossible.

11 (5) is the inverse of (2), not the contrapositive.

13 (3) follows from (1) and the converse of (2), but not from (1) and (2).
15 Counterexample: $2 + 4 = 6$, which is not odd
17 Counterexample: $x = 1$. Then $5^2 \neq 17$.
19 Ambiguous case. Two distinct triangles with given data are possible. (Fig. 1.12.19)

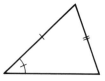

Figure 1.12.19

21 $(180° - \angle 1) + (180° - \angle 2) + (180° - \angle 3) = 180°$, or $\angle 1 + \angle 2 + \angle 3 = 360°$

Problems 2.5

5 $a + b + c + d + e = (a + b + c + d) + e$
7 $-4, 2, -\pi, 0, \sqrt{5}$ **9** $7, 3, 0, \frac{2}{3}, \frac{1}{2}$
11 -2 **13** -1
15 -33 **17** $a \times b \times c = (a \times b) \times c$
23 $3, -\frac{1}{4}, -\frac{4}{3}, 1$, does not exist.   **25** -29
27 -117 **29** 329
31 No   **33** No
35 No; $a - 0 = a$, but $0 - a \neq a$.   **37** Meaningless, 0, 1, 0, indeterminate
39 $-1, 0, 2$, none, 1, and 4   **41** $0, -2$, none, 0, 3
43 $x = -2, -3$
45 $x^2 - 5 = 0, (x + \sqrt{5})(x - \sqrt{5}) = 0, x = \sqrt{5}, -\sqrt{5}$
47 $(a + b) \times c = c \times (a + b) = (c \times a) + (c \times b) = (a \times c) + (b \times c)$

49 False. Put $a = b = c = 1$. Then $\dfrac{1}{2} \neq \dfrac{1}{1} + \dfrac{1}{1}$.

51 First, $x = -b/a$ is a solution. Second, if x_1 and x_2 are two solutions, $ax_1 + b = ax_2 + b$ or $ax_1 = ax_2$, or $x_1 = x_2$.
53 It is closed, commutative, associative. 0 is the identity. $a(\neq -1)$ has addiplicative inverse $-a/(1 + a)$.
55 (1) Given; (2) Theorem 1; (3) R7; (4) R9; (5) R8
57 (1) Given; (2) Theorem 1; (3) R7; (4) R9; (5) R8

Problems 2.7

31 Fails for $n = 1$, but II can be proved.
33 True for $n = 2$, but II is false. This statement is true for $n = 2, 4, 5, 7, 8, 10, 11, \ldots$. Can you state the general situation and prove it?
37 Counterexample: Any integer $\neq \pm 1, 0$. For then $1/a$ is not an integer.
39 No

Problems 2.10

3 In a/b, take a positive and $b = 1$.
5 Existence: $x = b - a$ satisfies the equation. If x_1 and x_2 satisfy the equation, then $a + x_1 = a + x_2$, or $x_1 = x_2$.
7 If $\dfrac{a/b}{c/d} = x$, then $(a/b) = x(c/d)$. Multiply both sides by bd. Then $ad = x(cb)$, or $x = (ad)/(bc)$.
9 Indirect proof. Suppose a is nearer to 0 than any other rational number. Then $a/2$ is a rational which is nearer to 0 than a.

11 $0.\overline{571428}$ **13** $0.\overline{4}$ **15** $3.\overline{1}$
17 $\frac{7}{9}$ **19** $1,214/99$ **21** $37,584/9,990 = 696/185$

23 There are only a finite number of remainders less than the divisor. Eventually one will repeat.

27 Write $4b^2 = a^2$. But now we cannot conclude that a is divisible by 4. Counterexample: $b = 3$, $a = 6$.

29 Write $2 - \sqrt{3} = \dfrac{a}{b}$, or $\sqrt{3} = 2 - \dfrac{a}{b} = \dfrac{2b - a}{a}$. This is impossible since $\sqrt{3}$ is irrational.

Problems 2.13

1 $5 > 2, 7 > -1, 4 > -4, -2 > -6, 0 > -3$

3 $a < c < b$ **5** 14, 12, 6, 2, 34   **7** (Fig. 2.13.7)

9 $++, --$ **11** IV, I, III, II, II   **13** 3, 2, 11, 11, 3

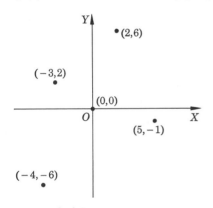

Figure 2.13.7

15 $|x_2 - x_1| = |x_1 - x_2|$

17 $|x_2 - x_1|^2 = (x_2 - x_1)^2 = (x_1 - x_2)^2$

19 $|x_2' - x_1'| = |(a + x_2) - (a + x_1)| = |x_2 - x_1|$

21 $K = C + 273°$

23 $|x_2' - x_1'| = |ax_2 - ax_1| = |a(x_2 - x_1)| = |a| \cdot |x_2 - x_1|$

25 $I = 12F$

27 $|x_2' - x_1'| = |(ax_2 + b) - (ax_1 + b)| = |a(x_2 - x_1)| = |a||x_2 - x_1|$

29 $F = \tfrac{9}{5} C + 32; \; -40°$

Problems 2.14

1 $13 + i$ **3** $-5 - 16i$ **5** $-3 - 12i$

7 $2 + 7i$ **9** $5 - 7i$ **11** $2 + 7i$

13 $5 + 2i$ **15** $-29 - 29i$ **17** $41 + i$

19 4   **21** 89   **23** $21 - 15i$

25 $-45 + 9i$ **27** -21 **29** i

31 $(49 - 10i)/61$ **33** $(11 + 52i)/25$ **35** $2 + i$

37 $(15 + 10i)/39$ **39** $(8 - 3i)/6$ **41** $x = -\tfrac{23}{26}, y = \tfrac{15}{26}$

43 $x = \tfrac{29}{17}, y = \tfrac{31}{17}$ **45** $x = 0, y = -\tfrac{3}{2}$ **55** $(5, -1)$

57 $(-10, 11)$ **59** $(\tfrac{10}{13}, \tfrac{11}{13})$

Problems 3.2

1 $-3a^2 + 2ab + 3c$ **3** $-5pq + 3q^3$

5 $12x^2y + 4xy^2 + 2xy + x^2 - 2y^2 + 5$ **7** $3x^3 - 5y^3 - 6y$

9 $x^2 - 2y^2 + 3x + 5a^2 + 3b^2 + 2y$ **11** $9x^2y - 3xy^2 + 10xy + 3x^3 + 2$

13 $5x^2 - 8y^2 - 2r^2 + 4s^2$ **15** $6x^4 - 5a^2 + 8xy$

17 $-5xy - x^2 - 3y^4 + 3y^2$ **19** $-a^2 - 13b^2 - 12ab$

Problems 3.3

1 $-6a^7b^6$ **3** $50x^5y^5wz^4$ **5** $8a^2 - 2ab - 28b^2$

7 $16x^2 + 16xy + 4y^2$ **9** $27p^3 + 27p^2q + 9pq^2 + q^3$ **11** $9x^2 - 16y^2$

13 $4x^6 + 11x^5 - 7x^4 - 13x^3 - 3x^2 + 2x + 6$

15 $-40x^6 - 13x^5 + 15x^4 + 2x^3 + 24x^2 + 3x$

17 $15a^7 - 38a^4b^2 + 5a^3b^3 + 7ab^4 - b^5 + 12a^4 - 28ab^2 + 4b^3$

19 $24x^6 - 7x^5y + 5x^4y^2 + 17x^3y^3 - 7x^2y^4 - 3xy^5 + y^6$

Problems 3.4

1 10, 21, 15, 1, 4 **3** $10 + 5 = 15$ **5** $1 + 4 + 6 + 4 + 1 = 16$

7 $32x^5 + 80x^4y + 80x^3y^2 + 40x^2y^3 + 10xy^4 + y^5$

9 $r^6 - 12r^5s + 60r^4s^2 - 160r^3s^3 + 240r^2s^4 - 192rs^5 + 64s^6$

11 $x^6 + 3x^5y + \tfrac{15}{4}x^4y^2 + \tfrac{5}{2}x^3y^3 + \tfrac{15}{16}x^2y^4 + \tfrac{3}{16}xy^5 + \tfrac{1}{64}y^6$

13 $8/x^3 + 12 + 6x^3 + x^6$ **15** 1.0510100501

17 $\binom{13}{5} = 1287$ **19** $\binom{7}{3} \cdot 2^3 \cdot (-3)^4 = 22{,}680$

21 6

23 $x^3 + y^3 + z^3 + 3(x^2y + x^2z + xy^2 + xz^2 + y^2z + yz^2) + 6xyz$

Problems 3.5

	Quotient	Remainder
1	$x^2 + 2x + 2$	0
3	$3x^2 - 2x + 6$	-4
5	$2x^2 + x + 5$	$2x - 5$
7	$4x^3 - x^2 + x + 5$	$-2x + 12$
9	$x^2 + 2$	$2x^2 + 5$
11	$-4x^3 + 2x^2 - x + 7$	$5x - 7$
13	$x^2 + 1$	0
15	$x^4 - x^2 + 1$	-2
17	$x^5 + x^4y + x^3y^2 + x^2y^3 + xy^4 + y^5$	0
19	$\tfrac{1}{3}x^2 - 2x + \tfrac{1}{2}$	$-\tfrac{1}{2}$

Problems 3.6

1 $x(y + 3) + 3(y + 3) = (x + 3)(y + 3)$

3 $x^2(x + 1) - (x + 1) = (x^2 - 1)(x + 1) = (x - 1)(x + 1)(x + 1)$

5 $s(x - y) + v(x - y) = (s + v)(x - y)$

7 $(x - 4)(x + 2)$ **9** $(x - 18)(x - 2)$ **11** $(x - 4)(x - 8)$

13 $(y - 3x)(y - 15x)$ **15** No such factors **17** $(3x + 1)(x + 4)$

19 $(3x + 1)(x - 4)$ **21** $(x + 3)(4x - 7)$ **23** $(5x + 6y)(x - 3y)$

25 $(2x + 3)(5x - 4)$ **27** $(4x + 9)(3x - 5)$ **29** $(3x + 25)(4x - 11)$

31 No such factors **33** $(3x + 4)(3x - 4)$ **35** $(x + \sqrt{7})(x - \sqrt{7})$

37 $(3x + 2i)(3x - 2i)$

39 $[(x + 3) + (x - 2)][(x + 3) - (x - 2)] = (2x + 1)(5)$

41 $(x^2 - y^2)(x^2 + y^2) = (x - y)(x + y)(x + iy)(x - iy)$

43 $[(x + 2) + (y + 3)][(x + 2) - (y + 3)] = (x + y + 5)(x - y - 1)$

45 $(3x - y)(9x^2 + 3xy + y^2)$ **47** $(x + 5)(x^2 - 5x + 25)$

49 $(a + 3b + 1)(a - 3b + 1)$ **51** $(x + 3)(x - 3)(x + 1)(x - 1)$

53 $x(2x + 3y)(x + 2y)$

55 $(2x + y)(x + 2y) - 3(2x + y) + 4(x + 2y) - 12 = (x + 2y - 3)(2x + y + 4)$

Problems 4.2

1 $\dfrac{x - 3}{x + 3}$ **3** $\dfrac{x + 4}{x + 5}$ **5** Does not simplify.

7 $\dfrac{2x + 1}{x + 4}$ **9** $\dfrac{3x - 2}{2x - 1}$ **11** $\dfrac{c + d}{4 - y}$

13 $\dfrac{x^2 + xy + y^2}{x + y}$

15 Does not simplify.

17 $\dfrac{x^2(a + 3)}{a - 4}$

19 Does not simplify.

21 Any x except 0, ⅜

23 Any x except $-1, 0, 1$

25 Any x except $x = 0$

Problems 4.3

1 $\dfrac{10x + 4}{x^2 - 1}$

3 $\dfrac{1}{2u - 3v}$

5 $\dfrac{2a - 5b}{a(a - b)}$

7 $\dfrac{5x + 3y}{xy(x + y)}$

9 $\dfrac{2x^2 - 13x}{(x + 4)(x + 1)(x - 4)}$

11 $\dfrac{-x^4 + 7x^2 + 8x}{(x^2 + 2x + 1)(x^2 + 4x + 4)}$

13 $\dfrac{3x^3 + 10x^2 + 14x + 15}{(x^2 + 3x + 3)(x^2 + x + 5)}$

15 $\dfrac{2x^3 - 3x^2 + 2x + 3}{x(x + 1)(x - 1)}$

17 $\dfrac{8x^2 + 41x + 33}{(x + 1)(x + 3)(x + 4)(x + 5)}$

19 $\dfrac{9x - 21 - 28i}{x^2 - 6x + 25}$

Problems 4.4

1 $\dfrac{x - 2}{3}$

3 $\dfrac{x + 1}{x - 1}$

5 $\dfrac{(x + 1)^2(x^2 - 1)}{(x + 2)^4}$

7 $\dfrac{(x + 3)(x - 4)(x + 5)}{(x + 4)(x + 1)}$

9 12

11 $x/2$

13 $\dfrac{(2x + 5)(x - 3)}{(4x + 1)(x + 5)}$

15 1

17 $\dfrac{y^2(x + y)}{x^2}$

19 p/r

Problems 4.5

1 $-\dfrac{4x}{(x - 2)(x - 1)}$

3 $\dfrac{4x + 7}{(x - 3)(x - 2)}$

5 1

7 $\dfrac{3x(12x^2 - 7)}{12x^2 - 5}$

9 $\dfrac{x^3 + 13x + 26}{x(x - 2)(x + 3)}$

11 $\dfrac{6x^2 - 54x + 22}{(x - 1)(x + 2)(x - 3)}$

13 $\dfrac{x + 3}{3x + 1}$

15 $\dfrac{(4x + 1)(2x + 1)(26x^2 - 20x + 4)}{x^2(3x + 1)(5x - 2)}$

17 $\dfrac{(x^3 + 4x^2 + x - 2)(x - 2)(x + 4)}{(x^3 + 4x^2 - 2x + 4)(x + 2)(x + 3)}$

19 $\dfrac{-2x^2 + 3x + 1 + 4xi}{x^2 - 2x + 5}$

Problems 5.2

1 4^{10}

3 5

5 $6^3/3^3 = 2^3 = 8$

7 -3^{21}

9 $49 + (4^2)(7) + 4^2 = 177$. The best method of solution is to multiply numerator and denominator by 4×7.

11 $x^4y^2 + 2x^3y^4 + y^6$. Multiply numerator and denominator by x^3y^4.

13 $\dfrac{a^2b^3 - c^5}{a^3bc^7}$. Multiply numerator and denominator by ab^2c^3.

15 $2y^3 + 5 - \dfrac{3}{y^3}$

17 $\dfrac{4x + 3x^2 - 5x^3}{3 + 4x^2}$

19 $\dfrac{3x^3 + 5x^2 + 4}{7x^6 + x^4}$

Problems 5.6

1 $4x^3 - 2x$

3 $x + 2x^{1/2}y^{1/2} + y$

5 $p - q$

7 $\dfrac{x^{5/2} + 4x}{2}$

9 $\dfrac{x^{7/3} + 2x^2 - 4x^{1/3}}{3x^{7/3} + 1}$. Multiply numerator and denominator by $x^{5/3}$.

11 $\dfrac{5a^{7/3} + 4a^{4/3}b^2}{5b + 3a^{7/3}b^{5/3}}$

13 $|x + 3|$

15 $|x + 1| + |x - 1|$. Counterexample $x = 0$

17 $\dfrac{|2x - 3|}{2x - 3}$

19 $\dfrac{x}{|x|}\sqrt{x^2 + 1}$

21 $x + 5$

23 $-x - 5$

25 $2x + 3$

27 $-2x - 3$

29 1

31 $-\sqrt{35}$

33 $12i$

35 -15

37 -4

39 $-24i$

41 $-\sqrt{5}$

43 0

45 $(7 - 12i)\sqrt{2}$

51 No, for $a^b \neq b^a$.

53 No

Problems 5.7

1 $3\sqrt{7}/7$

3 $5(\sqrt{2} + \sqrt{3})$

5 $\dfrac{4\sqrt{x - 1}}{x - 1}$

7 $-\sqrt{x - 2} + \sqrt{x + 1}$

9 $\dfrac{12\sqrt{2} + 4\sqrt{3}}{15}$

11 $\tfrac{1}{9}$

13 $\dfrac{5\sqrt{3} + 3\sqrt{5}}{15}$

15 $\dfrac{2x + x^{3/2} + 9}{9x - x^2}$

17 $\dfrac{4x^2}{1 - x}$

19 1.155

21 1.984

Problems 6.2

1 $A = B$

3 $A \supset B$

5 $A = B$

7 $A \subset B$

9 $A \supset B$

Problems 6.4

1 $-2 \pm i$

3 $-\tfrac{1}{3}$

5 $-\tfrac{5}{4}, \tfrac{7}{9}$

7 $\tfrac{9}{2}, -\tfrac{5}{3}$

9 $3 + i, 2 + i$

11 $(x + \tfrac{5}{2})^2 = \tfrac{21}{4}$

13 $(x - 4)^2 = 14$

15 $(2x - \tfrac{5}{2})^2 = \tfrac{13}{4}$

17 $3, 10$

19 $\tfrac{7}{4}, 3$

21 $-\dfrac{4 + 3i}{2 - i}, \dfrac{-1 + i}{2 - i}$

23 $k = \tfrac{9}{4}$

25 $k = \tfrac{49}{8}$

27 $\pm\tfrac{4}{3}$

29 $x^2 - 3x + 4 = 0$

31 $3x^2 - x + 2 = 0$

33 $x^2 - (3 - i)x + (2 + 5i) = 0$

35 $2\left(x + \dfrac{3 + i\sqrt{31}}{4}\right)\left(x + \dfrac{3 - i\sqrt{31}}{4}\right)$

37 $(x - 1 + i)(x - 1 - i)$

39 $(6x - 5)(9x + 7)$

41 9 and 5

43 $9x^2 + 5x + 3 = 0$

45 $x^2 - 5x - 66 = 0$

Problems 6.5

1 (Fig. 6.5.1)

3 (Fig. 6.5.3)

5 (Fig. 6.5.5)

7 (Fig. 6.5.7)

9 (Fig. 6.5.9)

11 (Fig. 6.5.11)

13 (Fig. 6.5.13)

15 (Fig. 6.5.15)

17 y increases as x approaches $\pm \infty$.

19 $y = 0$ when $x = r_1$ or r_2.

21 Since there are no real roots, y is never zero.

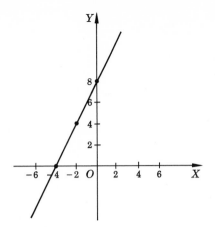

Figure 6.5.1

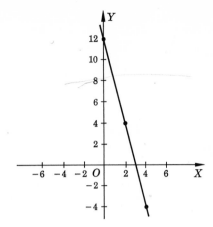

Figure 6.5.3

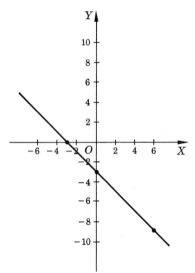

Figure 6.5.5

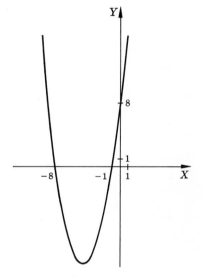

Figure 6.5.7

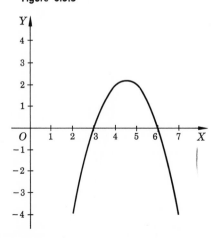

Figure 6.5.9

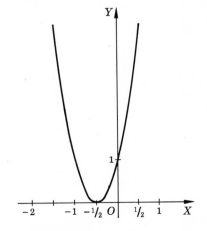

Figure 6.5.11

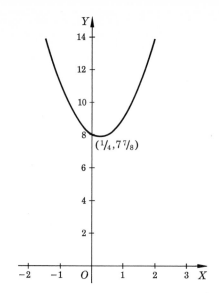

Figure 6.5.13

Figure 6.5.15

23 The highest or lowest point occurs when $\left(x + \dfrac{b}{2a}\right) = 0$. Hence $x = -\dfrac{b}{2a}$;
$$y = \frac{-b + 4ac}{4c}.$$

Problems 6.7

1	$\{1, -1\%\}$	**3**	$\{3, -\%\}$	**5**	$\{2, \%\}$
7	$\{3, -3\}$	**9**	$\{3\}$	**11**	$\{2\}$
13	$\emptyset$	**15**	$\{-5\}$	**17**	$\{0\}$
19	$\{9\}$				

Problems 6.10

1	(2,2) (Fig. 6.10.1)	**3**	(0,3) (Fig. 6.10.3)	**5**	(3,1) (Fig. 6.10.5)
7	No solution (Fig. 6.10.7)	**9**	Infinitely many solutions (Fig. 6.10.9)		
11	(1,−1,2)	**13**	(−3,1,2)	**15**	No solution
17	Line of solutions	**19**	Plane of solutions	**21**	$3x - y - 2 = 0$
23	$x - y = 0$	**25**	$x - 3y + 14 = 0$	**27**	$2x - 3y + 14 = 0$
29	A family of parallel lines				

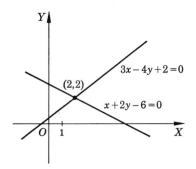

Figure 6.10.1

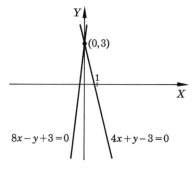

Figure 6.10.3

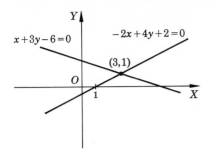

Figure 6.10.5

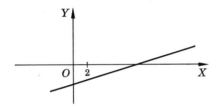

Figure 6.10.7

Figure 6.10.9

Problems 6.11

1 $(1,1)$, $(-3,-7)$ (Fig. 6.11.1)

3 $(-4,-9)$, $(3,12)$ (Fig. 6.11.3)

5 $(3,-8)$ (Fig. 6.11.5)

7 $\dfrac{-1 \pm i\sqrt{7}}{2}$, $\dfrac{-21 \pm i\sqrt{7}}{2}$ (Fig. 6.11.7)

9 $(5,-8)$, $(-\frac{3}{2},-\frac{45}{4})$ (Fig. 6.11.9)

11 $(1,2)$, $(-4,-3)$ (Fig. 6.11.11)

13 $(\pm\sqrt{3},1)$

15 $(14,-2)$, $(-2,14)$

17 $(1,1)$, $(-1,-1)$

19 $(1,2,-1)$

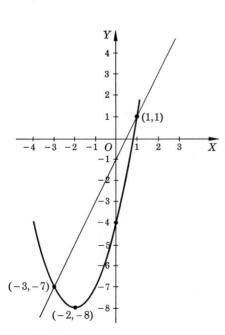

Figure 6.11.1

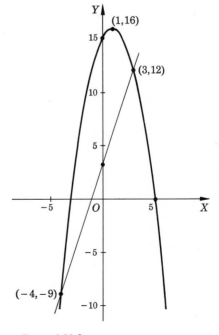

Figure 6.11.3

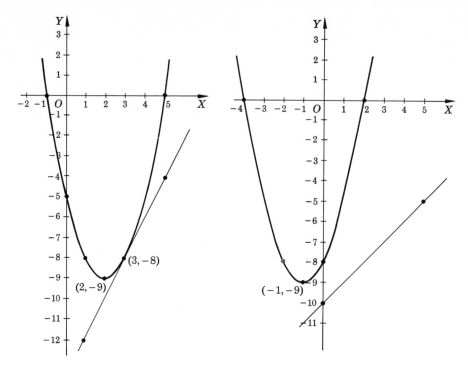

Figure 6.11.5

Figure 6.11.7

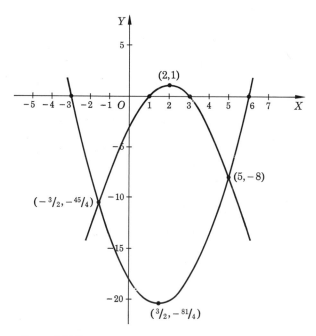

Figure 6.11.9

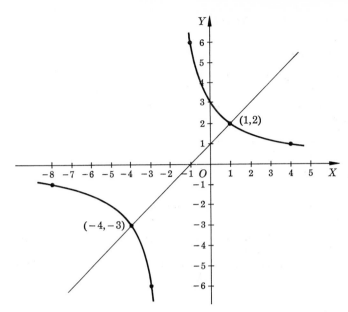

Figure 6.11.11

Problems 6.12

1	48	**3**	10, 40	**5** 576 ft²

7 14,625 **9** 7 and 9 bricks/min
11 Milkshake 30 cents, coffee 10 cents **13** Notar 30 cents, Green Grass 40 cents
15 One **17** 6 francs, 5 shillings, 10 marks
19 Neverstart 120 mi; Everknock 80 mi **21** 10g and 1g
23 Current 3 amp, resistance 5 ohms **25** 18 ft/sec/sec, 36 ft/sec/sec

Problems 6.13

1 $x' = 2x + 3$ **3** $x' = 2x - 80$; 80

5 From Eq. (2),

$$d'^2 = (x_2' - x_1')^2 + (y_2' - y_1')^2$$
$$= [(x_2 + a) - (x_1 + a)]^2 + [(y_2 + b) - (y_1 + b)]^2$$
$$= (x_2 - x_1)^2 + (y_2 - y_1)^2 = d^2$$

7 Given (x,y), there is a unique (x',y'). Given (x',y'), there is a unique (x,y), namely:
$x = x' - a$, $y = y' - b$.

9 From Eq. (2),

$$d'^2 = (ax_1 - ax_2)^2 + (ay_1 - ay_2)^2$$
$$= a^2[(x_1 - x_2)^2 + (y_1 - y_2)^2] = a^2 d^2$$

Hence $d' = |a|d$. If a rectangle has sides p and q, then $A = pq$. So

$$A' = (|a|p)(|a|q) = a^2 pq = a^2 A.$$

Two triangles whose corresponding sides have proportional lengths are similar.

11 Let $x' = x \neq 0$; $y' = y \neq 0$. Then $x = ax$, $y = ay$. So $a = 1$.
13 Equation (2) is unchanged.
15 Given (x,y), there is a unique (x',y'). Given (x',y') there is a unique (x,y), namely:
$x = -x'$, $y = y'$.
17 From Eq. (2),

$$d'^2 = [(ax_2 + by_2) - (ax_1 + by_1)]^2 + [(-bx_2 + ay_2) - (-bx_1 + ay_1)]^2$$
$$= (a^2 + b^2)[(x_2 - x_1)^2 + (y_2 - y_1)^2] = d^2$$

19 Let $x' = x \neq 0$, $y' = y \neq 0$. Then

$$\begin{cases} x = ax + by \\ y = -bx + ay \end{cases}$$

or $\begin{cases} x = ax + by \\ y = ay - bx \end{cases}$ or $\begin{cases} x^2 = ax^2 + bxy \\ y^2 = ay^2 - bxy \end{cases}$

So $x^2 + y^2 = a(x^2 + y^2)$ or $a = 1$, and then $b = 0$.

21 We must solve

$$\begin{cases} x = 2x + y \\ y = x + 2y \end{cases}$$ or $\begin{cases} 0 = x + y \\ 0 = x + y \end{cases}$

Any point on the line $x + y = 0$ is a solution.

23 We must solve

$$\begin{cases} x = a_1x + b_1y \\ y = a_2x + b_2y \end{cases}$$ or $\begin{cases} (a_1 - 1)x + b_1y = 0 \\ a_2x + (b_2 - 1)y = 0 \end{cases}$

From Theorem 9 there is the unique solution $(0,0)$ unless $(a_1 - 1)(b_2 - 1) - a_2b_1 = 0$ or $a_1b_2 - a_2b_1 - b_2 - a_1 + 1 = 0$

25 The equations can be solved uniquely for (x,y) in terms of (x',y') by virtue of Theorem 9.

Problems 7.3

1 $(-1,5)$ **3** $(-2,2)$ **5** $(1,3)$

7 $(-4,-1,3)$ **9** $(7,6,24)$ **11** $2i + 27j - 4k$

13 $(a,b) + (c,d) = (a + c, b + d)$
$(c,d) + (a,b) = (c + a, d + b)$
The right sides are equal from properties of real numbers, so the left sides are equal.

15 $(a,b,c) + (0,0,0) = (a,b,c); (0,0,0) + (a,b,c) = (a,b,c)$

17 12   **19** -8 **21** 0

23 3   **25** $\sqrt{5}$ **27** 1

29 $\sqrt{29}$

Problems 7.5

1 $\begin{pmatrix} 3 & 1 \\ 4 & 9 \end{pmatrix}$ **3** $\begin{pmatrix} 1 & -5 \\ 1 & -9 \end{pmatrix}$ **5** $\begin{pmatrix} 10 & 12 \\ 5 & 10 \end{pmatrix}$

11 19   **13** $\begin{pmatrix} 15 \\ 19 \end{pmatrix}$ **15** $\begin{pmatrix} -16 & 7 \\ 9 & 10 \end{pmatrix}$

17 $\begin{pmatrix} 22 & 18 \\ -22 & 20 \end{pmatrix}$ **19** $\begin{pmatrix} 13 & 22 \\ -3 & -18 \end{pmatrix}$ **21** $\begin{pmatrix} 13 & -3 \\ 8 & 10 \\ 9 & 25 \end{pmatrix}$

23 $\begin{pmatrix} 2 & -10 & 4 \\ 1 & -5 & 2 \\ 4 & -20 & 8 \end{pmatrix}$ **25** $\begin{pmatrix} 6 & 3 & -6 \\ 4 & -29 & 20 \\ 2 & -8 & 6 \end{pmatrix}$ **27** $\begin{pmatrix} 1 & 0 & 0 \\ 0 & 1 & 0 \\ 0 & 0 & 1 \end{pmatrix}$

29 $\begin{pmatrix} 0 & 0 \\ 0 & 0 \end{pmatrix}$. No. See Theorem 4, Sec. 2.4.

31 $\begin{pmatrix} 7 & 5 \\ 13 & 8 \end{pmatrix}$ $\begin{pmatrix} 12 & -3 & 7 \\ 2 & -1 & 1 \\ 7 & -2 & 4 \end{pmatrix}$ **33** AB is 2×2; BA is 3×3.

35 $\begin{pmatrix} 0 & 0 \\ 0 & 0 \end{pmatrix}$ **37** $\begin{pmatrix} 0 & 0 \\ 0 & 0 \end{pmatrix}$

39 $AX + D = 0$ where $A = \begin{pmatrix} 3 & 4 \\ 1 & -2 \end{pmatrix}$, $X = \begin{pmatrix} x \\ y \end{pmatrix}$, $D = \begin{pmatrix} -5 \\ 6 \end{pmatrix}$, $0 = \begin{pmatrix} 0 \\ 0 \end{pmatrix}$

41 $AX + D = 0$ where $A = \begin{pmatrix} 2 & -1 & 3 \\ 1 & 2 & 1 \\ 3 & 1 & -1 \end{pmatrix}$, $X = \begin{pmatrix} x \\ y \\ z \end{pmatrix}$

$D = \begin{pmatrix} -4 \\ 6 \\ -1 \end{pmatrix}$, $0 = \begin{pmatrix} 0 \\ 0 \\ 0 \end{pmatrix}$

43 $X' = AX$ where $X' = \begin{pmatrix} x' \\ y' \end{pmatrix}$, $A = \begin{pmatrix} a & b \\ c & d \end{pmatrix}$, $X = \begin{pmatrix} x \\ y \end{pmatrix}$

$X'' = BX'$ where $X'' = \begin{pmatrix} x'' \\ y'' \end{pmatrix}$, $B = \begin{pmatrix} p & q \\ r & s \end{pmatrix}$, $X' = \begin{pmatrix} x' \\ y' \end{pmatrix}$

Then $X'' = BX' = B(AX) = BAX$.

45 $x^2 + 2xy + y^2$

Problems 7.7

1 $\Delta = 10$, $A^{-1} = \dfrac{1}{10}\begin{pmatrix} 2 & -4 \\ 1 & 3 \end{pmatrix}$ **3** $\Delta = -13$, $A^{-1} = -\dfrac{1}{13}\begin{pmatrix} 1 & -2 \\ -5 & -3 \end{pmatrix}$

5 $\Delta = 0$; A^{-1} does not exist.   **7** $\Delta = 1$, $A^{-1} = \begin{pmatrix} 1 & 0 \\ 0 & 1 \end{pmatrix}$

9 $\Delta = 1$, $A^{-1} = \begin{pmatrix} \dfrac{1}{\sqrt{2}} & \dfrac{1}{\sqrt{2}} \\ -\dfrac{1}{\sqrt{2}} & \dfrac{1}{\sqrt{2}} \end{pmatrix}$

11 -18 **13** 49 **15** 0

17 -9 **19** $-3x - 5y + 14$

21 The second determinant becomes equal to

$\begin{vmatrix} a_1 & b_1 & c_1 \\ a_2 & b_2 & c_2 \\ a_3 & b_3 & c_3 \end{vmatrix} + k \begin{vmatrix} b_1 & b_1 & c_1 \\ b_2 & b_2 & c_2 \\ b_3 & b_3 & c_3 \end{vmatrix} = \begin{vmatrix} a_1 & b_1 & c_1 \\ a_2 & b_2 & c_2 \\ a_3 & b_3 & c_3 \end{vmatrix} + 0$

23 $\dfrac{1}{5}\begin{pmatrix} -2 & 2 & 3 \\ 3 & 2 & -2 \\ -3 & 3 & 2 \end{pmatrix}$ **25** $\begin{pmatrix} 2 & -3 & 11 \\ -2 & 4 & -13 \\ 1 & -2 & 7 \end{pmatrix}$

27 Does not exist.   **29** $\begin{pmatrix} 7 & -19 & 6 \\ -5 & 14 & -4 \\ 1 & -3 & 1 \end{pmatrix}$

Problems 7.8

1 $2, 1$ **3** $-2, 4$ **5** No solution

7 $1, 0, 2$ **9** $1, 1, 0$ **11** Line of solutions

13 $1, 1, 1$

15 $x = -\dfrac{\begin{vmatrix} d_1 & b_1 & c_1 \\ d_2 & b_2 & c_2 \\ d_3 & b_3 & c_3 \end{vmatrix}}{\begin{vmatrix} a_1 & b_1 & c_1 \\ a_2 & b_2 & c_2 \\ a_3 & b_3 & c_3 \end{vmatrix}}$ $y = -\dfrac{\begin{vmatrix} a_1 & d_1 & c_1 \\ a_2 & d_2 & c_2 \\ a_3 & d_3 & c_3 \end{vmatrix}}{\begin{vmatrix} a_1 & b_1 & c_1 \\ a_2 & b_2 & c_2 \\ a_3 & b_3 & c_3 \end{vmatrix}}$ $z = -\dfrac{\begin{vmatrix} a_1 & b_1 & d_1 \\ a_2 & b_2 & d_2 \\ a_3 & b_3 & d_3 \end{vmatrix}}{\begin{vmatrix} a_1 & b_1 & c_1 \\ a_2 & b_2 & c_2 \\ a_3 & b_3 & c_3 \end{vmatrix}}$

17 $x = -19k$, $y = 9k$, $z = 17k$ **19** $x = 16k$, $y = -3k$, $z = -7k$

21 $k(6, -7, 8)$ **23** $k(2, 16, -12)$

25 All three 2×2 determinants are zero.

27 $(13, -11, 5)$ **29** $(-28, -6, 13)$ **31** k

33 j **35** 0

Problems 7.9

1 $(25 \quad 56 \quad 200) \begin{pmatrix} 25 \\ 20 \\ 30 \end{pmatrix} = \$7,745$

$$
\begin{array}{ccc}
 & E & P & C
\end{array}
$$

3 $(5 \quad 20 \quad 85) \begin{pmatrix} 14.12 & 19.70 & 5.28 \\ 6.64 & 7.50 & 3.40 \\ 4.26 & 3.40 & 3.48 \end{pmatrix} \begin{array}{l} Herald \\ Tribune \\ Times \end{array}$

$$
\begin{array}{ccc}
\text{Editorial} & \text{Printing} & \text{Circulation} \\
= (\$565.50 & \$537.50 & \$390.20)
\end{array}
$$

5

$$
\begin{array}{c}
 \\
A \\
B \\
C
\end{array}
\begin{array}{ccc}
K & G & L/O \\
\begin{pmatrix} 40 & 20 & 7 \\ 30 & 35 & 2 \\ 27 & 25 & 15 \end{pmatrix}
\end{array}
\begin{pmatrix} 11 \\ 19 \\ 4 \end{pmatrix} \times 10^3 =
\begin{pmatrix} \$\ 848 \\ \$1,003 \\ \$\ 832 \end{pmatrix} \times 10^3
$$

7 (a)

$$
\begin{array}{cccc}
 & & P & E & S \\
(7 \quad 7 \quad 2000) & \begin{pmatrix} 25 & 15 & 10 \\ 20 & 20 & 15 \\ 0.10 & 0.08 & 0.06 \end{pmatrix}
\end{array}
$$

$$
\begin{array}{ccc}
\text{President} & \text{Executive} & \text{Salesman} \\
= (\$515 & \$405 & \$295)
\end{array}
$$

(b) $(515 \quad 405 \quad 295) \begin{pmatrix} 1 \\ 2 \\ 4 \end{pmatrix} = \$2,505$

9 (a)

$$
\begin{array}{ccc}
 & & \text{Revenue} & \text{Cost}
\end{array}
$$

$$
\begin{pmatrix} 40 & 20 & 7 \\ 30 & 35 & 2 \\ 27 & 25 & 15 \end{pmatrix}
\begin{pmatrix} 11 & 10.90 \\ 19 & 18.80 \\ 4 & 4.00 \end{pmatrix} \times 10^3 =
\begin{pmatrix} \$\ 848 & \$840 \\ \$1,003 & \$993 \\ \$\ 832 & \$824.3 \end{pmatrix} \times 10^3
$$

(b) $\begin{pmatrix} 848 & 840 \\ 1,003 & 993 \\ 832 & 824.3 \end{pmatrix} \begin{pmatrix} 1 \\ -1 \end{pmatrix} = \begin{pmatrix} \$\ 8 \\ \$10 \\ \$\ 7.70 \end{pmatrix} \times 10^3$

Problems 7.10

1 $P = \dfrac{1}{0.61} \begin{pmatrix} 0.8 & 0.3 \\ 0.1 & 0.8 \end{pmatrix} \begin{pmatrix} 5,600 \\ 5,400 \end{pmatrix} = \begin{pmatrix} \$10,000 \\ \$\ 8,000 \end{pmatrix}$

3 $C_E = \begin{pmatrix} 0.8 & -0.3 \\ -0.1 & 1 \end{pmatrix} \begin{pmatrix} 5,000 \\ 7,000 \end{pmatrix} = \begin{pmatrix} \$1,900 \\ \$6,300 \end{pmatrix}$

5 $P = \dfrac{1}{0.25} \begin{pmatrix} 0.62 & 0.38 & 0.36 \\ 0.24 & 0.51 & 0.22 \\ 0.34 & 0.41 & 0.52 \end{pmatrix} \begin{pmatrix} 104 \\ 4 \\ 25 \end{pmatrix} \times 10^9 = \begin{pmatrix} \$300 \times 10^9 \\ \$130 \times 10^9 \\ \$200 \times 10^9 \end{pmatrix}$

7 $P = \dfrac{1}{0.133} \begin{pmatrix} 0.26 & 0.14 & 0.09 \\ 0.08 & 0.35 & 0.13 \\ 0.07 & 0.14 & 0.28 \end{pmatrix} \begin{pmatrix} 2 \\ 48 \\ 23 \end{pmatrix} \times 10^9 = \begin{pmatrix} \$\ 70 \times 10^9 \\ \$150 \times 10^9 \\ \$100 \times 10^9 \end{pmatrix}$

9 $C_E = \begin{pmatrix} 1 & -0.9 & -0.8 \\ -0.1 & 0.6 & -0.6 \\ -0.1 & -0.2 & 0.7 \end{pmatrix} \begin{pmatrix} 19 \\ 31 \\ 21 \end{pmatrix} \times 10^9 = \begin{pmatrix} \$11.7 \times 10^9 \\ \$\ 8.3 \times 10^9 \\ \$\ 6.6 \times 10^9 \end{pmatrix}$

11 $P = \dfrac{1}{0.152} \begin{pmatrix} 0.34 & 0.22 & 0.08 \\ 0.15 & 0.41 & 0.08 \\ 0.07 & 0.09 & 0.24 \end{pmatrix} \begin{pmatrix} 6 \\ 34 \\ 4.5 \end{pmatrix} \times 10^9 = \begin{pmatrix} \$\ 65 \times 10^9 \\ \$100 \times 10^9 \\ \$\ 30 \times 10^9 \end{pmatrix}$

The consumption targets are impossible to attain.

Problems 8.3

1 $x > 3$ (Fig. 8.3.1)
5 $x > 2$ (Fig. 8.3.5)
9 $-1 < x < 5$ (Fig. 8.3.9)
13 $-3 < x < 3$ (Fig. 8.3.13)
17 $-\frac{5}{3} < x < 3$ (Fig. 8.3.17)

3 $x < -3$ (Fig. 8.3.3)
7 $x > 4$ or $x < -4$ (Fig. 8.3.7)
11 $x > 5$ or $x < -2$ (Fig. 8.3.11)
15 $-7 < x < 2$ (Fig. 8.3.15)
19 $1.9 < x < 2.1$ (Fig. 8.3.19)

Figure 8.3.1

Figure 8.3.3

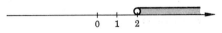

Figure 8.3.5

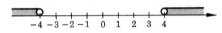

Figure 8.3.7

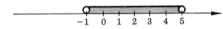

Figure 8.3.9

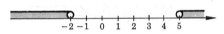

Figure 8.3.11

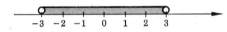

Figure 8.3.13

Figure 8.3.15

Figure 8.3.17

Figure 8.3.19

21 $|x - 4| < 3$
25 $|x - 2| > 3$
29 $x_1' - x_2' = a(x_1 - x_2)$, which is positive since each factor is positive.
45 $-1 < x - 3 < 1$, $3 < x + 1 < 5$, $7 < x + 5 < 9$. So $|x + 1| > 3$, $|x + 5| < 9$,
$$\left| \frac{x + 5}{x + 1} \right| < \frac{9}{3} = 3.$$

23 $|2x - 1| < 5$
27 $|4x - 1| > 6$

47 $-1 < x - 2 < 1$, $3 < x + 2 < 5$, $|x + 2| < 5$,
$|x^2 - 4| = |x - 2| \times |x + 2| < 1 \times 5 = 5$
49 $-0.01 < x - 2 < 0.01$, $3.99 < x + 2 < 4.01$, $|x + 2| < 4.01$.
$|x^2 - 4| < 0.01 \times 4.01 = 0.0401$

Problems 8.4

1 (Fig. 8.4.1)
5 (Fig. 8.4.5)
9 All x
13 (Fig. 8.4.13)
17 $\dfrac{x - 2}{(x + 2)(x - 1)} > 0$ (Fig. 8.4.17)

3 All x (Fig. 8.4.3)
7 (Fig. 8.4.7)
11 No x
15 (Fig. 8.4.15)
19 $\dfrac{(x + 1)(x + 13)}{(x + 4)(x - 5)} < 0$ (Fig. 8.4.19)

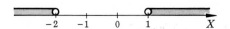

Figure 8.4.1

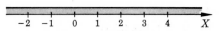

Figure 8.4.3

Figure 8.4.5

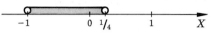

Figure 8.4.7

Figure 8.4.13

Figure 8.4.15

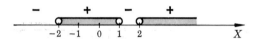

Figure 8.4.17

Figure 8.4.19

21 To prove II: For all $k \geq 5$: if $2^k > k^2$, then $2^{k+1} > (k+1)^2$.
By hypothesis, $2^k > k^2$. Hence $2^{k+1} > 2k^2$. We must prove that $2k^2 \geq (k+1)^2$, or $2k^2 \geq k^2 + 2k + 1$, or $k^2 - 2k - 1 \geq 0$ for $k \geq 5$. The roots of $k^2 - 2k - 1 = 0$ are $k = 1 \pm \sqrt{2}$; the larger of these is $1 + \sqrt{2} = 2.41$. Since $k^2 - 2k - 1 \geq 0$ for $k \geq 2.41$, the proof is complete. Alternatively divide $k^2 - 2k - 1 \geq 0$ by $k - 4$, which is positive for $k \geq 5$. The result is $k + 2 + \dfrac{7}{k-4} \geq 0$, which is true for $k \geq 5$.

23 I is true. II is false.　　　　**25** I is false.
27 I is false.　　　　　　　　　　**29** I is false.
31 We are given $(a+b)(a-b) > 0$ and $a > 0$, $b > 0$. Hence $a + b > 0$, and consequently $a - b > 0$.
33 $a > c$, $b > d$. Hence $a + b > c + d$.
35 Equality when $a \geq 0$ and $b \geq 0$.
37 $|(a-b)+b| \leq |a-b| + |b|$; $|a| \leq |a-b| + |b|$; $|a| - |b| \leq |a-b|$.
39 Equality when $a_1 = kb_1$; $a_2 = kb_2$.
41 From Cauchy's Inequality (Prob. 40)

$$\frac{(a_1b_1 + \cdots + a_nb_n)^2}{(a_1^2 + \cdots + a_n^2)(b_1^2 + \cdots + b_n^2)} \leq 1$$

Now rewrite in vector notation.

Problems 8.8

1 (Fig. 8.8.1)　　　　**3** (Fig. 8.8.3)　　　　**5** The whole plane
7 (Fig. 8.8.7)　　　　**9** (Fig. 8.8.9)　　　　**11** (Fig. 8.8.11)

13 (Fig. 8.8.13)   **15** The whole plane   **17** (Fig. 8.8.17)
19 (Fig. 8.8.19)   **21** (Fig. 8.8.21)   **23** (Fig. 8.8.23)
25 (Fig. 8.8.25)

27

r	0 $\cdots$ 0	1 $\cdots$ 1	2 $\cdots$ 2	3
p	10 $\cdots$ 18	10 $\cdots$ 15	9 $\cdots$ 12	9

when 180 sec left to play; impossible if 90 sec left to play

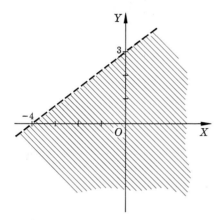

Figure 8.8.1

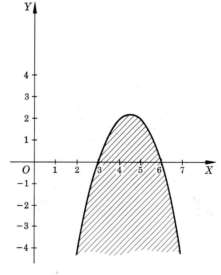

Figure 8.8.3

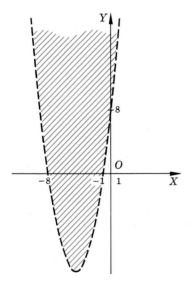

Figure 8.8.7

Figure 8.8.9

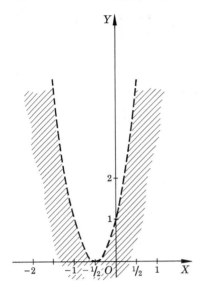

Figure 8.8.11

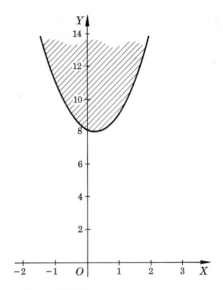

Figure 8.8.13

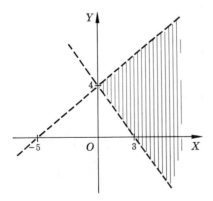

Figure 8.8.17

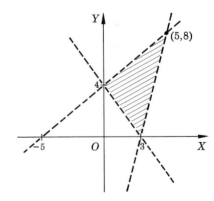

Figure 8.8.19

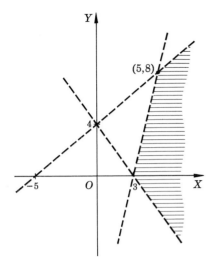

Figure 8.8.21

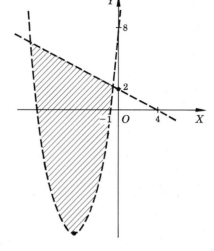

Figure 8.8.23

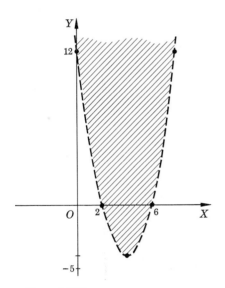

Figure 8.8.25

Problems 8.9

1 Cost $= 10a + 20b$. The minimum is attained at all points of segment QR (Fig. 8.20).
3 $I = r/5 + p/10$; $r = 0, p = 10$
5 H: $100x + 200y \geq 2,500$ (a) C $= 10x + 30y$; min $=$ \$250 at (25,0)
    P: $50x + 10y \geq 350$ (b) T $= kx + 2ky$; min $= 25k$ along segment PQ
    F: $30x \geq 150$ (c) W $= x + y$; min $= 15$ at (5,10)
           $x \geq 0; y \geq 0$ (Fig. 8.9.5)

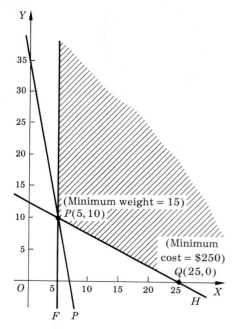

Figure 8.9.5

7 $12x + 25y \geq 435$ (a) Cost $= 5,000x + 12,000y$; min $=$ \$186,000 at (30,3)
$10x + 5y \geq 125$
$7x + 25y \geq 285$ (b) Time $= 40x + 50y$; min $= 950$ hr at (5,15)
$x \geq 0; y \geq 0$ (Fig. 8.9.7)

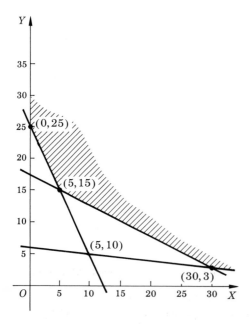

Figure 8.9.7

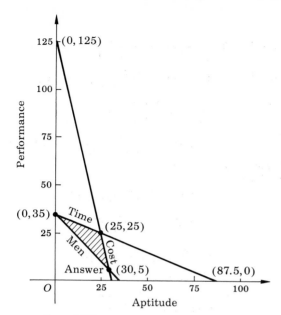

Figure 8.9.9

9 Cost: $20x + 5y \leq 625$ Errors $= 0.18x + 0.22y$
Tests: $x + y \geq 35$ Min $= 6.50$ at $(30,5)$
Time: $10x + 25y \leq 875$ (Fig. 8.9.9)
  $x \geq 0;\ y \geq 0$

11 No   **13** Max $= 14$; min $= -12$ **15** Max $= 20$; min $= 6$
17 $f(x,y)$ is constant; every point is both a max and a min.
19 $f(x,y)$ is constant on R.

Problems 9.1

1 Domain all reals; range all reals   **3** Domain all reals; range, $y \geq 1$
5 Domain, $-5 \leq x \leq 5$; range, $-5 \leq y \leq 5$ **7** Domain all reals; range all reals
9 Domain, $-1 \leq x \leq 1$; range, $-1 \leq y \leq 1$ **11** Domain all reals; range all reals
13 Domain, $x > 2$; range all reals   **15** Domain, $x \leq -1$; range, $y \geq 5$
17 Domain, $x = 1, 4$; range, $y = 4, 6$ **19** Domain, $x = 3, 7$; range, $y = 5$

Problems 9.2

1 Function   **3** Not a function   **5** X, Y
7 $X, \{y \mid y \geq 1\}$ **9** Domain $\{0, 3, 5\}$; range $\{1, 2\}$
11 Domain $\{1, 2\}$; range $\{5, 6\}$ **13** Domain $\{2, 3\}$; range $\{1\}$
15 $X, \{1, 2\}$ **17** $\{x \mid x$ is an integer$\}, \{0\}$
19 X, except $\{0, 1\}$; $\{y \mid y \leq -4\} \cup \{y \mid y > 0\}$
21 X, except $\{0, 2\}$; Y
23 X except $-2, 2$; Y except $0, \frac{1}{4}$
25 (a) $\{y \mid -1 \leq y \leq 0\}$; (b) $\{y \mid 0 \leq y\}$
27 $\frac{1}{8}, \frac{4}{8}, \frac{7}{8}, \frac{10}{8}, -\frac{2}{8}, -\frac{5}{8}$
29 Some answers are: The functions whose values are given by the rules:
  (a) $y = +\sqrt{25 - x^2}, -5 \leq x \leq 5, 0 \leq y \leq 5$
  (b) $y = -\sqrt{25 - x^2}, -5 \leq x \leq 5, -5 \leq y \leq 0$
  (c) $y = \begin{cases} +\sqrt{25 - x^2}, x = 0 \\ -\sqrt{25 - x^2}, 0 < |x| \leq 5 \end{cases}$
  Domain, $\{x \mid -5 \leq x \leq 5\}$; range, $\{5\} \cup \{y \mid -5 < y \leq 0\}$

Problems 9.3

1 $(f + g)(x) = \dfrac{2x + 1}{x(x + 1)}, x \neq 0, -1$

$(f - g)(x) = \dfrac{1}{x(x + 1)}, x \neq 0, -1$

$(fg)(x) = \dfrac{1}{x(x + 1)}, x \neq 0, -1$

$(f/g)(x) = \dfrac{x + 1}{x}, x \neq 0, -1$

3 $(f + g)(x) = \dfrac{x^2 + 3x + 3}{x + 1}, x \neq -1$

$(f - g)(x) = \dfrac{-(x^2 + 3x + 1)}{x + 1}, x \neq -1$

$(fg)(x) = \dfrac{x + 2}{x + 1}, x \neq -1$

$(f/g)(x) = \dfrac{1}{(x + 1)(x + 2)}, x \neq -1, -2$

5 $(f + g)(x) = x^2 + 5x + 6$, all reals
$(f - g)(x) = -(x^2 + 3x + 2)$, all reals
$(fg)(x) = (x + 2)^3$, all reals

$(f/g)(x) = \dfrac{1}{x + 2}, x \neq -2$

7 $x \neq \pm 1$:

$(f + g)(x) = \dfrac{x}{(x + 1)(x - 1)}$

$(f - g)(x) = \dfrac{x - 2}{(x + 1)(x - 1)}$

$(fg)(x) = \dfrac{1}{(x + 1)^2(x - 1)}$

$(f/g)(x) = x - 1$

9 $z = x^6 - 8x^3 + 5$, all reals

11 $g(f(x)) = x + 1$, all reals
$f(g(x)) = x + 1$, all reals

13 $g(f(x)) = \sqrt{|x|}$, all reals
$f(g(x)) = |\sqrt{x}|, x \geq 0$

15 $g(f(x)) = 1/x^6, x \neq 0$
$f(g(x)) = 1/x^6, x \neq 0$

17 (a) 2; (b) 2; (c) ¼

Problems 9.4

1 (Fig. 9.4.1)
7 (Fig. 9.4.7)
13 (Fig. 9.4.13)
19 (Fig. 9.4.19)
25 (Fig. 9.4.25)
31 (Fig. 9.4.31)
37 (Fig. 9.4.37)
43 Yes

3 (Fig. 9.4.3)
9 (Fig. 9.4.9)
15 (Fig. 9.4.15)
21 (Fig. 9.4.21)
27 (Fig. 9.4.27)
33 (Fig. 9.4.33)
39 (Fig. 9.4.39)

5 (Fig. 9.4.5)
11 (Fig. 9.4.11)
17 (Fig. 9.4.17)
23 (Fig. 9.4.23)
29 (Fig. 9.4.29)
35 (Fig. 9.4.35)
41 (Fig. 9.4.41)

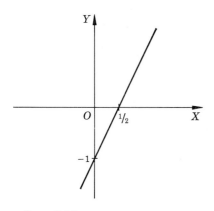

Figure 9.4.1

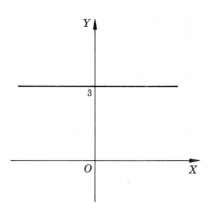

Figure 9.4.3

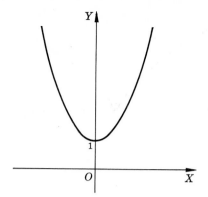

Figure 9.4.5

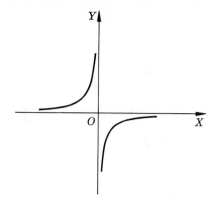

Figure 9.4.7

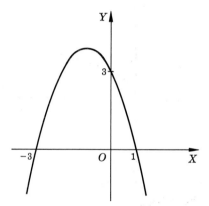

Figure 9.4.9

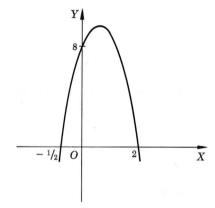

Figure 9.4.11

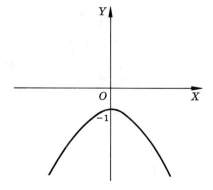

Figure 9.4.13

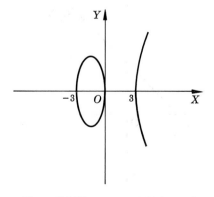

Figure 9.4.15

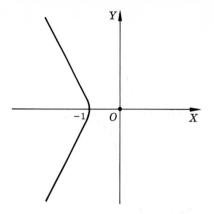

Figure 9.4.17

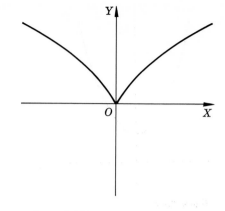

Figure 9.4.19

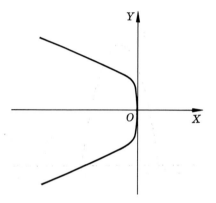

Figure 9.4.21

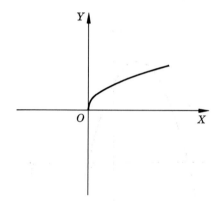

Figure 9.4.23

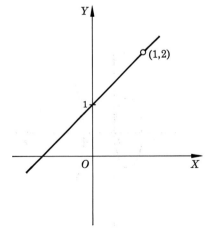

Figure 9.4.25

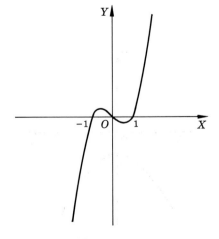

Figure 9.4.27

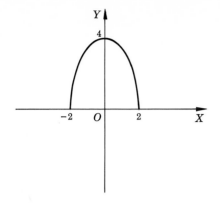

Figure 9.4.29

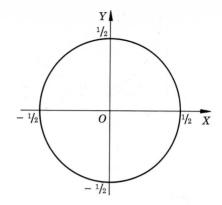

Figure 9.4.31

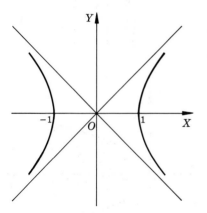

Figure 9.4.33

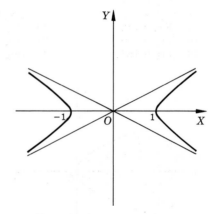

Figure 9.4.35

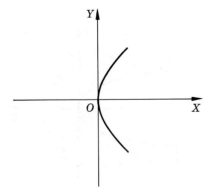

Figure 9.4.37

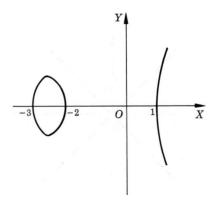

Figure 9.4.39

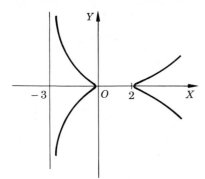

Figure 9.4.41

Problems 9.5

1	(Fig. 9.5.1)	**3**	(Fig. 9.5.3)	**5**	(Fig. 9.5.5)
7	(Fig. 9.5.7)	**9**	(Fig. 9.5.9)	**11**	(Fig. 9.5.11)
13	(Fig. 9.5.13)	**15**	(Fig. 9.5.15)	**17**	(Fig. 9.5.17)
19	(Fig. 9.5.19)	**21**	(Fig. 9.5.21)	**23**	(Fig. 9.5.23)
25	(Fig. 9.5.25)	**27**	(Fig. 9.5.27)		

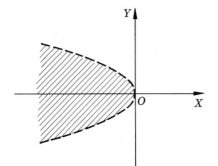

Figure 9.5.1

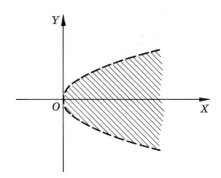

Figure 9.5.3

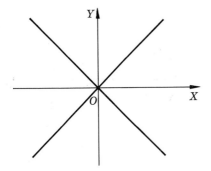

Figure 9.5.5

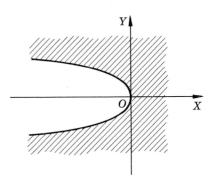

Figure 9.5.7

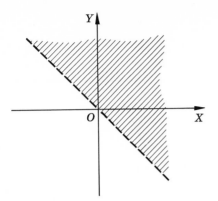

Figure 9.5.9

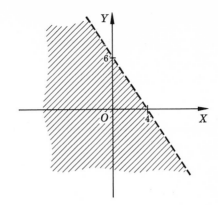

Figure 9.5.11

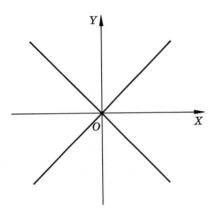

Figure 9.5.13

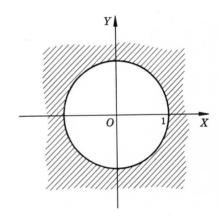

Figure 9.5.15

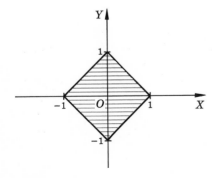

Figure 9.5.17

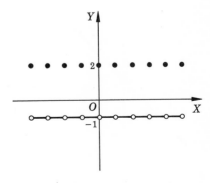

Figure 9.5.19

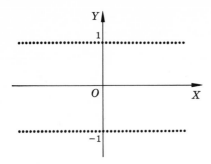

Figure 9.5.21

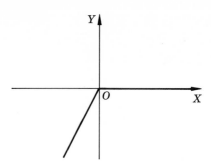

Figure 9.5.23

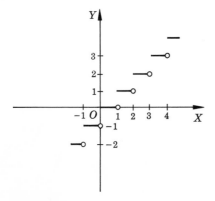

Figure 9.5.25

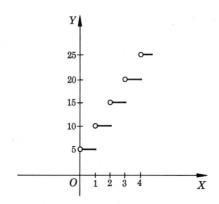

Figure 9.5.27

Problems 9.6

1 $f^{-1}(x) = x - 1$; domain all reals; range all reals (Fig. 9.6.1)

3 $f^{-1}(x) = \dfrac{x}{a} - \dfrac{b}{a}$; domain all reals; range all reals (Fig. 9.6.3)

5 $f^{-1}(x) = x$; domain all reals; range all reals (Fig. 9.6.5)

7 No inverse function (Fig. 9.6.7)

9 $f^{-1}(x) = x^{1/3}$; domain all reals; range all reals (Fig. 9.6.9)

11 $f^{-1}(x) = \sqrt{1 - x^2}$; domain, $0 \le x \le 1$; range, $0 \le y \le 1$ (Fig. 9.6.11)

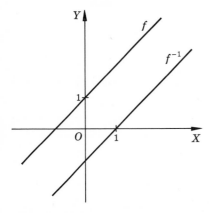

Figure 9.6.1

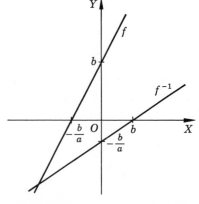

Figure 9.6.3

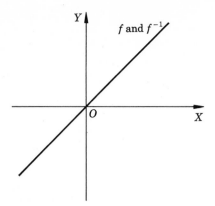

Figure 9.6.5

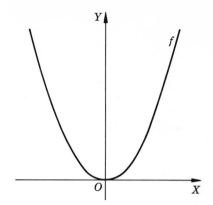

Figure 9.6.7

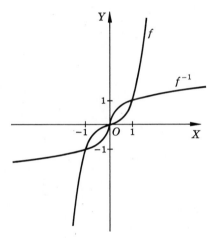

Figure 9.6.9

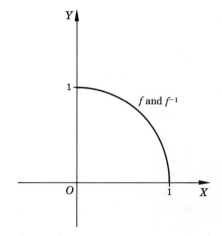

Figure 9.6.11

Problems 9.7

1 $f(x) = \dfrac{4x - 4}{3}$; d_f reals; r_f reals

3 $f(x) = \sqrt{2x^2 - 1}$; d_f, $|x| \geq \tfrac{1}{2}\sqrt{2}$; r_f nonnegative reals

5 $f(x) = \dfrac{1}{1 + x^2}$; d_f reals; r_f, $0 < y \leq 1$

7 $f(x) = -|x|$; d_f reals; r_f nonpositive reals

9 $f(t) = -16t^2 + v_0 t$; d_f, $0 < t < v_0/16$; r_f, $0 \leq s \leq v_0^2/64$

11 $f(x) = 1 + x^2$; d_f, $x \geq 0$; r_f, $y \geq 1$
[Or $f(x) = 5$; d_f, $x \geq 0$; r_f 5. Etc.]

13 $f(x) = 2 + x^7$; d_f reals; r_f, $y \geq 2$
[Or $f(x) = x + 1$, d_f reals; r_f reals. Etc.]

Problems 9.8

1 $q = 7$, $p = 3$ (Fig. 9.8.1)

3 $q = 12$, $p = 22$ (Fig. 9.8.3)

5 $q = 20$, $p = 10$ (Fig. 9.8.5)

7 $q = 2$, $p = 58$ (Fig. 9.8.7)

9 $q = 25$, $p = 20$ (Fig. 9.8.9)

11 $q = 0.4$, $p = 30$ (Fig. 9.8.11)

13 $q = 6$, $p = 70$ (Fig. 9.8.13)

15 $q = 24$, $p = 6$ (Fig. 9.8.15)

17 $q = 8$, $p = 30$ (Fig. 9.8.17)

19 $q = 5$, $p = 25$ (Fig. 9.8.19)

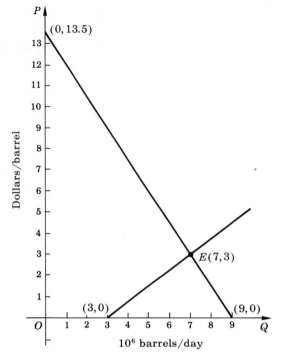

Figure 9.8.1

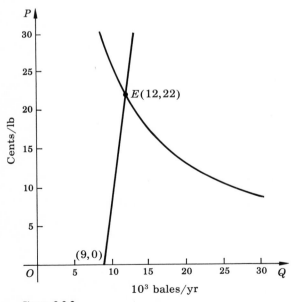

Figure 9.8.3

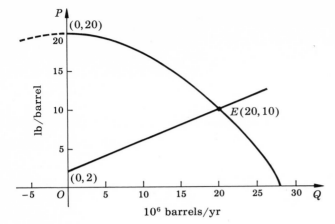

Figure 9.8.5

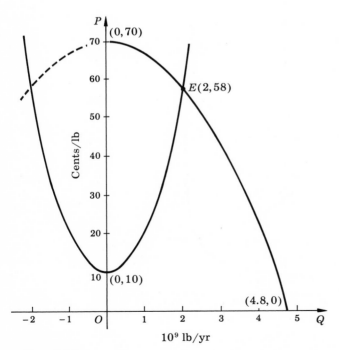

Figure 9.8.7

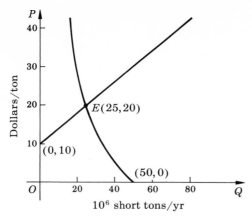

Figure 9.8.9

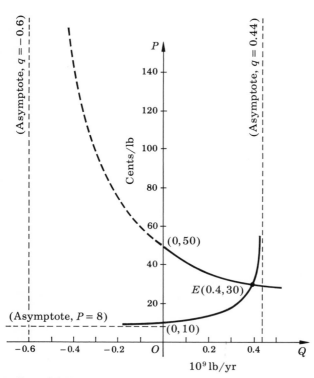

Figure 9.8.11

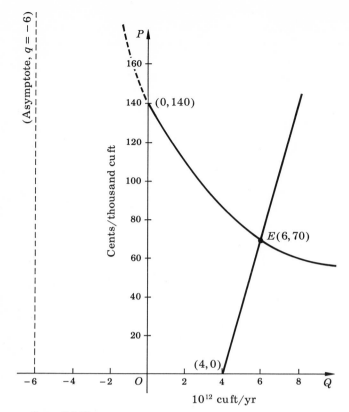

Figure 9.8.13

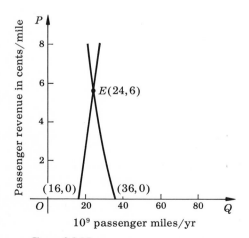

Figure 9.8.15

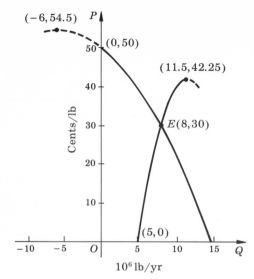

Figure 9.8.17

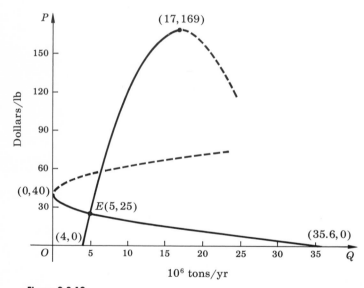

Figure 9.8.19

Problems 9.9

1 $E' = (6\frac{2}{3}, 3\frac{1}{2})$. Revenue before shift = 21×10^6; after shift = $18\frac{1}{3} \times 10^6$ (Fig. 9.9.1)

3 $E' = (10, 26.4)$. Revenue before and after shift = $1{,}261.92 \times 10^3$ (Fig. 9.9.3)

5 $E' = (29, 13.6)$. Revenue before shift = £200×10^6; after shift = £394.4×10^6 (Fig. 9.9.5)

7 $D: q^2 = \dfrac{70 - p}{3}$; $S: p = 12q^2 + 10 + x$

(*a*) $p = 58 + 9 = 67$. From D, $q^2 = (70 - 67)/3 = 1$; from S, $67 = 12 + 10 + x$; $x = 45$

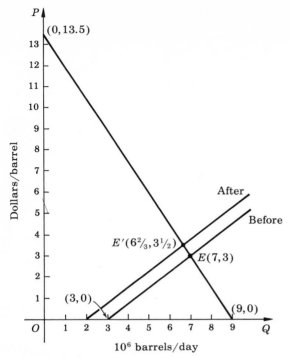

Figure 9.9.1

(b) $p = 58 - 3 = 55$. From D, $q^2 = (70 - 55)/3 = 5$; from S, $55 = 60 + 10 + x$;
$x = -15$

(c) The equilibrium point in (a) is $E_a(1,67)$. The equilibrium point in (b) is $E_b(\sqrt{5},55)$.
The total revenue before the shifts is $\$116 \times 10^7$. The total revenue after shift

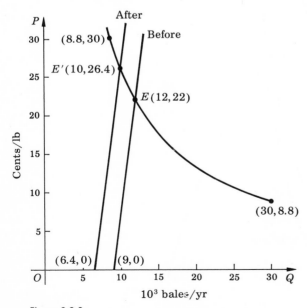

Figure 9.9.3

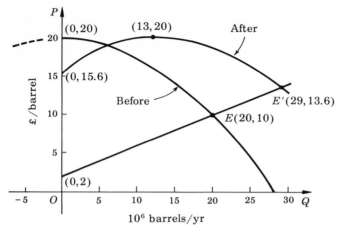

Figure 9.9.5

(a) is $(67 - 45) \times 1 \times 10^7 = \$22 \times 10^7$. The total revenue after shift (b) is $(55 + 15) \times \sqrt{5} \times 10^7 = \$156.52 \times 10^7$. (Fig. 9.9.7)

9 After shift:

$$D: q = \frac{1,000}{p + 20} - \frac{120}{19}; \quad S: q = \frac{5p - 50}{2}$$

Putting right sides equal, you can show that
$19p^2 + 238p - 10,440 = 0$ or $(p - 18)(19p + 580) = 0$
After-shift equilibrium point is $E'(20,18)$. Total revenue before shift $= \$500 \times 10^6$; after shift $= \$360 \times 10^6$ (Fig. 9.9.9)

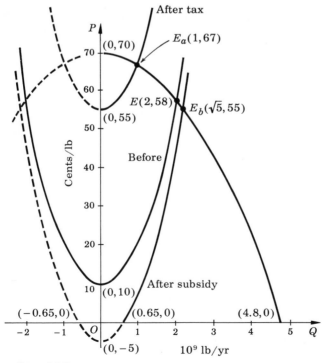

Figure 9.9.7

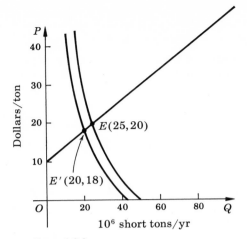

Figure 9.9.9

11 After shift:

$$D: q = \frac{150}{5p} - \frac{3}{5} + x; \quad S: q = \frac{11}{25} - \frac{22}{25(p - 8)} - y$$

Letting $q = 0.3$ and $p = 40$, we find $x = \frac{3}{20}$, $y = \frac{9}{80}$. Total revenue before and after shift $= \$12 \times 10^7$ (Fig. 9.9.11)

13 S: $3p - 2q = 36$
$q = 8\frac{1}{19}$, $p = 17\frac{7}{19}$

15 S: $p - 3q + 9 = 0$
$q = 5$, $p = 6$

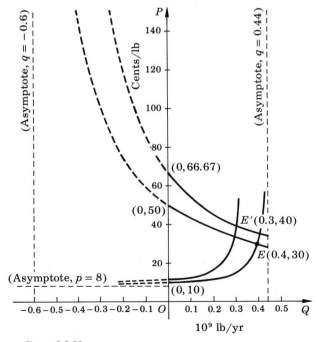

Figure 9.9.11

Problems 10.4

1 (b)   3 (a); (b); (c)   5 (a)
7 (a); (b)   9 (a); (b); (c)   11 (c)

13 (a) Polynomial, all reals   15 (a) Rational, $x \neq -2, 3$ 17 (a) Explicit, $x \geq 0$
   (b) Explicit, $x \geq 0$ (b) Explicit, all reals   (b) Rational, $y \neq 0$
   (c) Rational, $x \neq 0$ (c) Explicit, $-1 \leq x \leq 1$ (c) Explicit, $\theta \geq -1$

Problems 10.5

1 Polynomial; domain X; range Y; zeros $\{-1, 0, 1\}$ (Fig. 10.5.1)
3 Explicit algebraic; domain $-1 \leq x \leq 0$ or $x \geq 1$; range $y \geq 0$; zeros $\{-1, 0, 1\}$;
   end-points $\{-1, 0, 1\}$ (Fig. 10.5.3)

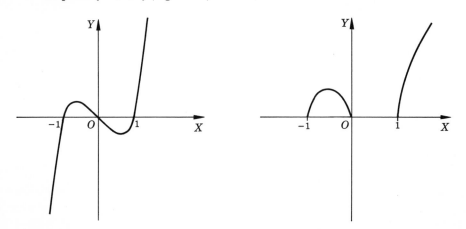

Figure 10.5.1   **Figure 10.5.3**

5 Polynomial; domain X; range (complete square) $y \leq \frac{1}{4}$; zeros $\{-1, 0 \text{ (double)}, 1\}$
   (Fig. 10.5.5)
7 Explicit algebraic; domain $|x| \leq 1$; range $0 \leq y \leq \frac{1}{2}$; zeros $\{-1, 0, 1\}$; end-points
   $\{-1, 1\}$ (Fig. 10.5.7)
9 Polynomial; domain X; range Y; zero $\{-1\}$ (Fig. 10.5.9)
11 Explicit algebraic; domain $x \geq -1$; range $y \geq 0$; zero $\{-1\}$; end-point $\{-1\}$
   (Fig. 10.5.11)
13 Explicit algebraic; domain X; range Y; zero $\{-1\}$; asymptote $y = x$ (cannot be found
   by methods in the text) (Fig. 10.5.13)

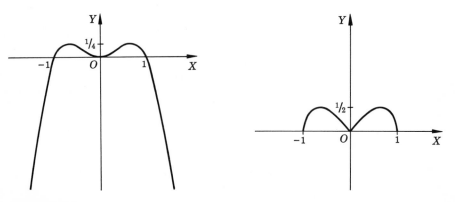

Figure 10.5.5   **Figure 10.5.7**

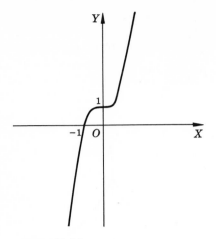

Figure 10.5.9

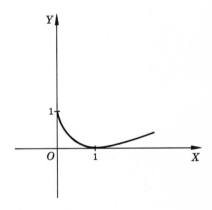

Figure 10.5.11

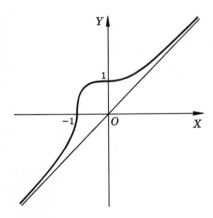

Figure 10.5.13

Figure 10.5.15

15 Explicit algebraic; domain $0 \leq x$; range $0 \leq y$; zero $\{1\}$; end-point $\{0\}$ (Fig. 10.5.15)
17 Rational; domain X; range $0 < y \leq 2$ (Fig. 10.5.17)
19 Explicit algebraic; domain $0 \leq x < 2$; range $y \geq 0$; zero $\{0\}$; end-point $\{0\}$; vertical asymptote $x = 2$ (Fig. 10.5.19)

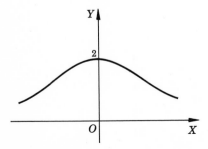

Figure 10.5.17

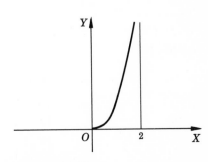

Figure 10.5.19

21 Explicit algebraic; domain $-3 \leq x \leq 0$ or $x \geq 1$; range $y \leq 0$; zeros $\{-3, 0, 1\}$; end-points $\{-3, 0, 1\}$ (Fig. 10.5.21)
23 Implicit algebraic; domain $-1 \leq x \leq 1$; range $-1 \leq y \leq 1$; zeros $\{-1, 1\}$; hypocycloid (Fig. 10.5.23)
25 Implicit algebraic; domain X; range Y; zero $\{0\}$ (Fig. 10.5.25)
27 The line $y = -x$ *and* the hyperbola $xy + 1 = 0$ (Fig. 10.5.27)
29 n even: domain X; range $y \geq 0$; zero $\{0\}$
n odd: domain X; range Y; zero $\{0\}$

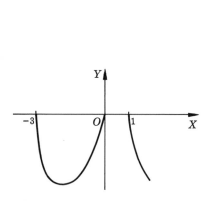

Figure 10.5.21

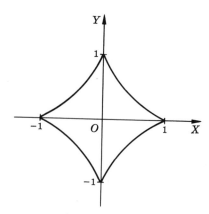

Figure 10.5.23

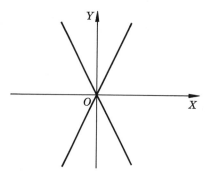

Figure 10.5.25

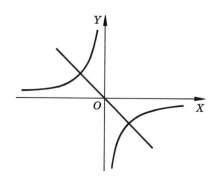

Figure 10.5.27

Problems 10.6

1 $x^2 - 5x + 6$	**3** $x - i$	**5** $x^4 - 6x^3 + 11x^2 - 6x$
7 $x^3 - (1 + i)x^2 + ix$	**9** $x^3 - 4x^2 + 5x - 2$	**11** $x^3 + x^2 - 2x$
13 $x^3 - 2x^2 - 3x + 6$	**15** $(x^2 - 2ax + a^2 + b^2)(x - r)$	
17 $x^3 - (a + 2)x^2 + (2a + 1)x - a$	**19** 3	
21 48	**23** 7	**25** $\pm 2, \pm 2i$
27 $5, \dfrac{1}{2} \pm i\dfrac{\sqrt{3}}{2}$	**29** 5	**31** 3

Problems 10.7

1 $P(1) = 8$
$P(-1) = 10$

3 $P(1) = 5$
$P(-1) = 1$

5 $P(\frac{1}{3}) = \frac{13}{9}$
$P(\frac{1}{2}) = \frac{3}{2}$

7 $q = 4x^2 + 6x + 3$
$r = 4$

9 $q = x^2 + 4x - 2$
$r = -10$

11 $q = x^3 - x^2 + x + 2$
$r = -8$

13 $q = x + \frac{7}{2}$
$r = -2\frac{1}{4}$

15 $q = x^2 + \frac{25}{3}x - \frac{2}{9}$
$r = \frac{160}{27}$

17
$$
\begin{array}{rrrr|r}
5 & 8 & -1 & 6 & \underline{-2} \\
 & -10 & 4 & -6 & \\
\hline
5 & -2 & 3 & 0 &
\end{array}
$$

19
$$
\begin{array}{rrrrr|r}
4 & 6 & -1 & 2 & -8 & \underline{-2} \\
 & -8 & 4 & -6 & 8 & \\
\hline
4 & -2 & 3 & -4 & 0 &
\end{array}
$$

21
$$
\begin{array}{rrrrr|r}
6 & -5 & 4 & -13 & -3 & \underline{\frac{3}{2}} \\
 & 9 & 6 & 15 & 3 & \\
\hline
6 & 4 & 10 & 2 & 0 &
\end{array}
$$

25 0

27 -150

29 $-7, 5$

Problems 10.8

1 $\{1, \frac{1}{2}\}$

3 $\{6, \frac{2}{3}\}$

5 $\{-\frac{3}{8}, \frac{5}{2}, 0\}$

7 $\{4, 5\}$

9 $\{\pm\frac{1}{2}\}$

11 $\{2, -3, -\frac{4}{3}\}$

13 $\{0 \text{ (double)}, \pm i\}$

15 $\{-\frac{1}{3}, \frac{1}{2}\}$

17 $\{0, 4, -\frac{5}{2}\}$

19 $\{0 \text{ (quadruple)}, \pm 1\}$

21 $\{-\frac{1}{2} \pm \frac{1}{2}\sqrt{13}, -\frac{1}{2} \pm \frac{1}{2}\sqrt{17}\}$

Problems 10.9

1 $\{-\frac{4}{3}, -\frac{5}{4}, 2\}$

3 $\{-4, 2, 3\}$

5 $\{1, -1 \pm \sqrt{7}\}$

7 $\{-\frac{8}{3}, \frac{5}{4}, 3, 4\}$

9 $\{-3, -2, \frac{1}{2}, 2\}$

11 $\{\frac{1}{2}, 1, -1 \pm i\sqrt{2}\}$

Problems 10.10

1 -3.9

3 0.3

5 -1.4

7 0.3

9 1.2

11 ± 1.4

13 -0.1

15 ± 1

Problems 10.11

1 $2 \text{ cis } \dfrac{\pi}{6}$

3 $2 \text{ cis } \dfrac{5\pi}{6}$

5 $17 \text{ cis } \dfrac{\pi}{2}$

7 $13 \text{ cis } 0$

9 $5 \text{ cis } \text{Tan}^{-1}\left(-\frac{4}{3}\right)$

11 1

13 $-2\sqrt{3} - 2i$

15 $5 - 5\sqrt{3}\,i$

17 $\frac{3}{2}\sqrt{2} - \frac{3}{2}\sqrt{2}\,i$

19 $(-6 - 8\sqrt{2}) + (4 + 8\sqrt{2} + 6\sqrt{3})i$

21 $6 \text{ cis } 120°$

23 $21 \text{ cis }\left(-\dfrac{\pi}{12}\right)$

25 $\frac{1}{2} \text{ cis } 120°$

27 $25 \text{ cis } \pi$

29 $32 \text{ cis } 315°$

31 $32 \text{ cis } \dfrac{5\pi}{3}$

33 $2 \text{ cis } 90, 2 \text{ cis } 210°, 2 \text{ cis } 330°$

35 $1, i, -1, -i$

37 $3 \text{ cis } \dfrac{\pi}{3}, 3 \text{ cis } \pi, 3 \text{ cis } \dfrac{5\pi}{3}$

39 $2 \text{ cis } \dfrac{2k\pi}{5}; k = 0, 1, 2, 3, 4$

41 $(\frac{4}{3})^{1/3} \text{ cis } \dfrac{2k\pi}{3}; k = 0, 1, 2$

Problems 11.1

1 $a^x \times a^y = a^{x+y}$ **3** 2.7   **5** 0.16
7 $256/27$ **9** 9   **11** 21
13 120   **15** a^{16x} **17** $a^{11x/6} \cdot b^{2x}$
19 $10^{0.5x-1.5}$ **21** (Fig. 11.1.21)   **23** (Fig. 11.1.23)
25 (Fig. 11.1.25)

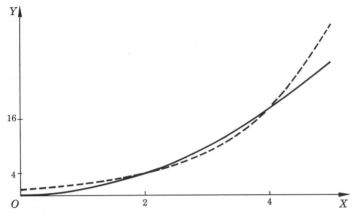

Figure 11.1.21

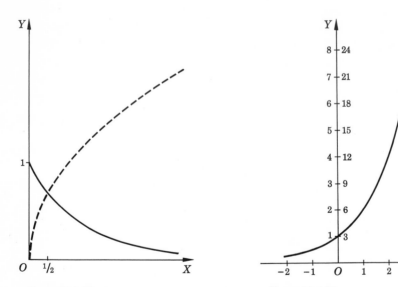

Figure 11.1.23   **Figure 11.1.25**

Problems 11.2

1 1.0513   **3** 6.0496   **5** 14.7782
7 0.13534   **9** 0.01291   **11** (Fig. 11.2.11)
13 (Fig. 11.2.13)   **15** (Fig. 11.2.15)   **17** (Fig. 11.2.17)
19 (Fig. 11.2.19)

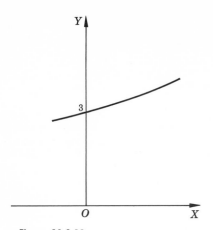

Figure 11.2.11

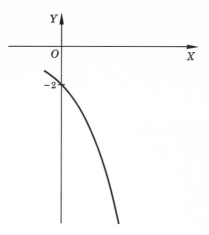

Figure 11.2.13

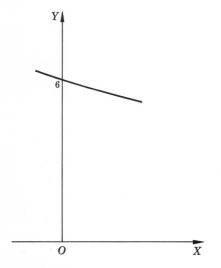

Figure 11.2.15

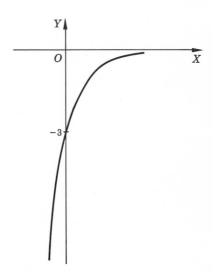

Figure 11.2.17

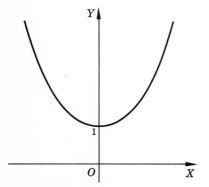

Figure 11.2.19

Problems 11.3

1	43.07	**3**	3.32	**5**	126
7	43.1	**9**	3.32	**11**	126
13	3	**15**	5	**17**	6

19 $a^{x \log_a x} = [a^{\log_a x}]^x = x^x$ **21** $a^{x(1+\log_a b)}$

23 $e = 2.718$. This suggests that $(1 + 1/n)^n$ approaches e as n gets large.

Problems 11.4

1	(Fig. 11.4.1)	**3**	(Fig. 11.4.3)	**5**	(Fig. 11.4.5)
7	(Fig. 11.4.7)	**9**	(Fig. 11.4.9)	**11**	(Fig. 11.4.11)
13	(Fig. 11.4.13)	**15**	1.53	**17**	-2
19	0, 0.7$-$				

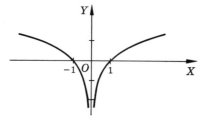

Figure 11.4.1

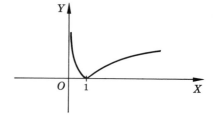

Figure 11.4.3

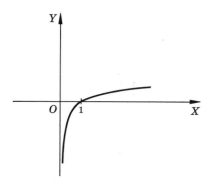

Figure 11.4.5

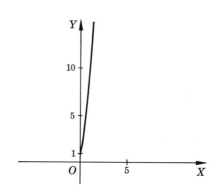

Figure 11.4.7

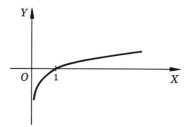

Figure 11.4.9

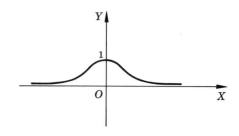

Figure 11.4.11

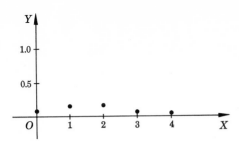

Figure 11.4.13

Problems 11.5

1 $t = 10 \log_e \tfrac{3}{5}$

3 $t = 5 \log_e \left(\dfrac{3.4}{2.5} \right)$

5 $k = \tfrac{1}{3} \, 2^{0.6}$

7 4.814 km

9 $r = \log_e \tfrac{1}{16}$

11 $\tfrac{1}{32}$ g

13 $t_1 = 3 \times 10^{-16}$

Problems 11.6

1 10,000

3 $y = 3x - 1$

5 $y = 2x^3$

7 Semilog

9 Log–log

11 Log–log

13 Log–log

15 Rectilinear

Problems 11.7

1 $U = (4\tfrac{1}{2}, 6\tfrac{3}{4})$

3 The logarithmic demand curve is a straight line making an angle of 45° with each axis. Every point has unit elasticity.

5 $U = (16.3, 13.3)$ **7** $U = (2.8, 46.7)$

9 U does not exist. Demand is inelastic at every point.

Problems 12.2

1 $\{HH, HT, TH, TT\}$

3 $\{H1, H2, H3, H4, H5, H6, T1, T2, T3, T4, T5, T6\}$

5 $\{1, 2, 3, 4\}$ **7** $\{1, 2, 3, 4, 5, 6, 7, 8\}$ **9** $1, 2, 3, \ldots, 20\}$

11 $\{1, 2, 3, 4, 5\}$. We assume no interest in sides marked 1 other than the mark "1".

13 $\{GFP, GFP', GF'P, G'FP, GF'P', G'FP', G'F'P, G'F'P'\}$

15 $\{MFF, FMF, FFM\}$

17 Permissible: $(a), (c), (d)$ not permissible: (b), not exhaustive; (e), not mutually exclusive

19 $(a), (b), (c), (d)$

Problems 12.4

1 $\{H, T\}, P(H) = P(T) = \tfrac{1}{2}$

3 $\{1, 2, 3, 4\}, P(i) = \tfrac{1}{4}, i = 1, 2, 3, 4$

5 $\{1, 2, 3\}, P(1) = \tfrac{1}{6}, P(2) = \tfrac{2}{6}, P(3) = \tfrac{3}{6}$

7 $\{1, 2, \ldots, 8\}, P(i) = \tfrac{1}{8}, i = 1, 2, \ldots, 8$

9 $\{1, 2, \ldots, 12\}, P(i) = \tfrac{1}{12}, i = 1, 2, \ldots, 12$

11 $\{1, 2, \ldots, 20\}, P(i) = \tfrac{1}{20}, i = 1, 2, \ldots, 20$

Problems 12.5

1 $P(\text{odd}) = P(\text{even}) = \frac{1}{2}$ **3** $P(\text{odd}) = P(\text{even}) = \frac{1}{2}$

5 $P(\text{odd}) = P(\text{even}) = \frac{1}{2}$ **7** $P(1) = \frac{1}{12}$, $P(\text{prime}) = \frac{5}{12}$, $P(\text{composite}) = \frac{6}{12}$

9 $P(\text{multiple of 3}) = \frac{6}{20}$, $P(\text{otherwise}) = \frac{14}{20}$

11 $P(\text{multiple of 6}) = \frac{3}{20}$, $P(\text{multiple of 7}) = \frac{2}{20}$, $P(\text{otherwise}) = \frac{15}{20}$

13 $P(c) = P(d) = P(h) = P(s) = \frac{1}{4}$

15 $P(i) = \frac{1}{13}$, $i = 2, 3, \ldots, 10$; $P(J) = P(Q) = P(K) = P(A) = \frac{1}{13}$

17 $P(\text{red}) = \frac{5}{12}$, $P(\text{green}) = \frac{4}{12}$, $P(\text{black}) = \frac{3}{12}$

19 $P(\text{odd}) = \frac{7}{12}$, $P(\text{even}) = \frac{5}{12}$

Problems 12.6

1 $U = U_1 \times U_1$ where $U_1 = \{1, 2, 3, 4\}$
$P(ij) = \frac{1}{16}$ $i, j = 1, 2, 3, 4$

3 $U = U_1 \times U_1$ where $U_1 = \{1, 2, \ldots, 12\}$
$P(ij) = \frac{1}{144}$ $i, j = 1, 2, \ldots, 12$

5 $U = \{1, 2, 3, 4\} \times \{1, 2, 3, 4, 5, 6, 7\,8\}$
$P(ij) = \frac{1}{32}$ $i = 1, 2, 3, 4, j = 1, 2, \ldots, 8$

7 $U = \{1, 2, 3, 4\} \times \{1, 2, \ldots, 20\}$
$P(ij) = \frac{1}{80}$ $i = 1, 2, 3, 4; j = 1, 2, \ldots, 20$

9 $U = \{1, 2, \ldots, 8\} \times \{1, 2, \ldots, 20\}$
$P(ij) = \frac{1}{160}$ $i = 1, 2, \ldots, 8; j = 1, 2, \ldots, 20$

11 $U = \{1, 2, 3\} \times \{1, 2, 3\}$
$P(11) = \frac{2}{16}$ $P(12) = \frac{4}{16}$ $P(13) = \frac{2}{16}$
$P(21) = \frac{1}{16}$ $P(22) = \frac{2}{16}$ $P(23) = \frac{1}{16}$
$P(31) = \frac{1}{16}$ $P(32) = \frac{2}{16}$ $P(33) = \frac{1}{16}$

13 $U = \{1, 2\} \times \{1, 2\}$
$P(11) = \frac{121}{144}$ $P(12) = \frac{11}{144}$
$P(21) = \frac{11}{144}$ $P(22) = \frac{1}{144}$

15 $U = \{1, 2, \ldots, n\} \times \{1, 2, \ldots, n\}$
$P(ij) = \dfrac{1}{n^2}$ $i, j = 1, 2, \ldots, n$

A "deck" of cards marked $1, 2, \ldots, n$

17 $U = \{HT\} \times \{HT\} \times \{HT\}$
$= \{HHH, HHT, HTH, THH, HTT, THT, TTH, TTT\}$
$P(HHH) = P(HHT) = \cdots = P(TTT) = \frac{1}{8}$

19 $U = \{1, 2\} \times \{1, 2\} \times \{1, 2\}$
$= \{111, 112, 121, 211, 122, 212, 221, 222\}$
$P(111) = \frac{27}{64}$ $P(112) = P(121) = P(211) = \frac{9}{64}$
$P(122) = P(212) = P(221) = \frac{3}{64}$ $P(222) = \frac{1}{64}$

21 $U = \{1, 2, \ldots, 52\} \times \{1, 2, \ldots, 52\}$
$P(ij) = (\frac{1}{52})^2$ $i, j = 1, 2, \ldots, 52$

23 $U = \{\text{red, black}\} \times \{\text{red, black}\}$
$P(rr) = \frac{16}{81}$ $P(rb) = \frac{20}{81}$
$P(br) = \frac{20}{81}$ $P(bb) = \frac{25}{81}$

25 $U = \{\text{red, black}\}$
$P(r) = \frac{4}{9}$ $P(b) = \frac{5}{9}$

27 $P(\text{all correct}) = (\frac{1}{5})^{50}$
$U = \{\text{correct, not correct}\} \times \cdots \times \{\text{correct, not correct}\}$, 50 times.   This nonuniform joint distribution has 2^{50} simple events.   No.

29 $\frac{49}{100} \times \frac{999}{1,000}$

31 $U = \{\text{I}s, \text{II}s, \text{II}g, \text{III}g\}$
$P(\text{I}s) = \frac{1}{3}$ $P(\text{II}s) = P(\text{II}g) = \frac{1}{6}$ $P(\text{III}g) = \frac{1}{3}$

Problems 12.7

1	$\frac{1}{2}$	**3**	$\frac{1}{6}$	**5**	$\frac{1}{4}$
7	$\frac{1}{10}$	**9**	$\frac{9}{80}$	**11**	$\frac{1}{4}$
13	$\frac{11}{72}$	**15**	$\frac{1}{8}$	**17**	$\frac{1}{8}$
19	$\frac{1}{8}$	**21**	$\frac{1}{8}$	**23**	$\frac{4}{21}$
25	$(\frac{1}{26})^4$	**27**	$(\frac{1}{365})^2$	**29**	$\frac{1}{64}$
31	$1/1{,}728$	**33**	$\frac{1}{9}$	**35**	$\frac{1}{2}$

37 (a) $\frac{1}{10}$, (b) $\frac{1}{4}$

Problems 12.8

1	$\frac{1}{2}$	**3**	$\frac{3}{4}$	**5**	$\frac{1}{2}$
7	$\frac{5}{8}$	**11**	$\frac{7}{8}$	**13**	$\frac{1}{12}$
15	$\frac{11}{12}$	**17**	$\frac{7}{36}$	**19**	$\frac{1}{6}$

21 $P(A \cap B) = \frac{1}{4}$ $P(A \cup B) = \frac{3}{4}$ **23** $\frac{35}{36}$

25 $\frac{8}{9}$ **27** $\frac{3}{7}$

29 $\frac{215}{216}$ **31** $\frac{21}{41}$

33 (a) $\frac{12}{25}$, (b) $\frac{17}{25}$ **35** (a) 1.00, (b) 0.80

Problems 12.9

1	$\frac{1}{2}$	**3**	$\frac{3}{7}$	**5**	$\frac{3}{4}$
7	$\frac{1}{4}$	**9**	$\frac{2}{3}$	**11**	$\frac{2}{3}$
13	$\frac{2}{5}$	**15**	$\frac{1}{3}$	**17**	$\frac{1}{2}$
19	$\frac{3}{5}$				

Problems 12.11

1	(a) 20, (b) 380	**3**	5,040	**5**	(a) 720, (b) 46,656
7	125	**9**	(a) 90, (b) 45	**11**	3,780

13 $11 \cdot 10 \cdot 9 \cdot 7 \cdot 5 - 7 \cdot 6 \cdot 5 \cdot 4 = 33{,}810$

15 $12 \cdot 11 \cdot 10 \cdot 9 \cdot 8 \cdot 7 \cdot 6$ **17** $15!/4!$ **19** $24^2 \cdot 15 \cdot 10^4$

21 $4 + 4^2 + 4^3 + 4^4 - 27 = 3 \cdot 4^3 + 2 \cdot 4^2 + 5 = 313$

23 $\dfrac{(n-1)!}{2}$ **25** $45 \cdot 28 \cdot 15 \cdot 6$ **27** 30

29 m^n

Problems 12.12

3	1	**5**	(a) 120, (b) 120	**7**	$20!/3!$
9	8	**11**	19	**13**	(a) 12, (b) 24
15	15	**17**	39	**19**	$\binom{30}{6}!$

21 (a) 120, (b) 252, (c) 126   **23** 22

25 $\binom{40}{5} + 60 \binom{40}{4} + 1{,}770 \binom{40}{3}$ **27** $4n^4$

29 $3\binom{33}{3} + 33^3$

Problems 12.14

1 (a) $\frac{1}{4}$, (b) $\frac{9}{64}$, (c) $\frac{5}{32}$ **3** $\dfrac{53}{5^5}$ **5** (a) $\frac{25}{216}$, (b) $\dfrac{171}{6^4}$

7 $\dfrac{30}{6^5}$ **9** $[1 - (\frac{1}{2})^{500} - 500(\frac{1}{2})^{500}]$

11 $\displaystyle\sum_{i=2}^{5} \binom{500}{i}(\frac{1}{2})^{500}$ **13** $\dfrac{11}{3^5}$ **15** $\dfrac{200}{3^5}$

17 $\frac{3}{16}$ **19** $1 - \dfrac{131}{6^5}$ **21** $\frac{1}{4}$

23 $\binom{25}{x}(\frac{3}{4})^{25-x}(\frac{1}{4})^x$

25 $\frac{17}{30}$

27 $\dfrac{\dfrac{1}{\binom{64}{2}}\cdot 7\cdot 7\cdot 2}{\dfrac{1}{\binom{64}{2}}\cdot\binom{64}{2}}=\dfrac{7}{144}$

29 $(a)\dfrac{\binom{52}{1}\binom{48}{1}\binom{44}{1}\binom{40}{1}\binom{36}{1}}{\binom{52}{5}}$, $(b)\dfrac{\binom{4}{2}\binom{48}{3}+\binom{4}{3}\binom{48}{2}+\binom{4}{4}\binom{48}{1}}{\binom{52}{5}}$

31 $(a)\ \frac{5}{12}$, $(b)\dfrac{\binom{11}{1}+2\binom{11}{2}+\binom{11}{3}}{6^{11}}$

33 $\dfrac{15!\cdot 14\cdot 13\cdot 12\cdot 2}{19!}$

35 $\frac{2}{9}$

Problems 12.15

1 (a) Reject, (b) accept, (c) accept, (d) accept, (e) accept
3 Accept
7 Reject
5 Quit playing, but don't shoot him.
9 Reject

Problems 12.16

1 $3.50
3 (a) $(\frac{1}{2}+\frac{1}{4}+\frac{1}{8}+\cdots)=1
$\quad$ (b) $2(\frac{1}{2})+4(\frac{1}{4})+8(\frac{1}{8})+\cdots=\infty$
5 ($1\times\frac{7}{22})+($1.50\times\frac{35}{66})+($2\times\frac{5}{33})=$1.42$
7 0
9 $m=2.1;\ \sigma^2=1.09;\ \sigma=\sqrt{1.09}$
11 $m=9;\ \sigma^2=7.2;\ \sigma=\sqrt{7.2}$
13 3, 2
15 $m=\dfrac{n+1}{2};\ \sigma^2=\dfrac{n^2-1}{12}$

Problems 12.17

1 (a) and (c)
3 If $a\le b$ or if $a\le d$, a satisfies the requirements. If $a>b$ and $a>d$ and $b\ne d$, the smaller of b and d satisfies the requirements. If $a>b$ and $a>d$ and $b=d$, both b and d satisfy the requirements.
5 Strictly determined. Player 1 plays row 1; Player 2 plays column 1; value = 3.
7 $p=q=\frac{1}{3};\ 1-p=1-q=\frac{2}{3};\ v=0$. Fair
9 $p=\frac{1}{5};\ 1-p=\frac{4}{5};\ q=\frac{2}{5};\ 1-q=\frac{3}{5};\ v=\frac{13}{5}$

11

Offense	Defense R	Defense P
R	3	8
P	20	0

The offense should run four-fifths of the time. The defense should set a running defense eight-twenty-fifths of the time. The expected gain is $6\frac{2}{5}$ yd.

13

	1	2
1	2	3
2	3	4

Strictly determined. Player 1 shows 2 fingers; Player 2 shows 1 finger; value = 3.

		Enemy	
		Do not mine	Mine
15	Do not	0	-100
We sweep			
	Sweep	-10	-5

We sweep with probability $^{20}/_{21}$. Enemy lays mines with probability $^{2}/_{21}$. Expected cost to us $= \$^{200}/_{21}$ million.

17 If we do not sweep, our expected net cost is \$5 million. If we do sweep, our expected net cost is \$9.75 million. *Ans.*: Do no sweeping.

19 $(0 \times p_1) + (-1 \times p_2) + (1 \times p_3) = (1 \times p_1) + (0 \times p_2) + (-1 \times p_3) = (-1 \times p_1) + (1 \times p_2) + (0 \times p_3)$, where $p_1 + p_2 + p_3 = 1$. *Ans.*: $p_1 = p_2 = p_3 = \frac{1}{3}$; $v = 0$. Same for Player 2.

Index

Index

DATE DUE

4-5-67 New Books		
MAY 24 '67		
SEP 27 '67		
MAY 8 '68		
JUL 3 '68		
JUL 10 '68		
AUG 7 '68		
AUG 21 '68		
JAN 15 '69		
MAY 19 '69		
Sept-17 69		
AUG 19 70		
NOV 15 1977		
MAY 13 1987		
DEC 17 1990		
GAYLORD		PRINTED IN U.S.A.

The Symbols Used in This Book

and the pages on which they are defined

7	{ }	notation for a set
7	∅	the empty set
7	{3}	the set of which 3 is the only element
7	{2, 4, 7, 9}	the set whose elements are 2, 4, 7, and 9
7	$X = \{x \mid x \text{ is a real number}\}$	the set of real numbers
7	$X \times Y$	the set of ordered pairs of real numbers
8	⊆	inclusion for sets
8	⊂	proper inclusion for sets
11	U	universal set
12	p_x, q_x	open sentences with variable x
12	P, Q	truth sets of p_x and q_x, respectively
12	$p_x \leftrightarrow q_x$	equivalence of p_x and q_x
14	A'	complement of the set A
14	∪	union of sets
15	∩	intersection of sets
18	$p_x \rightarrow q_x$	implication of q_x by p_x
18	$\forall_x$	for all x
42	$\lvert a \rvert$	absolute value
59	$1.\overline{142857}$	repeating decimal
63	<	less than
63	>	greater than
64	≤	less than or equal to
64	≥	greater than or equal to
65	(x, y)	coordinates of a point in the plane
69	$a + bi$	complex number
69	i	unit imaginary number, $i^2 = -1$
83	$n!$	n factorial
83	$0!$	zero factorial
83	$\binom{n}{r}$	binomial coefficient
112	$\sqrt[n]{a} = a^{1/n}$	positive nth root of a
112	$\sqrt[q]{a^p} = a^{p/q}$	positive qth root of a^p
159	(a,b) or (a,b,c)	row vector
159	$\begin{pmatrix} a_1 & b_1 \\ a_2 & b_2 \end{pmatrix}$	matrix
159	$\begin{pmatrix} a_1 & b_1 & c_1 \\ a_2 & b_2 & c_2 \end{pmatrix}$	matrix